AF380548

Operator Theory: Advances and
Applications
Vol. 158

Editor:
I. Gohberg

Selected Topics in Complex Analysis

The S. Ya. Khavinson Memorial Volume

V. Ya. Eiderman
M. V. Samokhin
Editors

Birkhäuser Verlag
Basel · Boston · Berlin

Editors:

Vladimir Ya. Eiderman
Mikhail V. Samokhin
Moscow State Civil Engineering University
Department of Higher Mathematics
Yaroslavskoye shosse, 26
129337 Moscow
Russia
e-mail: eiderman@orc.ru
         msamokh@mgsu.ru

2000 Mathematics Subject Classification 01-02, 01-06, 01A70, 30-06, 31-06, 32-06, 47-06

A CIP catalogue record for this book is available from the
Library of Congress, Washington D.C., USA

Bibliographic information published by Die Deutsche Bibliothek
Die Deutsche Bibliothek lists this publication in the Deutsche Nationalbibliografie; detailed
bibliographic data is available in the Internet at <http://dnb.ddb.de>.

ISBN 3-7643-7251-6 Birkhäuser Verlag, Basel – Boston – Berlin

© 2005 Birkhäuser Verlag, P.O. Box 133, CH-4010 Basel, Switzerland
Part of Springer Science+Business Media
Printed on acid-free paper produced from chlorine-free pulp. TCF ∞
Cover design: Heinz Hiltbrunner, Basel
Printed in Germany
ISBN 10: 3-7643-7251-6
ISBN 13: 978-3-7643-7251-4
9 8 7 6 5 4 3 2 1   www.birkhauser.ch

Dedicated to the memory of

Semyon Yakovlevich Khavinson

Semyon Yakovlevich Khavinson
1927–2003

Contents

Operator Theory:
Advances and Applications, Vol. 158, 1–22

Semyon Yakovlevich Khavinson
(May 17, 1927 – January 30, 2003)

V.P. Havin

1. Life

S.Ya. Khavinson was born in Moscow. His mother, M.B. Zucker, was a medicine professor, his father, Ya.S. Khavinson, a well-known journalist; in 1940–1943 he was the head of TASS (Telegraph Agency of the Soviet Union), and for a long time the editor of the journal "World Economy and International Relations". Brought up in an atmosphere of the ubiquitous communist ideology, S.Ya. made a long way to independent thinking (the sobering process was completely over when I met him for the first time in 1957). Moved by romantic infatuation with aviation and the trend of the epoch, after highschool in 1943, he entered MAI (Moscow Institute of Aviation) where a new infatuation prevailed over the first one, and S.Ya. was forever captivated by mathematics. He left MAI for the math department of MGU (Moscow State University). This change was not an easy undertaking, but he succeeded in catching up with the much more advanced mathematical program and became a pure math student of the third year of MGU, the institution of a very high level of teaching and research.

This episode confronted him for the first time with the difference in teaching mathematics to pure and applied mathematicians, the problem he had to cope with all his life: after he graduated from MGU in 1949, he taught in Yelets (a small town in Lipetsk region) and then in Vladimir where his students were future highschool teachers. From 1956 on he taught at MISI (Moscow Institute of Civil Engineering) where he became the head of a huge "chair of higher mathematics" (some 80 members!). All his life he had to combine intense and fruitful activity of "a pure mathematician" proving new theorems, with the everyday worries and headaches of a pedagogue and organizer of a complicated teaching process for masses of students which did not consider math as their primordial task, but had to overcome it according to the Russian tradition of educating engineers of any speciality. And S.Ya. contrived to successfully agree both jobs, being at the same time a bright and passionate complex analyst with an impressive output of results and ideas, and a teacher of mathematics for non-professionals, fully devoted to this difficult duty.

S.Ya. started his scientific work as a student of MGU under the guidance of A.I. Markushevich who also acted as S.Ya.'s adviser during his extramural graduate student years in Yelets and Vladimir. He defended his first ("the candidate" $\cong$ PhD) thesis in 1953. The second ("the doctoral") was defended in 1962 (the title was "Duality method in extremal and approximation problems of function theory").

The starting point of his work were the duality relations (see Section 2 below) first applied to some concrete problems by M.G. Krein (1938) and S.M. Nikolski (1946). S.Ya. made the duality method a powerful tool of Complex Analysis. (The same approach was found by Rogosinski & Shapiro in 1953, but the first articles of S.Ya. devoted to this method appeared in 1949 and 1951). A more detailed survey of his results follows (see Sections 2–6).

As I have mentioned, apart from pure mathematics, S.Ya. worked as the leader of an unusually numerous collective, the chair of higher mathematics of MISI. The common practice is to split such collectives into several smaller and more manageable units. But due to some personal features of S.Ya., his wise, tactful ways to contact people, everybody preferred to have him as a leader, and it was eventually decided to let the chair as it was and not split it. The energy of S.Ya. made the chair and its analysis seminar one of the centers of mathematical work in Moscow. S.Ya. was a successful adviser of ten graduate students in math; all of them had defended their theses, a result not very common for a non-mathematical teaching institution. Two of former students of S.Ya., V.Ya. Eiderman and M.V. Samokhin, defended their second ("doctoral") theses and are full professors.

S.Ya. wrote a number of textbooks for his students, future engineers, and also for mathematicians working in various universities of Russia who regularly come to MISI for their sabbaticals and take special courses in math to maintain their shape. S.Ya. was famous as a brilliant, thorough and very clear lecturer. His general tendency in his lectures and texts was to convey the very essence, "the truth of the subject" to his listeners and readers, minimizing all kinds of technicalities and doctrinalities. The permissible degree of this minimization will forever remain a theme of innumerable controversies, and I remember – with great pleasure – some fervent disputes with S.Ya. on this point. (They are reflected in his inscription on his gift to me, the book "Lectures on intregral calculus": "To dear Vitya Havin for the repetition of the live integral calculus which he somewhat forgot composing his german-pedantical booklets on this subject, from the loving author. 30.11.1976"; the books S.Ya. alluded at were intended for pure mathematicians). In any case, books of S.Ya. are nicely written and deservedly popular.

* * *

Life of S.Ya. Khavinson was not always easy. He had to live through some very bitter moments. But for his friends and colleagues he invariably remained the same, with his kindness, readiness to help, unusual width of interests (one of them was poetry: he could recite his favorite poems for hours). But I guess the main pivot of his life enabling him to weather all kinds of personal crises was his mathematics. His legacy is rich, both in content and volume (some 170 publications). Its detailed survey would take a book. My modest aim in subsequent pages is just to order Khavinson's results according to their main guidelines and briefly orient the reader in their highlights.

2. Linear Programming in Complex Analysis

The mathematical career of S.Ya. Khavinson started at the end of forties, in less than twenty years after the publication of S. Banach's "Théorie des opérations linéaires" which opened the "Sturm und Drang" period of Functional Analysis. Its methods and ideas were merging with classical, or "concrete", analysis. Complex Analysis was probably the last area to be conquered by the new approach, and Khavinson was in the first ranks of its proponents. His impressive contribution to this merging process was the application of the duality of linear spaces to classical extremal problems of Complex Analysis. Its eternal and ubiquitous theme, maximizing or minimizing a functional on a class of analytic functions, was given a completely new interpretation in Khavinson's work.

For a more specific discussion we need some notation: $H^\infty(G)$ will denote the space of all functions analytic and bounded in the domain G; the norm of an $f \in H^\infty(G)$ is $\|f\|_\infty = \sup_G |f|$; $\mathbb{D}$ will stand for the open unit disc.

2.1. A result of E. Landau

For a function f analytic in $\mathbb{D}$ put

$$c_k(f) = \frac{f^{(k)}(0)}{k!},$$

the k-th Taylor coefficient of f. For a given N Landau computed

$$\max\left\{ \left| \sum_{k=0}^{N} c_k(f) \right| \ : \ f \in H^\infty(\mathbb{D}), \quad \|f\|_\infty \le 1 \right\}.$$

In 1913 he found a quite explicit expression of this quantity (as a function of N) and exhibited the maximizing f. This example is generic for the problems studied by Khavinson for many years, starting from 1949. A short and very clear exposition of Landau's result is in §10 of [Kh111, Kh123][1] where one can find historical information and a list of Landau's successors. Note that in his early publication [Kh5] Khavinson obtained an elegant generalization of Landau's estimate (with $f^{(k)}(a)/k!$, in place of c_k, $a \in \mathbb{D}$). The gist of [Kh5] was a purely quantitative aspect of the problem whereas Khavinson's efforts in his subsequent works were mainly

---

[1] References [Kh1], [Kh2], etc., see in the section "S.Ya. Khavinson: Bibliography", following the present paper.

concentrated on "qualitative" questions. Nevertheless we believe that Landau's result and his analogs combined with Functional Analysis were among the principal incentives which had ignited Khavinson's interest in extremal problems in spaces of analytic functions.

2.2. Khavinson's approach to extremal problems: a sketch

We will have to confine our description to a rather particular situation in order to just allude at the possibilities of the method. The modest aim of this section is to convey *the flavor* of Khavinson's ideas. They first appeared in [Kh1], [Kh6] to be largely developed in various directions in [Kh11], [Kh55], [Kh26], [Kh49], [Kh152], [Kh161] (we quote only cornerstone papers where references to shorter publications are available). For the first reading we recommend a very nicely written exposition in [Kh123].

The duality of extremal problems, the main theme of this section, was implicit in Landau's argument as in many other classical works (see the information, e.g., in [Kh123]). But it appeared there as an experimental fact attached to a concrete situation. Its presence could be discerned only a posteriori. The explanation of the true underlying mechanism is due to Khavinson. It is rooted in quite general and abstract facts.

2.2.1. Abstract scheme. Given a complex linear normed space X and its subspace X_0 we denote by X^* the conjugate space of X and by $X_0^\perp$ the polar set of X_0,

$$X_0^\perp = \{f \in X^* : f|X_0 = 0\}.$$

Fix an $f \in X^*$ and put $f_0 = f|X_0$. Then

$$\|f - g\|_{X^*} \geq \|f_0\|_{X_0^*} \quad \text{for any } g \in X_0^\perp,$$

and, by Hahn-Banach, the equality occurs for a $g_0 \in X_0^\perp$. Thus

$$\min\{\|f - g\|_{X^*} : g \in X_0^\perp\} = \sup\{|f(x)| : x \in X_0, \ \|x\|_X \leq 1\}. \qquad (2.1)$$

Note that we write "min", not just "inf" to the left, the infimum being attained for a $g_0 \in X_0^\perp$, a best approximant of f by the elements of $X_0^\perp$.

The general duality relation (2.1) applies to a vast class of concrete extremal problems for analytic functions. As a zero order approximation to rigorous statements imagine a space Y of functions analytic in a plane domain G. Due to the maximum modulus principle and its numerous analogs Y can be very often identified (isometrically) with a subspace X_0 of a normed space X consisting of functions defined on the boundary Γ of G. A linear functional F on Y can be viewed as an element f_0 of X_0^*, and we may apply the "minimax" (or, to be more precise, the "minisup") relation (2.1) to the problem of maximizing F on the unit ball of Y. At the first glance, (2.1) does not yield any maximizing element in its right side (and even does not guarantee its existence). Equality (2.1) just transforms the maximization problem into a dual best approximation problem (see the left side of (2.1)). But the very *coincidence* of these two problems is a powerful tool of analysis of the extremal elements in both sides of (2.1).

2.2.2. Relation (2.1) and extremal problems in $H^\infty(G)$. To illustrate these vague ideas let Y be the space $H^\infty(G)$ of all functions analytic and bounded in G with the usual uniform norm:

$$\|x\|_{H^\infty(G)} = \|x\|_\infty = \sup_G |x|, \quad x \in H^\infty(G).$$

We assume G is bounded and finitely connected,

$$\Gamma = \Gamma_1 \cup \cdots \cup \Gamma_n,$$

where the connected components Γ_j of the boundary Γ are rectifiable Jordan loops. Any $x \in H^\infty(G)$ has angular boundary values $\hat{x}(\xi)$ at s-almost all $\xi \in \Gamma$ (w.r. to the length s on Γ). Moreover,

$$\|x\|_{H^\infty(G)} = \operatorname*{vrai\,sup}_{\xi \in \Gamma} |\hat{x}(\xi)|.$$

Identifying $x \in Y = H^\infty(G)$ with $\hat{x}$ we make Y a subspace X_0 of $X = L^\infty(\Gamma)$; we denote this subspace by $H^\infty(\Gamma)$.

2.2.3. Digression. Now we have to devote a paragraph to the Smirnov class $E^1(G)$ whose participation in this context is unavoidable. It can be defined as the set of all functions φ analytic in G possessing angular boundary values $\hat{\varphi}(\xi)$ at almost all $\xi \in \Gamma$ and such that $\hat{\varphi} \in L^1(\Gamma)$ $(= L^1(\Gamma, s))$, and φ is representable by its Cauchy integral:

$$\varphi(a) = \frac{1}{2\pi i} \int_\Gamma \frac{\hat{\varphi}(\xi)\, d\xi}{\xi - a}, \quad a \in G.$$

The mapping $\varphi \mapsto \hat{\varphi}$ of $E^1(G)$ into $L^1(\Gamma)$ is one-to-one, and putting

$$\|\varphi\|_{E^1(G)} = \|\hat{\varphi}\|_{L^1(\Gamma)},$$

we turn $E^1(G)$ to a normed space which we identify with the subspace $E^1(\Gamma)$ of $L^1(\Gamma)$, the set of all functions $\hat{\varphi}$ for $\varphi \in E^1(G)$.

We are now almost ready to explain the role played by $E^1(G)$ for our theme, but first let us consider the subspace $C_A(G)$ of $H^\infty(G)$ consisting of all elements of $H^\infty(G)$ continuously extendable to $G \cup \Gamma$; by $C_A(\Gamma)$ we denote the subspace of $L^\infty(\Gamma)$ formed by all functions $\hat{\varphi}$ with $\varphi \in C_A(G)$. We need the following identity:

$$E^1(\Gamma) = \left\{ \omega \in L^1(\Gamma) : \int_\Gamma \omega(\xi)\psi(\xi)\, d\xi = 0 \ \text{ for any } \psi \in H^\infty(\Gamma) \right\}. \tag{2.2}$$

The next assertion is a generalization of an important result due to F. and M. Riesz who discovered it for $G = \mathbb{D}$:

Theorem. *Let μ be a complex Borel measure on Γ. The following are equivalent:*
1) $\int_\Gamma x\, d\mu = 0$ *for any $x \in C_A(\Gamma)$;*

2) μ *is s-absolutely continuous, and $\dfrac{d\mu}{ds} \in E^1(\Gamma)$.*

A concise survey of the Smirnov classes is in [Kh123, p. 57–61].

2.2.4. Let us return to our main theme, that is to a linear functional $F \in (H^\infty(G))^*$ and maximizing of $|F(y)|$ w.r. to $y \in H^\infty(G)$, $\|y\|_\infty \le 1$; the maximum is $\|F\|_{(H^\infty(G))^*}$.

Many concrete problems of Complex Analysis (in particular, of Geometric Function Theory, remember the Riemann mapping theorem) suggest the following questions:

1) Is there an extremal element $y^* \in H^\infty(G)$ such that

$$\|y^*\|_\infty \le 1, \quad |F(y^*)| = \|F\|_{(H^\infty(G))^*}?$$

2) Suppose y^* does exist, is it unique (up to a constant unimodular factor)?

3) Is y^* unimodular on Γ (i.e., $|\hat{y^*}| = 1$ a.e. on Γ)?

The rich experience of concrete problems suggests that question 3) is natural and, generically, the answer should be "yes".

Note that in these questions we are dealing with "qualitative" problems: the quantitative problem of computing $\|F\|_{(H^\infty(G))^*}$ is not primordial. The interest of these questions is determined by the following experimental fact: for many functionals F the maximizer y^* turns out to possess some useful properties (e.g., to map G onto $\mathbb{D}$); this is what makes its existence, uniqueness and description very attractive aims.

Of course, the answers to questions 1)–3) depend on F. Identifying $H^\infty(G)$ with $X_0 = H^\infty(\Gamma) \subset L^\infty(\Gamma) = X$ (see **2.2.2**) we may perceive F as $f_0 = f|H^\infty(\Gamma)$ for an $f \in (L^\infty(\Gamma))^*$ which is, generally speaking, a mysterious object. But for really interesting functionals F there is no loss in assuming that f is generated by *a function* $\omega \in L^1(\Gamma)$:

$$f(x) = \int_\Gamma \omega(\xi)x(\xi)\,d\xi, \quad x \in L^\infty(\Gamma). \tag{2.3}$$

As an illustration we may use $\omega(\xi) = (\xi - a)^{-2}/2\pi i$ for a fixed $a \in G$; the respective F is just

$$y \mapsto y'(a), \quad y \in H^\infty(G).$$

A direct application of relation (2.1) to our present pair $X = L^\infty(\Gamma)$, $X_0 = H^\infty(\Gamma)$ is hindered by a complicated nature of X^*. To avoid the analysis of $X_0^\perp$ we replace the pair $(X, X_0) = (L^\infty(\Gamma), H^\infty(\Gamma))$ by $(C(\Gamma), C_A(\Gamma))$, see **2.2.3**; $C(\Gamma)$ is the usual space of all complex continuous functions on Γ, $C_A(\Gamma)$ is its subspace consisting of boundary traces of functions continuous in $G \cup \Gamma$ and analytic in G. Unlike $(L^\infty(\Gamma))^*$ the space $(C(\Gamma))^*$ can be conveniently and isometrically identified as $M(\Gamma)$, the space of all Borel complex measures on Γ whereas $(C_A(\Gamma))^\perp$ becomes

$$\{\mu \in M(\Gamma) : d\mu = f(\xi)\,d\xi \text{ for an } f \in E^1(\Gamma)\},$$

see Theorem in **2.2.3**.

Now we apply (2.1) to $X = C(\Gamma)$, $X_0 = C_A(\Gamma)$ and to the functional $x \mapsto \int_\Gamma x(\xi)\omega(\xi)\,d\xi$ generated on $C(\Gamma)$ by measure $\nu \in M(\Gamma)$, $d\nu = \omega(\xi)\,d\xi$. We get a function $\omega_0 \in E^1(\Gamma)$ such that

$$\|f_0|C_A\|_{C_A^*} = \int_\Gamma |\omega - \omega_0|\, ds = \min\left\{\int_\Gamma |\omega - \tilde\omega|\, ds : \tilde\omega \in E^1(\Gamma)\right\}. \qquad (2.4)$$

From now on we forget about $C(\Gamma)$, $C_A(\Gamma)$ and return to the functional (2.3) on $L^\infty(\Gamma)$ and its restriction f_0 onto $H^\infty(\Gamma)$. By the Montel theorem there exist a sequence $(x_n)_{n=1}^\infty$ and a function x^* in $H^\infty(G)$ such that

$$\|x_n\|_\infty \leq 1, \quad x_n \xrightarrow[n\to\infty]{} x^* \text{ pointwise in } G, \quad |f(x_n)| \xrightarrow[n\to\infty]{} \|f_0\|_{(H^\infty(\Gamma))^*}.$$

It is not hard to see that $\lim_{n\to\infty} f(x_n) = f(x^*)$ (this is obvious for the functional $y \mapsto y'(a)$, $y \in H^\infty(G)$). Hence

$$|f(x^*)| = |f_0(x^*)| = \|f_0\|_{(H^\infty(\Gamma))^*},$$

and the answer to the first of the three questions in **2.2.4** is positive. Now,

$$|f_0(x^*)| = \|f_0\|_{(H^\infty(\Gamma))^*} \geq \|f_0|C_A(\Gamma)\|_{(C_A(\Gamma))^*} = \int_\Gamma |\omega - \omega_0|\, ds$$

$$\geq \int_\Gamma |x^*|\,|\omega - \omega_0|\, ds \geq \left|\int_\Gamma x^*(\xi)(\omega(\xi) - \omega_0(\xi))\, d\xi\right|$$

$$= \left|\int_\Gamma x^*(\xi)\,\omega(\xi)\, d\xi\right| = |f_0(x^*)|$$

(we have used (2.4), the inclusion $\omega_0 \in E^1(G)$, and (2.2)). Thus the couple (ω_0, x^*) of extremal functions satisfies *the integral* relation

$$\left|\int_\Gamma x^*(\xi)\,(\omega(\xi) - \omega_0(\xi))\, d\xi\right| = \int_\Gamma |x^*(\xi)|\,|\omega(\xi) - \omega_0(\xi)|\, ds. \qquad (2.5)$$

Rewrite $d\xi$ in the left integral as $\xi'_s\, ds$ and recall the equality $|\xi'_s| = 1$ a.e. on Γ. Then from the integral relation (2.5) we obtain *a pointwise* relation connecting ω_0 and x^* : for an $\alpha \in \mathbb{R}$

$$x^*(\omega - \omega_0)\xi'_s = |x^*|\,|\omega - \omega_0|\, e^{i\alpha} \quad \text{a.e. on } \Gamma, \qquad (2.6)$$

a source of precious information on *both* functions ω_0 and x^*.

We exclude the trivial case $\omega \in E^1(\Gamma)$ corresponding to the zero functional f. Then the length of the set

$$a = \{\xi \in \Gamma : \omega(\xi) - \omega_0(\xi) \neq 0\}$$

is positive, and, by (2.6),

$$|x^*| = 1 \quad \text{a.e. on } a.$$

If, say, ω coincides on Γ with a function analytic in $\mathbb{C} \setminus$ (a compact set in G) and vanishing at infinity (as is the case with $\omega(\xi) = (\xi - b)^{-2}/2\pi i$, $b \in G$), then

$s(\Gamma \setminus a) = 0$, and $|x^*| = 1$ a.e. on Γ (cf. question 3)). In any case (2.6) can be given the following form a.e. on Γ :

$$x^*(\omega - \omega_0)\xi'_s = e^{i\alpha}\,|\omega - \omega_0|$$

whence x^* is uniquely defined on a up to a constant factor $e^{i\beta}$, $\beta \in \mathbb{R}$. By the boundary uniqueness theorem for $H^\infty(G)$, it is uniquely defined (with the same accuracy) in the whole of G, whence we have a positive answer to question 2) in **2.2.4**.

Function ω_0, the best $L^1(\Gamma)$-approximant of ω in $E^1(\Gamma)$, is, generally speaking, not unique. It is unique if G is simply connected. This case can be reduced, by conformal mapping, to $G = \mathbb{D}$ where (2.6) becomes

$$x^*(e^{it})(\omega(e^{it}) - \omega_0(e^{it}))ie^{it} = e^{i\alpha}|\omega(e^{it}) - \omega_0(e^{it})| \qquad (2.7)$$

for almost all $t \in \mathbb{R}$. Denote the unit circle by $\mathbb{T}$ and suppose $\omega_1 \in E^1(\mathbb{T}) = H^1(\mathbb{T})$, and

$$\int_0^{2\pi} |\omega(e^{it}) - \omega_0(e^{it})|\, dt = \int_0^{2\pi} |\omega(e^{it}) - \omega_1(e^{it})|\, dt.$$

Then (2.7) holds with ω_1 in place of ω_0, so that $Q(\xi) = e^{-i\alpha}x^*(\xi)(\omega_1(\xi) - \omega_0(\xi))i\xi$ *is real a.e. on* $\mathbb{T}$. But Q as a function of the variable $\xi \in \mathbb{D}$ belongs to the Hardy class $H^1(\mathbb{D})$ and is representable by its Poisson integral. Therefore $Q \equiv \mathrm{const}$; but $Q(0) = 0$ whence $\omega_1 = \omega_0$ in $\mathbb{D}$ and almost everywhere on $\mathbb{T}$.

2.3. Concluding remarks

Here we have to stop our superficial survey of Khavinson's approach to dual extremal problems in Complex Analysis. Due to the volume limitation we cannot anymore afford an exposition as detailed as in **2.2**. We have to emphasize that we have not exploited the crucial pointwise relation (2.6) in its full depth. In fact in Khavinson's work it is applied to a much more thorough investigation of x^* which enabled him to trace the improvement of the boundary behavior of x^* depending on the properties of ω, the number of zeros of x^*, representation of x^* as a product of canonical factors (which becomes especially explicit for $G = \mathbb{D}$ and ω corresponding to some classical functionals F) and, last but not least, mapping properties of x^*. We restrict ourselves to a partial quotation from Theorem 7.1 of [Kh123] shedding new light on classical results by Ahlfors, Grunsky and Garabedian:

if ω coincides on Γ with a function meromorphic in a domain $\tilde{G}$ with exactly m poles in G, then either x^* is constant or maps G onto the k-sheeted unit disc where

$$n \le k \le n + m - 2.$$

Returning to the example of $\omega(\xi) = (\xi - a)^{-2}/2\pi i$, $a \in G$ (see **2.2.4**), we get $k = n$, a result due to Ahlfors.

In publications mentioned at the beginning of **2.2** the reader can find numerous results based on analogs of the scheme sketched in **2.2** and adjusted to different spaces of analytic functions (in particular to Smirnov classes $E^p(G)$, $1 \leq p \leq +\infty$, also with some weights on Γ). These results form the ramified and efficient *Khavinson's theory of extremal problems*. Our next section is devoted to one of its ramifications.

3. Extremal problems with supplementary restrictions

3.1. Introductory remarks

The norms of all spaces mentioned in Section 2 depend on the restrictions of functions on Γ, the boundary of G. They were mainly weighted L^p-norms on Γ, $1 \leq p \leq +\infty$; we call them *boundary norms*. The variant of the theory to be described in the present section is aimed at norms measuring the restriction of a function on a set including Γ *and a part of* G. This new setting entails considerable complications. It is, however, not a mere formal generalization. To the contrary, it is motivated by a class of very popular problems of Complex Analysis.

As a famous saying goes, an analytic function behaves as a live organism reacting as a whole to a slightest local pressure. If, say, $f \in H^\infty(G)$ is oppressed on a part $D \subset G$, so that

$$\sup_D |f| < \varepsilon \tag{3.1}$$

with a small $\varepsilon > 0$, then the effect of this oppression is felt throughout G, at *any* of its points. This phenomenon is observed not only for the uniform norm $f \mapsto \sup_D |f|$, but for many other norms as well. Its consequences are numerous uniqueness theorems, one of the main themes of Complex Analysis. This explains the permanent interest of complex analysts in global estimates of analytic functions satisfying quantitative restrictions analogous to the *local inequality* (3.1).

If we are working in a space of analytic functions with a boundary norm, then (3.1) is "a supplementary restriction" strengthening the general restriction $\|f\|_\Gamma < \infty$ whence the title of this section.

3.2. Examples. A duality relation

Let G be a domain as in **2.2.2** and D its compact subset carrying a non-negative Borel measure μ. As in **2.2.4** we consider the functional (2.3) generated in $H^\infty(G) \cong H^\infty(\Gamma)$ by an $\omega \in L^1(\Gamma)$. But unlike **2.2.4** where we were interested in $\sup\{|f(x)| : x \in H^\infty(\Gamma), \|x\|_\infty \leq 1\}$, we concentrate now on

$$\sup\{|f(x)| : x \in H^\infty(\Gamma), \ \|x\|_\infty \leq 1, \ \|x\|_{L^p(D,\mu)} \leq \varepsilon\}, \tag{3.2}$$

involving a supplementary condition, that is the smallness of x expressed in terms of the $L^p(\mu)$-norm on D and a small number ε. This problem is interesting even in the particular case of *a finite D* and $p = \infty$. On the other hand, the norms in (3.2) could be made much more general replacing $H^\infty(\Gamma)$ by $E^p(\Gamma)$ (= boundary traces of functions of a Smirnov class in G endowed with a weighted $L^p(\Gamma)$-norm) whereas the supplementary norm might include a number of derivatives of x (see [Kh55]). The supremum in (3.1) can be given a form quite similar to the right side of (2.1), since (3.2) is just the norm of $f_0 = f|H^\infty(\Gamma)$, with one important distinction: this time $H^\infty(\Gamma)$ is endowed with a stronger norm $\|\ \|_{\infty,p,\mu,\varepsilon}$,

$$x \mapsto \max(\|\hat{x}\|_{L^\infty(\Gamma)}, \ \|x\|_{L^p(D,\mu)}/\varepsilon). \tag{3.3}$$

(Note that this norm is comparable with $\|\ \|_{\infty,\Gamma}$ due to the compactness of D.) This new norm takes into account not only $\|\hat{x}\|_{\infty,\Gamma}$, but also the magnitude of $|x|$ on D. It is not a boundary norm (see the beginning of **3.1**). This fact complicates the answers to the main questions which remain essentially the same as in **2.2.4**. The general approach is nevertheless also the same as in **2.2**. It is still based on duality of extremal problems: the search of the supremum in (3.2) is again equivalent to a (duly modified) best approximation problem. We avoid here the most general statements of this kind, see [Kh55, Kh123], and illustrate the new situation by an analog of (2.4) related to (3.1), i.e., to $p = \infty$ in (3.2); we also omit generalizations involving norms of the derivatives of x on D. Avoiding abstract preliminaries underlying the general scheme of [Kh55] we start with the space $X_0 = C_A(\Gamma) \times C(D)$ which is a subspace of $X = C_\Gamma \times C(D)$ normed by

$$(x,y) \mapsto \max(\|x\|_{\infty,\Gamma}, \ (\|y\|_{\infty,D})/\varepsilon).$$

As in (2.4), for an arbitrary $F \in X^*$ we get

$$\|F|X_0\|_{X_0^*} = \min\{\|F - \Phi\|_{X^*} : \Phi \in X_0^\perp\}. \tag{3.4}$$

Now, $F \cong (\mu,\nu)$ where $\mu \in M(\Gamma)$, $\nu \in M(D)$, the spaces of complex Borel measures, resp., on Γ and D, and

$$\|F\|_{X^*} = \|\mu\|_{M(\Gamma)} + \varepsilon\|\nu\|_{M(D)}, \quad F(x,y) = \int_\Gamma x\, d\mu + \int_D y\, d\nu.$$

A pair $(\rho,\sigma) \in X^*$ belongs to $X_0^\perp$ iff

$$\int_\Gamma x\, d\rho + \int_D x\, d\sigma = 0 \quad \text{for any } x \in C_A(G). \tag{3.5}$$

Any compactly supported $\lambda \in M(\mathbb{C})$ generates its Cauchy potential C^λ,

$$C^\lambda(a) = \frac{1}{2\pi i} \int_{\operatorname{supp}\lambda} \frac{d\,\lambda(\xi)}{\xi - a}, \quad a \in \mathbb{C}\backslash\operatorname{supp}\lambda$$

and, by the Cauchy formula, in (3.5) we have $x(a) = C^{\hat{x}\,d\xi}(a)$, $a \in G$, whence (3.5) becomes

$$\int_\Gamma x(\xi)\, d\rho(\xi) - \int_\Gamma C^\sigma(\xi)x(\xi)\, d\xi = 0 \quad \text{for any } x \in C_A(G),$$

or (see **2.2.3**) $d\rho(\xi) - C^\sigma(\xi)\, d\xi = \varphi(\xi)\, d\xi$ on Γ for a $\varphi \in E^1(\Gamma)$.

We are only interested in functionals F with $d\mu(\xi) = \omega(\xi)\,d\xi,\ \ \nu = 0$, where $\omega \in L^1(\Gamma)$, and (3.4) takes the following form:

$$\|F|X_0\|_{X_0^*} = \int_\Gamma |\omega - \varphi^* - C^{\sigma^*}|\,ds + \varepsilon \int_D |d\sigma^*|$$

for a pair $(\varphi^*, \sigma^*) \in E^1(\Gamma) \times M(D)$. Following the pattern of **2.2.4** we get an $x^* \in H^\infty(G),\ \ \|x^*\|_\infty \le 1$, such that

$$\sup\left\{\left|\int_\Gamma \hat{x}(\xi)\omega(\xi)\,d\xi\right| : x \in H^\infty(G),\ \|x\|_\infty \le 1\right\}$$

$$= \left|\int_\Gamma x^*(\xi)\omega(\xi)\,d\xi\right| = \int_\Gamma \left|x^*(\xi)[\omega(\xi) - \varphi^*(\xi) - C^{\sigma^*}(\xi)]\right|\,ds + \varepsilon \int_D |d\sigma^*|. \quad (3.6)$$

Removing the integrals (as in **2.2.4**) we end up with *a pointwise* relation connecting x^*, φ^* and σ^*. Then we may turn to its analysis with the same fruitful results: dependence of local boundary properties of x^* on the quality of ω, representation of x^* as a product of standard factors, location and number of zeros of x^*, and geometric properties of x^*: in many important cases x^* still maps G onto a k-sheeted disk $\mathbb{D}$, and k can be estimated from above and from below. An interesting feature of the problem is its stability: x^* ignores the presence of the "supplementary term" $\sup_D |x|/\varepsilon$ in the norm (3.3) and behaves as if it were absent (i.e., $\varepsilon = +\infty$).

We have to emphasize that our discussion applies only to the case when $D \subset G$ is compact. The situation gets more complicated if D is not separated from Γ. Some cases when $D \cap \Gamma \ne \emptyset$ are treated in Theorem 6.3 of [Kh55].

3.3. A Tchebyshev-like phenomenon

A curious effect was observed in [Kh55]. Let us denote the norm (3.3) by $\|\ \|_{\infty,D,\varepsilon}$ stressing its dependence on D. For $G = \mathbb{D}$ and many compact sets $D \subset \mathbb{D}$ the maximization problem of **3.1**, i.e.,

$$\max\{|F(x)| : x \in H^\infty(\mathbb{D}),\ \|x\|_{\infty,D,\varepsilon} \le 1\},$$

is shown to be equivalent to the same maximization, but with D replaced by *a finite* subset D_N of its boundary: the maximizing functions x_D^* and $x_{D_N}^*$ coincide, $\#D_N = N$. For some classical functionals F and sets D the information on D_N and estimates of N are quite precise and generalize earlier results by M. Heins.

The existence of D_N reminds of the Tchebyshev theorem reducing the best polynomial approximation on an interval I by a polynomial to the same problem on a finite set $I_N \subset I$.

3.4. Some quantitative results

The general approach to extremal problems with supplementary restrictions is illustrated in [Kh55] by some classical situations (the Walsh problem concerning the three circles problem by Hadamard, the Milloux and Heins problems, the

Whittacker inequality etc.). The results are mainly qualitative (i.e., in the spirit of questions posed in **2.2.4**). But some are quantitative yielding very useful explicit inequalities. Consider, e.g., a particular case of the problem discussed in **3.2**: given $z_0 \in \mathbb{D}$ put $f(x) = x(z_0)$, $x \in H^\infty(\mathbb{D})$, $D = \{\alpha_1, \ldots, \alpha_n\}$, a finite set in $\mathbb{D}$. In §11 of [Kh55] the duality relation (3.6) is a source of an explicit estimate of $|x(z_0)|$ for $x \in H^\infty(\mathbb{D})$ satisfying $\|x\|_\infty \leq 1$, $|x(\alpha_j)| \leq \varepsilon_j$, $j = 1, \ldots, N$. This estimate, in its turn, implies an interesting quantitative refinement of the classical uniqueness theorem for $H^\infty(\mathbb{D})$:

if a sequence $(\alpha_j)_{j=1}^\infty$ in $\mathbb{D}$ satisfies

$$\sum_{j=1}^\infty (1 - |\alpha_j|) = \infty, \tag{3.7}$$

then $\{x \in H^\infty(\mathbb{D}) \ \& \ x(\alpha_j) = 0, \ j = 1, 2, \ldots\} \implies x \equiv 0$. In §12 of [Kh55] it is shown that this uniqueness result *is stable:* $x \in H^\infty(\mathbb{D})$ vanishes identically if it decays fast enough along the sequence $(\alpha_j)_{j=1}^\infty$, provided (3.7) is fulfilled, and the sequence is rarefied (in a sense). This theorem was one of the first results of this kind. It was developed in various directions by Ushakova, Lyubarski–Seip and Eiderman.

We conclude this section with a remark on the role of the right side of the duality relation (3.6). It seems to be a purely ancillary tool, since our main concern was the maximizer x^* figuring in the *left* side of (3.6). But the right side is important as well. It contains an interesting extra term $\varepsilon \int |d\sigma^*|$ responsible for the magnitude of the charge σ^* generating the Cauchy potential C^{σ^*} approximating (in cooperation with $\varphi^* \in E^1(\Gamma)$) the given $L^1(\Gamma)$-function ω. Thus we arrive in a compulsory way to an approximation problem taking into account not only the accuracy of approximation (expressed by the first integral in the right side of (3.6)), but also "the price of approximation" expressed by $\int_D |d\sigma^*|$. Following S. Khavinson we return to this theme in Section 6 of this survey.

4. Spaces of analytic functions in multiply connected domains

4.1. Classes E^p

As in Section 2, G will denote a bounded domain in $\mathbb{C}$ with the boundary $\Gamma = \Gamma_1 \cup \cdots \cup \Gamma_m$, Γ_j being disjoint Jordan rectifiable loops. We denote by G_j the unbounded component of $\mathbb{C} \setminus \Gamma_j$, $j = 2, \ldots, m$; G_1 is the bounded component of $\mathbb{C} \setminus \Gamma_1$, so that

$$G = G_1 \cap G_2 \cap \cdots \cap G_m.$$

For a simply connected G (i.e., for $m = 1$) the Smirnov classes $E^p(G)$ appeared in Smirnov's works (see [6, 7] and references therein). A set of (equivalent) definitions of the class $E^p(G)$ (for any m) is collected on pp. 132–133 of [5]. The shortest is due to Khavinson & Tumarkin: $f \in E^p(G)$, $p > 0$, if

$$\sup_j \int_{bG_j} |f(\xi)|^p \, ds < +\infty \tag{4.1}$$

for an increasing sequence of subdomains $G_j \Subset G$ with rectifiable boundaries bG_j (like G itself) covering G. The rest of definitions (due to Smirnov and Keldysh & Lavrentiev for a simply connected G) are not so simple. In fact there exists a universal sequence (G_j) such that (4.1) holds for any $f \in E^p(G)$. Thus the definition looks very much like a formal reproduction of the definition of the Hardy class $H^p(\mathbb{D}) = E^p(\mathbb{D})$. As it seems, the main incentive for the initial development of the theory of spaces E^p was the desire to test the new possibilities opened in the 20ies and 30ies by the advent of the Lebesgue theory of integration created at the beginning of the 20th century. But the classes E^p turned out to be a necessary tool in many areas of Complex Analysis; these classes *had to be discovered* to satisfy essential needs of this discipline.

One of the impressive examples is Khavinson's theory of extremal problems described in Sections 2 and 3. The definitions of the spaces $C_A(G)$ and $H^\infty(G)$ are simple and self-imposing and do not contain any integration. They preexisted Lebesgue integration and Functional Analysis forming a natural context for many classical problems. But the investigation even of the most popular maximization problems in $C_A(G)$ and $H^\infty(G)$ is *equivalent* to a best approximation problem just in $E^1(G) \cong E^1(\Gamma)$. The spaces $E^1(\Gamma)$ emerge in a compulsory way in the general schemes developed by Khavinson.

But the theory of the Smirnov classes was not ripe enough when he started: its state was satisfactory only for simply connected domains G. The theory has been completed in a series of joint works by S. Khavinson and G. Tumarkin. A brief description of this series is the aim of the present section.

In [Kh18–Kh20] the definitions by Smirnov and Keldysh & Lavrentiev were complemented by some equivalent definitions (for $m = 1$) and generalized to $m > 1$. These publications also contain some results for non rectifiable Γ's.

Special attention is paid to the decompositions

$$f = f_1 + \cdots + f_m$$

where f_j are analytic in G_j (and vanish at infinity for $j \geq 2$) and to the equivalence

$$f \in E^p(G) \iff f_j \in E^p(G_j), \quad j = 1, 2, \ldots, m$$

which becomes especially delicate for a non rectifiable Γ.

The representability of $f \in E^p(G)$ by the Cauchy and Green formulas (with integrals taken over Γ) is considered in [Kh19]. Another important theme is the characterization of "polar sets" $(E^p(\Gamma))^\perp$ in $L^q(\Gamma)$, i.e., the formula

$$E^p(\Gamma) = \left\{ \omega \in L^q(\Gamma) : \int_\Gamma \omega(\xi) x(\xi) \, d\xi = 0 \ \text{ for any } x \in E^p(\Gamma) \right\}, \quad 1/p + 1/q = 1$$

(not excluding $p = 1$ or ∞, see [Kh33]).

4.2. Class $D(G)$. Smirnov domains

This important class is often denoted by $N^+(G)$ (to emphasize its close relation to the Nevanlinna class $N(G)$). It was discovered by Smirnov (see [7, 5]). This class plays an outstanding role: it appears in some *necessary and sufficient* conditions of validity of the maximum modulus principle, in polynomial approximation, and in the theory of conformal mapping. Khavinson and Tumarkin studied this class in general domains and found a complete description of removable singularities for this class. Removable sets turned out to be just the sets of zero logarithmic capacity [Kh17]. This fact improves earlier results by Parreau and Rudin.

V.I. Smirnov introduced a class S of Jordan domains G with a rectifiable boundary ([7]) which is now called "the Smirnov class of domains": $G \in S$ iff $\log |\omega'|$ is representable in $\mathbb{D}$ by the Poisson integral, where ω is a conformal mapping of $\mathbb{D}$ onto G. Smirnov showed that some familiar properties of functions analytic in "nice" domains hold *exactly* in domains of the class S. At the beginning it was not clear whether S coincides with the class of all Jordan domains with a rectifiable boundary. A negative answer was obtain by Keldysh & Lavrentiev (see [6] and [5] for further development).

In [Kh22] Khavinson and Tumarkin generalized Smirnov's theory to multiply connected domains and studied classes $E^p(G)$ and $D(G)$ in domains $G \in S$.

The culmination of the series of joint works by Khavinson and Tumarkin is [Kh26] where their results on classes of analytic functions in multiply connected domains are applied to a systematic study of dual extremal problems in the spirit of our Section 2 for very general weighted boundary norms.

4.3. Factorization problems

A prominent role in Complex Analysis and many of its applications to Operator Theory and Harmonic Analysis is played by the so-called inner-outer factorization discovered by Nevanlinna, Szegö, and Smirnov (see historical information in [5]). Any function f of the Hardy class $H^p(\mathbb{D})$ is representable as the product of two canonical factors O and I,

$$f = IO.$$

The first factor I is called "inner" and the second is "outer" (terminology of Beurling, nowadays adopted by all complex analysts). The outer factor is defined by the formula

$$O(a) = \exp \frac{1}{2\pi} \int\limits_{|\xi|=1} \log |f(\xi)| \frac{\xi + a}{\xi - a} \cdot \frac{d\xi}{i\xi}, \quad |a| < 1,$$

so that O does not vanish in $\mathbb{D}$, and $|\hat{O}| = |\hat{f}|$ a.e. on the unit circle $\{|a| = 1\}$. The inner factor is bounded in $\mathbb{D}$, and $|\hat{I}| = 1$ a.e. on $\{|a| = 1\}$. In its turn, I can be split into two more inner factors, a Blaschke product B responsible for the zeros of f in $\mathbb{D}$, and the so-called "singular inner factor" S which does not vanish in $\mathbb{D}$. The representation

$$f = BSO \tag{4.2}$$

is in fact applicable to any $f \in D(\mathbb{D})$, the Smirnov class, and may serve as the definition of that class. It is often called the canonical parameterization of $D(\mathbb{D})$, since it involves certain (infinite dimensional) free parameters, a supple and convenient tool to characterize important subclasses of $D(\mathbb{D})$.

One of Khavinson's favorite themes was the search of generalizations of (4.2) to multiply connected domains. This case included serious difficulties absent in the simply connected domains G where one can easily transplant (4.2) from the disc by a conformal mapping, or proceed directly in G just copying the construction of the factors for the disc. Let us briefly discuss two problems arising for the multiply connected G.

1. We may try to copy the construction of outer factor in (4.2) starting with the solution of the Dirichlet problem for G with the boundary data $\log|f|\,\|\Gamma$. This can be done replacing the Poisson kernel for the disc by the normal derivative of the Green function for G, and we get a real harmonic function u in G satisfying the condition

$$u|\Gamma = \log|f|\,\|\Gamma \quad \text{a.e. on } \Gamma$$

(we mean angular boundary values and drop $\hat{}$). The next step would be to define O as $\exp(u + iv)$ where v is the harmonic conjugate of u. Unfortunately, this idea does not work, since, generally speaking, v *is multivalued* unlike the case of the disc or a simply connected G.

2. Another obstacle is the lack of a natural "Blaschke factor"

$$\left(a \mapsto \frac{a - b}{1 - a\bar{b}}, \ |b| < 1, \ \text{for } G = \mathbb{D} \right) :$$

any function analytic in $G \cup \Gamma$ and unimodular on Γ has at least n zeros in G.

Sections 4.1–4.2 dealt with the aspects of the theory of function spaces in multiply connected domains which were *subordinated* to the needs of dual extremal problems. In [Kh23] Khavinson and Tumarkin reverse this order and apply the duality (as in Section 2) to construct (or rather to prove the existence) of a factor O_1 which makes the product $\exp(u + iv)O_1$ (see 1. above) one valued whereas its modulus is still $|f|$ a.e. on Γ.

As to the second difficulty and canonical factorization in general, we recommend the article [Kh128] where a factorization theory is exposed. It is applicable to classes of analytic functions on compact Riemann surfaces with border. This article sums up a series of results by Khavinson and other analysts (including Dmitry, Khavinson junior).

Another important source of information on factorization is the treatise [Kh110]. It is, unfortunately, not easily available, being a Xerox edition of a book which could by no means be published normally due to the peculiar publishing policy in the USSR of the sixties, seventies, and early eighties. The same fate, alas, struck the texts [Kh110–Kh112], [Kh115–Kh116]. Luckily, two of them ([Kh111–Kh112]) are translated into English and published by the AMS [Kh123].

5. Analytic capacity

5.1. Definition

Let K be a compact set in $\mathbb{C}$, G be the unbounded component of its complement,

$$H_0^\infty(G) := \{x \in H^\infty(G) : x(\infty) = 0\}$$
$$= \{x \in H^\infty(G) : x(\xi) = c_1(x)/\xi + O(1/\xi^2), \ \xi \to \infty\}.$$

The analytic capacity of K is, by definition,

$$\gamma(K) := \sup\{|c_1(x)| : x \in H_0^\infty(G), \ \|x\|_\infty \le 1\}$$
$$= \sup\left\{\left|\frac{1}{2\pi}\int_{bG} x(\xi)\,d\xi\right| : x \in H_0^\infty(G), \ \|x\|_\infty \le 1\right\}.$$

In other words, $\gamma(K)$ is the norm of the functional $x \mapsto c_1(x)$ on the space $H_0^\infty(G)$. *The term* "analytic capacity " was coined in 1958 by V.D. Erokhin, but *the notion* appeared earlier in [1], and from that moment on and until now it remains one of the most frequent and popular objects of Complex Analysis. It plays a capital role in the theory of removable singularities of analytic functions, and in the theory of conformal mapping. A powerful impetus to the interest of analytic capacity was a breakthrough in rational approximation by Mergelyan and Vitushkin. In particular, Vitushkin showed that analytic capacity plays a crucial role in criteria of approximation by rational functions. Last years are also marked by great events in the study of analytic capacity (see a concise survey in [Kh161] where one can find further references). The efforts were mainly concentrated on quantitative aspects (relations of $\gamma(K)$ with more palpable ("metric") characteristics of K, say, its length, semiadditivity of γ, etc). Khavinson's interests, not ignoring the quantitative problems (e.g., relations between γ and similar functionals adjusted to Smirnov's classes E^p, see below), were, nevertheless, directed at qualitative aspects (existence, uniqueness and special properties of extremal functions appearing implicitly in the very definition of γ and its dual equivalents). Khavinson's results in this area form a considerable contribution to the subject. His main publications on the theme of the present section are [Kh47, Kh123, Kh116, Kh152, Kh161].

5.2. The Schwarz lemma in arbitrary domains

Present in any textbook on Complex Analysis, the Schwarz lemma asserts that $\max\{|x'(0)| : x \in H^\infty(\mathbb{D}), x(0) = 0, \|x\|_\infty \le 1\}$ is attained only by $x(\xi) \equiv c\xi$, for a $c \in \mathbb{C}$, $|c| = 1$. A natural generalization of the problem is to maximize $|x'(\xi_0)|$ for $x \in H^\infty(G)$, $x(\xi_0) = 0$, $\|x\|_\infty \le 1$, where G is a domain and ξ_0 its fixed point. For a multiply connected G with a rectifiable boundary the maximizing function $x^*_{\xi_0,G}$ was studied by Ahlfors [1] and Garabedian [2]. Returning to the situation of **5.1** (i.e., of *an arbitrary* compact set $K \subset \mathbb{C}$ and the unbounded component G of $\mathbb{C} \setminus K$) we may connect $\gamma(K)$ with the Schwarz lemma for G and $\xi_0 = \infty$, the substitute for $x'(\infty)$ being $c_1(x)$. The extremal function $x^*_{\infty,G}$ is called *the Ahlfors* *function for G*, if it is normalized by

$$x^*_{\infty,G}(\xi) = \frac{\gamma(K)}{\xi} + \frac{c_2}{\xi^2} + \cdots \quad \text{for } |\xi| \gg 1.$$

Its existence is obvious. In the case of $K = K_1 \cup K_2 \cup \ldots \cup K_m$ where K_j are disjoint closed bounded Jordan domains with rectifiable boundaries Γ_j the uniqueness of $x^*_{\infty,G}$ was proved in [1] and [2]. It was also shown that $x^*_{\infty,G}$ maps G onto the m-sheeted unit disc.

Khavinson was the first to generalize and develop these results for *an arbitrary* compact set K. They can be deduced for a multiply connected G from the general scheme of Section 2 involving the duality relations

$$\gamma(K) = \frac{1}{2\pi i} \int_\Gamma x^*_{\infty,G}(\xi)\, d\xi$$

$$= \max\left\{ \left| \frac{1}{2\pi} \int_\Gamma x(\xi)\, d\xi \right| : x \in H_0^\infty(G),\ \|x\|_\infty \le 1 \right\} \tag{5.1}$$

$$= \min\left\{ \frac{1}{2\pi} \int_\Gamma |1 + \varphi(\xi)|\, ds : \varphi \in E^1(G),\ \varphi(\xi) = O(1/\xi) \text{ as } \xi \to \infty \right\}$$

and the pointwise relation connecting $x^*_{\infty,G}$ with the minimizer φ^*_G in the last term of (5.1):

$$x^*_{\infty,G}(\xi)(1 + \varphi^*(\xi))\, d\xi = |x^*_{\infty,G}(\xi)|\, |1 + \varphi^*_G(\xi)|\, ds \tag{5.2}$$

a.e. on $\Gamma = \Gamma_1 \cup \cdots \cup \Gamma_m$. An important role is played by the so-called Garabedian function L_G of G,

$$L_G := 1 + \varphi^*.$$

To adapt this approach to a general compact set K we need, first of all, an appropriate definition of the Smirnov class $E^1(G)$ (classes $E^p(G)$ with $p > 1$ will be also needed). We won't reproduce the definition given in [Kh47] and only mention that it includes a sequence $(G_n)_{n=1}^\infty$ of finitely connected domains with rectifiable boundaries exhausting G from within. The duality relations (5.1), (5.2) are preserved in an approximate form (no integration over the boundary of G is possible); a Garabedian function L_G, a limit of L_{G_n}, can still be associated with G. It does not vanish and, moreover, $L_G = \exp \psi_G$, where ψ_G is analytic in G. The approximate form of (5.1) and (5.2) is sufficient to prove the uniqueness of $x^*_{\infty,G}$ and to derive many of its properties. E.g., $x^*_{\infty,G}(G)$ covers $\mathbb{D}$ with the exception of a set of zero analytic capacity [Kh116, p. 73]; the zeros of $x^*_{\infty,G}$, except infinity, all lie in the convex hull of K.

5.3. Khavinson's measure. Bounded analytic functions and Cauchy potentials

Applying the duality relation (5.1) to G_n and passing to the limit as $n \to \infty$, Khavinson proves in [Kh47] the existence of a *unique non-negative* measure

$\mu^* = \mu_G^*$ concentrated on K and such that

$$x_{\infty,G}^*(a)L_G(a) = \int_K \frac{d\mu^*(\xi)}{\xi - a} \quad (= C^{\mu^*}(a)), \quad a \in G. \tag{5.3}$$

I call μ_G^* *the Khavinson measure* of G. Note that $\mu_G^*(K) = \gamma(K)$.

Khavinson's measure is important for representing bounded analytic functions as Cauchy potentials. Let us recall that for many sets K *any* $x \in H_0^\infty(G)$ is the Cauchy potential C^μ of a complex Borel measure μ_x on K:

$$x(a) = \int_K \frac{d\mu_x(\xi)}{\xi - a} = C^{\mu_x}(a), \quad a \in G. \tag{5.4}$$

This is true, for instance, whenever the Painlevé length of Γ, the boundary of G, is finite. But in general *not every* $x \in H_0^\infty(G)$ is a Cauchy potential (5.4). But a theorem by Khavinson asserts that for any $x \in H_0^\infty(G)$ the product xL_G is *always* (i.e., for *any* K) a Cauchy potential:

$$xL_G = C^{\tilde{\mu}_x} \quad \text{in } G$$

for a complex Borel measure $\tilde{\mu}_x$ on K. Moreover, $\tilde{\mu}_x$ is absolutely continuous w.r. to μ_G^*, and

$$\left| \frac{d\tilde{\mu}_x}{d\mu^*} \right| \leq \|x\|_\infty, \quad x \in H_0^\infty(G).$$

This estimate can be given a local form. Suppose m is a non-negative number, $E \subset \Gamma$ (= the boundary of G), $x \in H_0^\infty(G)$, and $\overline{\lim}_b |x| \leq m$ for any $b \in E$. Then

$$\int_E |d\tilde{\mu}_x| \leq m\mu_G^*(E)$$

[Kh116, pp. 63–66].

Generally speaking, Khavinson's measure μ_G^* is, in a sense, parallel to the harmonic measure ω_G of G on Γ (for the point at infinity), these two measures being proportional when K is connected (a continuum); relations of μ_G^* with the analytic capacity are similar to relations of ω_G with the logarithmic capacity.

5.4. E^p-capacities and analytic capacity

Put $E_0^p(G) := \{x \in E^p(G) : x(\infty) = 0\}$. Each class $E_0^p(G)$ generates a new kind of analytic capacity of K, i.e., a number $\gamma_p(K)$ defined as the norm of the same functional $x \mapsto c_1(x)$ w.r. to the E^p-norm in $E_0^p(G)$. The corresponding maximizer $x_{\infty,G,p}^*$ satisfies a duality relation analogous to (2.4) and involving a minimizer $\varphi_{G,p}^* \in E^q(G)$, $q = p/(p-1)$. It is important that any such relation *characterizes* the pair $(x_{\infty,G,p}^*, \varphi_{G,p}^*)$; the same is true for $(x_{\infty,G}^*, \varphi_G^*)$. Comparing these duality relations (or rather their pointwise versions) for different values of p, S. Khavinson arrived at equalities connecting $\gamma(K)$ (see **5.1**) with $\gamma_p(K)$ [Kh47, p. 9]. It may happen that $H^\infty(G)$ (or $E^p(G)$) is trivial, that is, consists of the zero function, in which case K is said to be removable for $H^\infty(G)$ (resp. $E^p(G)$).

An important theorem due to Khavinson says (roughly speaking) that the classes $H^\infty(G)$ and $E^p(G)$, $p > 1$, are trivial or nontrivial simultaneously (for $p = 2$ this statement is literally true, but for $1 < p \neq 2$ there are some subtle points which we won't discuss here). In order to be more precise we should write $E^p((G_n))$, not $E^p(G)$ and not forget the sequence (G_n) defining the class; the capacities $\gamma_p(K)$, however, depend only on G.

The problem of simultaneous triviality of $E^p(G)$ and $H^\infty(G)$ was of special interest in the sixties due to the (then open) Denjoy problem posed in 1909: is it possible for a $K \subset \Gamma$ (= a rectifiable simple arc) to be removable for $H^\infty(\mathbb{C} \setminus K)$ if length of K is positive? It was known long ago that $\gamma_2(K) > 0$ would imply $\gamma(K) > 0$ (Garabedian), whereas Khavinson's results showed that $\gamma_p(K) > 0$ for a $p > 1$ would suffice as well. On the other hand $\gamma_p(K) > 0$ would easily follow from the (then also unknown) L^p-continuity of the singular integral operator with the Cauchy kernel on Γ (see, e.g., [Kh46]). And, in fact, the negative answer to Denjoy's question turned out to be an almost direct corollary of Calderón's famous theorem on the L^p-continuity of the Cauchy singular integral operator on smooth arcs (1977), see the discussions in [Kh116, §5] and [4].

5.5. Some set functions related to analytic capacity

In the theory of analytic capacity γ and neighboring problems some other set functions naturally arise. It is, first of all, the Cauchy capacity γ^C. Denote by $M(K)$ the set of all Borel complex measures supported on K, and put

$$\gamma^C(K) = \sup\{|\mu(K)| : \mu \in M(K), \ \|C^\mu\|_{\infty,G} \leq 1\}. \tag{5.5}$$

The "real" and "positive" Cauchy capacities γ^R, γ^+ can be defined exactly as in (5.5) just replacing $M(K)$ by $M^R(K)$ and $M^+(K)$ (resp., real and non-negative measures on K). According to a remarkable result by Tolsa, the three set functions γ^C, γ^R, γ^+ are comparable with γ (but the coincidence of γ with γ^C is still an open problem; see a brief survey with references in [Kh161] where a systematic treatment of various relatives of γ can be found). Dual definitions of the three Cauchy capacities (and some others) are studied there summing up and developing further some earlier results of Khavinson. They are connected with general approximation problems "with size constraints" considered below in Section 6. Let us quote here just one typical, but also the simplest, result discussed in [Kh161]:

$$\gamma(K) = \inf\left\{ \varliminf_{n} \int |L_G| \, |d\nu_n| \right\}, \tag{5.6}$$

infimum being taken over all sequences (ν_n) of finite linear combinations of unit point masses in G such that

$$\lim_{n \to \infty} \max_{K} |C^{\nu_n} - 1| = 0.$$

If the Painlevé length of K is finite, then L_G (the Garabedian function) in (5.6) can be changed by one. Similar formulas hold for γ^R, γ^+ as well.

The study of various kinds of capacities and corresponding extremal problems in [Kh161] is based on general duality relations like in Section 2, but this time they have to be adjusted to approximation by the elements of a conical wedge, not just a linear subspace. The necessary abstract result is due to Garkavi; see also [Kh158].

The results we are discussing yield new removability criteria for a compact set K w.r. to the space $H^\infty(G)$. E.g., this space turns out to be trivial iff for any $\varepsilon > 0$ there exists a complex Borel measure ν with a compact support in G such that $\operatorname{var} \nu < \varepsilon$, and

$$C^\nu \equiv 1 \text{ near } K$$

(a big Cauchy potential can be induced on K by a negligible charge ν located off K). This fact is just a simple representative of a series of deeper results of the same kind in [Kh161].

An interesting new turn to this theme was given in [Kh152]. This article contains, by the way, an excellent survey of the state of affairs with the analytic capacity by 1999. Its new ingredients were the so-called Golubev sums

$$\sum_{j=1}^{N} (C^{\mu_j})^{(j)}$$

where μ_j denote measures with compact support (on these sums, including $N = \infty$, see [3]). In [Kh152] some analogs of the Cauchy capacity related to the Golubev sums are studied (among many other things).

6. Approximation

Approximation problems are present (be it implicitly) in any of the preceding sections where they mostly appeared *as a tool* (e.g., in dual counterparts of maximization problems). But approximation also appears explicitly, as the main theme, in many of Khavinson's works. His contribution to approximation theory is, first of all, marked by a special setting aimed not only *at the deviation* of the approximant from the function to be approximated, but also at *the size* of the approximant measured in terms of a separate norm (say, directly depending on the moduli of the coefficients of linear combinations which are the approximants). This "size" can be interpreted as the price to be paid for the accuracy of approximation.

This setting is very clearly described and illustrated by concrete examples in [Kh108] and [Kh121] where references to other publications of S. Khavinson on this subject can be found. Let us briefly discuss this setting in a very general form.

Consider a continuous seminorm p in a linear topological space E with the conjugate space E^*, and a seminorm p_1 on $\mathbb{R}^n$ (if E is real; if it is complex, then p_1 is defined on $\mathbb{C}^n$). For a linearly independent family $(x_1, x_2, \ldots, x_n)$ in E and a vector $y \in E$ put

$$\alpha = \inf_{\lambda_1, \ldots, \lambda_n} \left[p\left(g - \sum_{k=1}^{n} \lambda_k x_k \right) + p_1(\lambda_1, \ldots, \lambda_n) \right].$$

The last term is "the price of approximation" of y by the elements of the linear span of $x_1, x_2, \ldots, x_n$. It turns out that

$$\alpha = \beta$$

where

$$\beta = \sup\Big\{ |f(y)| : f \in E^*, \ |f(x)| \le p(x) \ \forall x \in E$$

$$\& \ \Big| \sum_{k=1}^{n} \lambda_k f(x_k) \Big| \le p_1(\lambda_1, \ldots, \lambda_n) \text{ for arbitrary scalars } \lambda_1, \ldots, \lambda_n \Big\}.$$

This abstract fact (and some neighboring facts, as, e.g., the existence of extremizers for α and β and their interrelations) can be given a lot of concrete embodiments making it a fact of classical (say, polynomial) approximation, or a moment problem (in the spirit of the Tchebyshev-Markov problem) or Complex Analysis. It is shown (among other things) in [Kh108] that the usual Tchebyshev polynomial of best approximation (corresponding to $p_1 \equiv 0$) preserves its best approximation property (i.e., minimizes α) if $p_1 = \eta p_0$ where p_0 is a given seminorm on $\mathbb{R}^n$ (or $\mathbb{C}^n$) and $\eta > 0$ is small enough. This article contains a list of elegant examples of concrete situations where the general theory is a source of explicit and sharp results concerning the moment problem, polynomial approximation, and sharp interpolation inequalities for some classes of analytic functions.

Another important turn of this theme is based on the notions of $O(p)$- and $o(p)$-completeness of a family of vectors. Let E be a (say, real) vector normed space and p a continuous seminorm in the space $\mathbb{R}_0^\infty$ of all finite sequences

$$(\lambda_1, \lambda_2, \ldots, \lambda_n, 0, 0, 0 \ldots), \quad \lambda_j \in \mathbb{R}.$$

A sequence $(x_k)_{k=1}^\infty$ of vectors in E is said to be $O(p)$-complete in E if for any $x \in E$ there exists a constant $C(x) > 0$ such that for any $\varepsilon > 0$ there exists a $\lambda \in \mathbb{R}_0^\infty$ satisfying

$$\Big\| x - \sum_{k=1}^{n} \lambda_k x_k \Big\|_E < \varepsilon, \quad p(\lambda) \le C(x). \tag{6.1}$$

Replacing $C(x)$ by ε in (6.1) we get an $o(p)$-complete sequence $(x_k)_{k=1}^\infty$. A systematic theory of $O(p)$- and $o(p)$-completeness is built in [Kh79] including some stability problems ("an $O(p)$-complete system remains $O(p)$-complete under small perturbations"). We again refer the reader to the list of publications in [Kh108, Kh121], but we cannot resist temptation to quote separately an elegant application of the theory given in [Kh71, Kh96] where the Weierstrass theorem on the uniform polynomial approximation on $[0, 1]$ (and, in fact, the Lavrentiev theorem on approximation on nowhere dense plane compacta) is complemented by exhaustive information on the rate of growth of the coefficients of approximating polynomials.

Another aspect of approximation which attracted S. Khavinson is related to the 13th Hilbert problem on representation of a continuous function of several variables by linear compositions of functions of a lesser number of variables. From the series of problems studied by S. Khavinson in this area we only mention here the passage from functions of two variables to a more difficult case of three or more variables. To make this passage possible S. Khavinson had to overcome essential difficulties discovering underway a number of new effects. His results in this area are summed up in his monograph [Kh148].

We conclude by just naming two more works devoted to "the pure approximation theory": [Kh21] is an essential development of earlier works by Jackson and M. Krein on the uniqueness of the polynomial of best approximation in L^1, and [Kh162], the last publication of S. Khavinson, is devoted to the uniform approximation by elements of an interval ($= \{x \in C(T) : a \le x \le b\}$ where a, b are given continuous functions on a compact space T); the results are applied to the uniform approximation of continuous functions of several variables by tensor products.

References

[1] L. Ahlfors, *Bounded analytic functions*, Duke Math. J., **1** (1947), 1–11.

[2] P. Garabedian, *Schwartz's lemma and the Szegö kernel function*, Trans. Amer. Math. Soc., **67** (1949), no. 1, 1–35.

[3] V.P. Havin, *Golubev series and analyticity on continua*, Lecture Notes in Math., **1043**, Springer, 1984, pp. 670–672.

[4] J. Marshall, *Removable sets for bounded analytic functions*, Lecture Notes in Math., **1043**, Springer, 1984, pp. 485–490.

[5] N.K. Nikolski, V.P. Havin, *Results of V.I. Smirnov on complex analysis and their posterior development*, In: V.I. Smirnov, "Selected Works. Complex Analysis. Mathematical Diffraction Theory", Leningrad State Univ., 1988, pp. 111–145. (Russian)

[6] I.I. Privalov, *Boundary properties of analytic functions*, Gostekhizdat, 1950. (Russian)

[7] V.I. Smirnov, *Selected Works. Complex Analysis. Mathematical Diffraction Theory*, Leningrad State Univ., 1988, pp. 82–110. (Russian)

V.P. Havin
Department of Mathematics and Mechanics
St. Petersburg State University
Bibliotechnaya pl. 2
Stary Petergof
St. Petersburg
198904, Russia
e-mail: `havin@VH1621.spb.edu`

Operator Theory:
Advances and Applications, Vol. 158, 23–32
© 2005 Birkhäuser Verlag Basel/Switzerland

S.Ya. Khavinson: Bibliography

[1] *On an extremal problem of the theory of analytic functions*, Uspekhi Matem. Nauk, **4** (1949), no. 4 (32), 158–159. (Russian)

[2] *Fourier coefficients of boundary values of analytic functions*, In: I.I. Privalov, "Boundary properties of analytic functions", Gostekhizdat, 1950, pp. 170–172. (Russian); German transl., *Fourier-Koeffizienten der Randwerte analytischer Funktionen*, In: I. Privalov, "Randeigenschaften analytischer Funktionen", Deutschen Verlag der Wissenschaften, Berlin, 1956.

[3] *On weak convergence of boundary values of analytic functions*, In: I.I. Privalov, "Boundary properties of analytic functions", Gostekhizdat, 1950, pp. 284–293. (Russian); German transl., *Sätze über Eindeutigkeit analytischer Funktionen auf dem Rande*, In: I. Privalov, "Randeigenschaften analytischer Funktionen", Deutscher Verlag der Wissenschaften, Berlin, 1956.

[4] *Boundary uniqueness theorems*, In: I.I. Privalov, "Boundary properties of analytic functions", Gostekhizdat, 1950, pp. 298–333. (Russian); German transl., In: I. Privalov, "Randeigenschaften analytischer Funktionen", Deutscher Verlag der Wissenschaften, Berlin, 1956.

[5] *An estimate of Taylor sums of bounded analytic functions in a circle*, Doklady Akad. Nauk SSSR **80** (1951), no. 3, 333–336. (Russian)

[6] *On certain extremal problems in the theory of analytic functions*, Uchenye Zapiski Moskov. Gos. Univ., Ser. Matematica, **148** (1951), no. 4, 133–143. (Russian)

[7] *On extremal properties of functions mapping a region on a multy-sheeted circle*, Doklady Akad. Nauk SSSR, **88** (1953), no. 6, 957–959. (Russian)

[8] *On some extremal problems in theory of analytic functions*, Thesis for a candidate's degree, Moscow, 1953, pp. 1–140. (Russian)

[9] *On some nonlinear extremal problems for bounded analytic functions*, Doklady Akad. Nauk SSSR, **92** (1953), no. 2, 243–245. (Russian)

[10] *Extremal problems for certain classes of analytic functions*, Doklady Akad. Nauk SSSR, **101** (1955), no. 3, 421–424. (Russian)

[11] *Extremal problems for certain classes of analytic functions in finitely connected regions*, Mat. Sb. **36(78)** (1955), no. 3, 445–478. (Russian); English transl., Amer. Math. Soc. Transl. (2), **5** (1957), 1–33.

[12] *On uniqueness of a polynomial of best approximation in the metric of the space L1*, Doklady Akad. Nauk SSSR **105** (1955), no. 6, 1159–1161. (Russian)

[13] *The Carathéodory-Fejér problem for analytic functions in multiply connected domains*, Vladimir Gos. Ped. Inst. Uchen. Zap., **2** (1956), 43–50. (Russian)

24   S.Ya. Khavinson: Bibliography

[14] *On Lebesgue theorem for Riemann integral*, Vladimir Gos. Ped. Inst. Uchen. Zap., **2** (1956), 51–54. (Russian)

[15] *P.L. Chebyshev systems and uniqueness of the polynomial for best approximation*, Trudy 3 Vsesoyuzn. Matem. S'ezda, vol. 1, 1956, p. 130. (Russian)

[16] *On the dimension of polyhedron of best approximation in L_1-space*, Trudy MISI, **19**, Moscow Inst. of Civil Engin., Moscow, 1957, pp. 18–29. (Russian)

[17] (with G.C. Tumarkin), *On the removing of singularities for analytic functions of a certain class (class D)*, Uspekhi Mat. Nauk, **12** (1957), no. 4(76), 193–199. (Russian)

[18] (with G.C. Tumarkin), *On the definition of analytic functions of class E_p in the multiply connected domains*, Uspekhi Mat. Nauk, **13** (1958), no. 1(79), 201–206. (Russian)

[19] (with G.C. Tumarkin), *Classes of analytic functions in multiply connected regions represented by Cauchy-Green formulas*, Uspekhi Mat. Nauk **13** (1958), no. 2(80), 215–221. (Russian)

[20] (with G.C. Tumarkin), *An expansion theorem for analytic functions of class E_p in multiply connected domains*, Uspekhi Mat. Nauk, **13** (1958), no. 2(80), 223–228. (Russian)

[21] *On uniqueness of functions of best approximation in the metric of the space L_1*, Izv. Akad. Nauk SSSR, Ser. Mat., **22** (1958), no. 2, 243–270. (Russian)

[22] (with G.C. Tumarkin), *Analytic functions on multiply connected regions of the class of V.I. Smirnov (class S)*, Izv. Akad. Nauk SSSR, Ser. Mat., **22** (1958), no. 3, 379–386. (Russian)

[23] (with G.C. Tumarkin), *Existence in multiply connected regions of single-valued analytic functions with a given modulus of boundary values*, Izv. Akad. Nauk SSSR, Ser. Mat., **22** (1958), no. 4, 543–562. (Russian)

[24] (with G.C. Tumarkin), *Conditions for the representability of a harmonic function by Green's formula in a multiply connected region*, Mat. Sb., **44(86)** (1958), no. 2, 225–234. (Russian)

[25] (with G.C. Tumarkin), *Properties of extremum functions in extremum problems for certain classes of analytic functions with a weighted metric*, Dokl. Akad. Nauk SSSR, **119** (1958), no. 2, 215–218. (Russian)

[26] (with G.C. Tumarkin), *Investigation of the properties of extremal functions by mean of duality relations in extremal problems for classes of analytic functions in multiply connected domains*, Mat. Sb. **46(88)** (1958), no. 2, 195–228. (Russian)

[27] *The radii of univalence, starlikeness and convexity of a class of analytic functions in multiply connected domains*, Izv. Vys. Uchebn. Zaved., Matematika, 1958, no. 3(4), 233–240. (Russian)

[28] (with G.C. Tumarkin), *Power series and their generalizations. Monogeneity problem. Boundary properties*, Matematika v SSSR za 40 let, vol. 1, Fizmatgiz, Moscow, 1958, pp. 407–444. (Russian)

[29] *The analytic capacity of plane sets, some classes of analytic functions and the extremum function in Schwarz's lemma for arbitrary regions*, Dokl. Akad. Nauk SSSR, **128** (1959), no. 5, 896–898. (Russian)

[30] *The analytic capacity of sets as related to mass distributions*, Dokl. Akad. Nauk SSSR, **128** (1959), no. 6, 1129–1131. (Russian)

[31] *On a property of theoretic-functional null-sets*, Nauchnye Doklady Vyssh. Shkoly, Ser. Fizmat, 1959, no. 3, 71–77. (Russian)

[32] *Extremal problems with additional conditions*, Tezisy dokl. 1 Megvuz. konf. po konstruktivnoi teorii funktsii, Leningrad State Univ., Leningrad, 1959. (Russian)

[33] (with G.C. Tumarkin), *Mutual ortogonality of boundary values of certain classes of analytic functions in multiply connected domains*, Uspekhi Mat. Nauk **14** (1959), no. 3, 173–180. (Russian)

[34] (with G.C. Tumarkin), *Classes of analytic functions on multiply connected domains*, Issledovaniya po sovremennym problemam teorii funktsii komplexnogo peremennoogo (Ed. A. I. Markushevich), Gosudarstv Izdat. Fiz.-Mat. Lit., Moscow, 1960, pp. 45–77. (Russian); French transl., *Classes de fonctions analytiques dans de domaines multiplement connexes*, Fonctions d'une variable complexe. Problèmes contemporains, Gauthier-Villars, Paris, 1962, pp. 37–71.

[35] (with G.C. Tumarkin), *Qualitative properties of solutions of certain type of extremal problems*, Issledovaniya po sovremennym problemam teorii funktsii komplexnogo peremennoogo (Ed. A. I. Markushevich), Gosudarstv Izdat. Fiz.-Mat. Lit., Moscow, 1960, pp. 77–95. (Russian); French transl., *Propriétés qualitative de solutions des problèmes extremaux de certains types*, Fonctions d'une variable complexe. Problèmes contemporains, Gauthier-Villars, Paris, 1962.

[36] *Extremal problems with additional conditions for analytic functions*, Tezisy dokl. 5 Vsesoyuzn. konf. po teorii funktsii komplexnogo peremennoogo, Yerevan State Univ., Yerevan, 1960. (Russian)

[37] *On stability property for theoretic-functional null-sets*, Trudy MISI, kaf. Vyssh. Matem., Moscow Inst. of Civil Engin., Moscow, 1960, pp. 127–136. (Russian)

[38] *An extremal problem for analytic functions of the class H_2*, Trudy MISI, kaf. Vyssh. Matem., Moscow Inst. of Civil Engin., Moscow, 1960, pp. 137–141. (Russian)

[39] *Duality method in extremal and approximation problems of function theory*, Thesis for a Doctor's degree, Moscow, 1960. (Russian)

[40] *On a class of extremal problems for polynomials*, Dokl. Akad. Nauk SSSR, **130** (1960), no. 5, 997–1000. (Russian); English transl., Soviet Math. Dokl., **1** (1960), 137–140.

[41] (with Iu. E. Alenicyn) *The radius of p-valence for bounded analytic functions in multiply connected domains*, Mat. Sb. **52(94)** (1960), 653–657. (Russian)

[42] *Approximation on sets of zero analytic capacity*, Dokl. Akad. Nauk SSSR, **131** (1960), no. 1, 44–46. (Russian); English transl., Soviet Math. Dokl., **1** (1960), 205–207.

[43] *On the radius of holomorphy of functions inverse to functions which are analytic and bounded in multiply connected regions*, Izv. Vys. Uchebn. Zaved., Matematika, 1960, no. 5 (18), 190–194. (Russian)

[44] *Extremum problems for functions satisfying some supplementary restrictions inside the region and the application of these problems to questions of approximation*, Dokl. Akad. Nauk SSSR, **135** (1960), no. 2, 270–273. (Russian); English transl., Soviet Math. Dokl., **1**, 1960, 1263–1266.

[45] *Some problems concerning the completeness of systems*, Dokl. Akad. Nauk SSSR, **137** (1961), no. 4, 793–796. (Russian); English transl., Soviet Math. Dokl., **2** (1961), 358–361.

[46] (with V.P. Havin) *Some estimates of analytic capacity*, Dokl. Akad. Nauk SSSR, **138** (1961), no. 4, 789–792. (Russian)

[47] *Analytic capacity of sets, joint nontriviality of various classes of analytic functions and the Schwarz lemma in arbitrary domains*, Mat. Sb., **54(96)** (1961), no. 1, 3–50. (Russian); English transl., American Math. Soc. Transl. (2), **43** (1964).

[48] *On approximation with account taken of the size of the coefficients of the approximants*, Trudy Mat. Inst. Steklov, **60** (1961), 304–324. (Russian)

[49] *On two classes of extremal problems for polynomials and moments*, Izv. Akad. Nauk SSSR, Ser. Mat., **25** (1961), no. 4, 557–590. (Russian)

[50] *Extremal and approximation problems with additional conditions*, Studies of Modern Problems of constructive Theory of Functions, Fizmatgiz, Moscow, 1961, pp. 347–352. (Russian)

[51] (with A.I. Markushevich and G.C.Tumarkin) *Some current problems of the theory of boundary properties*, Issledovaniya po soremennym problemam teorii funktsii komplexnogo peremennogo, **2**, Fizmatgiz, 1961, pp. 100–110. (Russian)

[52] *Some solved and unsolved problems connected with analytic capacity of sets*, Tezisy dokladov nauchno-tekhicheskoi konferentsii MISI, Moscow Inst. of Civil Engin., Moscow, 1962. (Russian)

[53] (with A.I. Markushevich and G.C.Tumarkin) *The current status and problems of the theory of boundary properties of analytic functions*, Proc. Fourth All-Union Math. Congr. (Leningrad, 1961), vol. 2, Izdat. Nauka, Leningrad, 1964, pp. 661–667. (Russian)

[54] *Some remarks on integral of Cauchy-Stieltjes type*, Litovsk. Mat. Sb., **2** (1962), no. 2, 281–288. (Russian)

[55] *The theory of extremal problems for bounded analytic functions satisfying additional conditions inside the domain*, Uspekhi Mat. Nauk, **18** (1963), no. 2 (110), 25–98. (Russian)

[56] *On the removal of singularities*, Litovsk. Mat. Sb., **3** (1963), no. 1, 271–287. (Russian)

[57] *Analytic functions of bounded type*, Mathematical Analysis. 1963, Akad. Nauk SSSR Inst. Nauchn. Informacii, Moscow, 1965, pp. 5–80. (Russian)

[58] *On some extremal problems of the theory of analytic functions*, American Math. Soc. Transl. (2), **32** (1963).

[59] *Approximations with reasoning the coefficients of polynomials*, American Math. Soc. Transl. (2), **44** (1965).

[60] *On the Rudin-Carleson theorem*, Dokl. Akad. Nauk SSSR, **165** (1965), no. 2, 497–499. (Russian)

[61] *Representation and approximation of functions on thin sets*, Contemporary Problems in Theory Anal. Functions (Internat. Conf., Erevan, 1965), Nauka, Moscow, 1966, pp. 314–318. (Russian)

[62] *An approximation problem of Walsh*, Izv. Akad. Nauk Armjan. SSR, Ser. Mat., **1** (1966), no. 4, 231–237. (Russian)

[63] *Approximation by elements of convex sets*, Dokl. Akad. Nauk SSSR, **172** (1967), no. 2, 294–297. (Russian)

[64] *On approximation by elements of convex sets and an Walsh problem*, Trudy 4 Vsesoyuznogo matem. s'ezda, Nauka, 1967. (Russian)

[65] *Boundary values of arbitrary functions*, Uspekhi Mat. Nauk, **22** (1967), no. 1 (133), 133–136. (Russian)

[66] (with A.N. Kochetkov) *The removal of singularities of analytic functions of certain classes*, Litovsk. Mat. Sb., **9** (1969), no. 1, 181–192. (Russian)

[67] *Indefinite integral. Definite integrals (lecture synopsis)*, Moscow Inst. of Civil Engineering, Moscow, 1969. (Russian)

[68] *Calculation of definite integrals (lecture synopsis)*, Moscow Inst. of Civil Engineering, Moscow, 1969. (Russian)

[69] *Additional chapters of integral calculus (lecture synopsis)*, Moscow Inst. of Civil Engineering, Moscow, 1969. (Russian)

[70] *Lectures on approximation by elements of convex sets*, Akad. Nauk Kazakh. SSR, 1969, pp. 1–72. (Russian)

[71] *Allowable magnitudes for the coefficients of the polynomials in the uniform approximation of continuous functions*, Mat. Zametki, **6** (1969), no. 5, 619–625. (Russian)

[72] *A Chebyshev theorem for the approximation of a function of two variables by sums* $\varphi(x)+\psi(y)$, Izv. Akad. Nauk SSSR, Ser. Mat., **33** (1969), no. 3, 650–666. (Russian)

[73] *Survey of some results in approximation theory with taking into account the values of approximating polynomials*, Primenenie funktsion. analiza v teorii pribligenii, Kalinin State Univ., Kalinin, 1970. (Russian)

[74] *Green, Gauss, Ostrogradsky and Stokes formulas (lecture synopsis)*, Moscow Inst. of Civil Engineering, Moscow, 1970, pp. 3–74. (Russian)

[75] (with M.I. Skanavi) *Foundations of vector analysis (lecture synopsis)*, Vysshaya Shkola, Moscow, 1971, pp. 1–70. (Russian)

[76] (with M.I. Skanavi) *Field theory and its applications to hydromechanics (lecture synopsis)*, Vysshaya Shkola, Moscow, 1971, pp. 1–76. (Russian)

[77] (with L.Ya. Tslaf) *A.F. Bermant – scientist and educator*, Matematika. Sbornik nauchno-metodich. statei, no. 1, Vysshaya shkola, 1971. (Russian)

[78] *Some approximation theorems involving the values of the coefficients of the approximating functions*, Dokl. Akad. Nauk SSSR, **196** (1971), no. 6, 1283–1286. (Russian)

[79] *A notion of completeness that takes into account the magnitude of the coefficients of the approximating polynomials*, Izv. Akad. Nauk Armian. SSR, Ser. Mat., **6** (1971), no. 2–3, 221–234. (Russian)

[80] (with Z.S. Romanova) *Approximation properties of subspaces of finite dimention in the space* L_1, Mat. Sb., **89(131)** (1972), no. 1, 115–165. (Russian)

[81] *In memory of M.I. Skanavi*, Matematika. Sbornik nauchno-metodich. statei, no. 3, Vysshaya shkola, 1973. (Russian)

[82] *The representation of extremal functions in the classes* E_p *by the functions of Green and Neumann*, Mat. Zametki, **16** (1974), no. 5, 707–716. (Russian)

[83] *The distribution of zeros of extremal functions in the space E_p in finitely connected domains*, Mat. Zametki, **16** (1974), no. 6, 879–885. (Russian)

[84] (with V.V. Ryžkov) *On application of educational TV for the purpose of improvement of mathematical training of engineers*, Uchebnoe televidenie, Yerevan State Univ., 1974. (Russian)

[85] *Methodical description of a course of TV-lectures on mathematics (for first-year studetns)*, Manuscript, 1974, 35 p. (Russian)

[86] *Introduction to analysis. Text-book for extramural training by TV*, Manuscript, 1974, 70 p. (Russian)

[87] *Extremal functions in problems on the distortion of lengths and areas under transformations by bounded analytic functions*, Nekotorye voprosy matematiki i mekhaniki sploshnoi sredy. Moskov. Ing.-Stroitel. Inst., Sb. Trudov, **130**. Moscow, 1975, pp. 3–6. (Russian)

[88] *A universal potential for compact sets of capacity zero, and other approximation characteristics of sparse sets*, Izv. Akad. Nauk Armjan. SSR, Ser. Mat., **10** (1975), no. 4, 356–372. (Russian)

[89] *On simultaneous introduction of all types of definite integrals*, Matematika. Sbornik nauchno-metodich. statei, no. 5, Vysshaya shkola, 1975.

[90] *Removable singularities of analytic functions of the V.I. Smirnov class*, Some problems in modern function theory. (Proc. Conf. Modern Problems of Geometric Theory of Functions, Inst. Math. Acad. Sci. USSR, Novosibirsk, 1976), Akad. Nauk SSSR, Sibirsk. Otdel. Inst. Mat., Novosibirsk, 1976, pp. 160–166. (Russian)

[91] *Lectures on integral calculus*, Vysshaya shkola, 1976, pp. 1–198. (Russian)

[92] (with E.Sh. Chatskaya) *Chebyshev range of ideas in the best approximation theory. Part 1. Duality relations and criteria for best approximation elements*, Moskov. Inzh.-Stroit. Inst. (Fak. povyshen. kvalif), Moscow Inst. of Civil Engineering, Moscow, 1976, pp. 1–46. (Russian)

[93] *Extremal properties and sets of removable singularities of analytic functions*, Monograph, Manuscript, 1976, 660 p. (Russian)

[94] *An extremal problem for bounded analytic functions connected with L_p-metric*, Matematika v stroitel'stve, Moskov. Ing.-Stroitel. Inst., Sb. Trudov, **153**. Moscow Inst. of Civil Engineering, Moscow, 1977, pp. 7–10. (Russian)

[95] *Generalization of a H. Bohr theorem*, Matematika v stroitel'stve, Moskov. Ing.-Stroitel. Inst., Sb. Trudov, **153**. Moscow Inst. of Civil Engineering, Moscow, 1977, pp. 10–17. (Russian)

[96] (with O.A. Muradjan) *The values of coefficients of the polynomials in the Weierstrass approximation theorem*, Mat. Zametki, **22** (1977), no. 2, 269–276. (Russian)

[97] (with E.Sh. Chatskaya) *Chebyshev range of ideas in the best approximation theory. Part 2. Completeness of systems*, Moskov. Inzh.-Stroit. Inst. (Fak. povyshen. kvalif), Moscow Inst. of Civil Engineering, Moscow, 1977. (Russian)

[98] (with E.Sh. Chatskaya) *Chebyshev range of ideas in the best approximation theory. Part 3. Uniqueness and plurality of best approximation elements*, Moskov. Inzh.-Stroit. Inst. (Fak. povyshen. kvalif), Moscow Inst. of Civil Engineering, Moscow, 1977. (Russian)

[99] *A problem on the exact majorant*, In: "99 problems of function theory", Leningrad branch of Steklov Math. Inst., Leningrad, 1978. (Russian)

[100] (with G.S. Vardanyan, I.G. Filippov, B.P. Osilenker, I.Yu. Yarantseva, N.A. Medvedeva, L.A. Karazina) *Methodical instructions on the calculation task "Mathematical physics" for MISI students*, Moscow Inst. of Civil Engineering, Moscow, 1979. (Russian)

[101] (with A.I. Markushevich) *Duality in analytic functions theory*, Mathem. Encyclopaedia, vol. 2 (1978), pp. 42–43. (Russian)

[102] *It is interesting and timely*, Matematika v shkole, 1979, no. 3. (Russian)

[103] *On representation of analytic functions by Cauchy potential and Golubev series*, Problemy matematiki i mekhaniki deformiruemoi sredy. Moskov. Ing.-Stroitel. Inst., Sb. Trudov, **173**. Moscow Inst. of Civil Engineering, Moscow, 1979, pp. 105–113. (Russian)

[104] *On two new directions in approximation theory*, Monograph, Manuscript, 1980, 250 p. (Russian)

[105] (with V.Kh. Belyi) *Introduction to analysis. Part 2. Text-book for extramural training by TV*, Manuscript, 1980, 60 p. (Russian)

[106] (with V.Kh. Belyi) *Differential calculus. Text-book for extramural training by TV*, Manuscript, 1980, 60 p. (Russian)

[107] *A class of sharp estimates for polynomials, moments and analytic functions*, Uspekhi Mat. Nauk, **36** (1981), no. 5(221), 199-200. (Russian)

[108] *On a class of sharp inequalities for polynomials, moments and analytic functions*, Sibirsk. Mat. Zh., **22** (1981), no. 6, 188-207. (Russian)

[109] *The efficient solution of some extremal problems of moment theory*, Studies in the theory of functions of several real variables. Yaroslav. Gos. Univ., 1981, pp. 81–90, 111. (Russian)

[110] *Factorization of analytic functions in finitely connected domains*, Moskov. Inzh.-Stroit. Inst. (Fak. povysheniya kvalif), Inst. of Civil Engineering, Moscow, 1981, pp. 1–117. (Russian)

[111] *Foundations of the theory of extremal problems for bounded analytic functions and various generalizations of them*, Moskov. Inzh.-Stroit. Inst. (Fak. povyshen. kvalif), Inst. of Civil Engineering, Moscow, 1981, pp. 1–92. (Russian); English Transl., Amer. Math. Soc. Transl. (2), **129**, 1986, 1–61.

[112] *Foundations of the theory of extremal problems for bounded analytic functions with additional conditions*, Moskov. Inzh.-Stroit. Inst. (Fak. povyshen. kvalif), Inst. of Civil Engineering, Moscow, 1981, pp. 1–89. (Russian); English Transl., Amer. Math. Soc. Transl. (2), **129**, 1986, 63–114.

[113] *Correlation between teaching of mathematics and other branches of science during training of construction engineers*, Moskov. Inzh.-Stroit. Inst. (Fak. povyshen. kvalif), Moscow Inst. of Civil Engineering, Moscow, 1981. (Russian)

[114] (with V.E. Nazaretan) *Obituary: V.V. Zorin*, Matematika v shkole, 1981, no. 5. (Russian)

[115] *Foundation of the theory of removable sets of analytic functions*, Moskov. Inzh.-Stroit. Inst. (Fak. povyshen. kvalif), Inst. of Civil Engineering, Moscow, 1982, pp. 3–90. (Russian)

[116] *Additional problems of the theory of removable sets of analytic functions*, Moskov. Inzh.-Stroit. Inst. (Fak. povyshen. kvalif), Inst. of Civil Engineering, Moscow, 1982, pp. 3–97. (Russian)

[117] *Stability of extremal elements under perturbation of best approximation problems, and extremal problems in the moment theory*, Litovskii Matem. sbornik, **22** (1982), no. 4. (Russian)

[118] *On a strengthened extremal property of the Caratéodory-Fejer functions*, Analysis Mathematica, **9** (1983), no. 2, 99–111.

[119] *Exact majorant problem*, In: "999 problems on complex and linear analysis". Lecture Notes in Math., **1043**, Springer, 1984.

[120] *Representation of functions of two variables by the sums $\varphi(x) + \psi(y)$*, Izv. Vyssh. Uchebn. Zaved., Matem., 1985, no. 2, 66–73. (Russian)

[121] *Complete systems in Banach spaces*, Izv. Akad. Nauk Armyan. SSR, Ser. Mat., **20** (1985), no. 2, 89–111. (Russian); English transl., Soviet J. Contemp. Math. Anal., **20** (1985), no. 2, 1–26.

[122] *Characterization of compact sets, on which every continuous function is uniformly approximated by linear superpositions*, Tezisy dokladov Litovskogo Mat. Obschestva, Vilnius, 1985. (Russian)

[123] *Two Papers on Extremal Problems in Complex Analysis*, American Math. Soc. Transl. (2), **129**, Providence, Rhode Island, 1986.

[124] *Remarks on Wronsky determinants*, Voprosy mat., mekh. sploshnyh sred i primeneniya matem. metodov v stroitel'stve, Moscow Inst. of Civil Engin., Moscow, 1987, pp. 37–40. (Russian)

[125] *On representation and approximation of functions of two variables by linear superpositions*, Voprosy mat., mekh. sploshnyh sred i primeneniya matem. metodov v stroitel'stve, Moscow Inst. of Civil Engin., Moscow, 1987, pp. 89–92. (Russian)

[126] *Factorization on compact Riemann surfaces with a boundary*, Doklady rasshirennyh zased. Inst. Prikl. Matem. im. I.N. Vekua, **3** (1988), no. 1. (Russian)

[127] *Factorization of univalent analytic functions on compact bordered Riemann surfaces*, Zap. Nauchn. Sem. LOMI, **170** (1989), 285–313. (Russian); English transl., J. Soviet Math., **63** (1993), no. 2, 275–290.

[128] *Theory of factorization of single-valued analytic functions on compact Riemann surfaces with a boundary*, Uspekhi Mat. Nauk, **44** (1989), no. 4(268), 155–189. (Russian); English transl., Russian Math. Surveys, **44** (1989), no. 4, 113–156.

[129] (with T.S. Kuzina) *Introduction to analysis. Variables and functions*, Methodical instructions and tasks for a self-dependent work of students, Moscow Inst. of Civil Engin., Moscow, 1989, pp. 1–53. (Russian)

[130] (with T.S. Kuzina) *Limit of a variable. Limit of a function*, Methodical instructions and tasks for a self-dependent work of students, Moscow Inst. of Civil Engin., Moscow, 1989, pp. 1–64. (Russian)

[131] (with O.K. Aksentyan) *Structures of metric and normed spaces and their applications to civil engineering courses*, Text-book, Moscow Inst. of Civil Engin., Moscow, 1989. (Russian)

[132] (with O.K. Aksentyan) *Linear operators and their applications to civil engineering problems*, Text-book, Moscow Inst. of Civil Engin., Moscow, 1989. (Russian)

[133] *A method of removing the multivalence of analytic functions*, Izv. Vyssh. Uchebn. Zaved., Mat., 1990, no. 11, 64–72. (Russian); English transl., Sov. Math. (Izv. VUZ), **34** (1990), no. 11, 80–90.

[134] (with E.G. Kiryatski) *Compactness of families of univalent functions that are defined by the intersection of two hyperplanes*, Litovsk. Mat. Sb., **30** (1990), no. 2, 268–274. (Russian); Engl. transl., Lithuanian Math. J., **30** (1990), no. 2, 120–125.

[135] *Duality method in the theory of extremal problems for analytic functions*, Manuscript, 1988, 70 p. (Russian)

[136] *A.I. Markushevich as a mathematician*, Mat. v shkole, 1990, no. 1, 64. (Russian)

[137] (with B.Ya. Levin, S.N. Mergelyan, I.V. Ostrovskii and others) *Anatolii Asirovich Gol'dberg (on the occasion of his sixtieth birthday)*, Uspekhi Mat. Nauk, **45** (1990), no. 5 (275). (Russian)

[138] (with T.S. Kuzina) *Zeros of functions, wwhich are the extreme points of H_1-space on Riemann surface*, Voprosy mat., mekh. sploshnyh sred i primeneniya matem. metodov v stroitel'stve, Moscow Inst. of Civil Engin., Moscow, 1992, pp. 50–59. (Russian)

[139] *Automorphy and orthogonality*, Voprosy mat., mekh. sploshnyh sred i primeneniya matem. metodov v stroitel'stve, Moscow Inst. of Civil Engin., Moscow, 1992, pp. 59–63. (Russian)

[140] *On interpolational properties of outer functions in Kolmogorov superpositions*, Mat. Zametki, **55** (1994), no. 4, 104–113. (Russian); English transl., Mathem. Notes, **55** (1994), no. 3–4, 406–412.

[141] (with S.V. Konyagin, S.M. Nikolskii, S.B. Stechkin) *A.L. Garkavi (on the occasion of his seventieth birthday)*, Uspekhi Mat. Nauk, **50** (1995), no. 2(302), 230–231. (Russian); English transl., Russian Math. Surveys, **50** (1995), no. 2, 461–463. .

[142] *Certain approximate properties of linear superpositions*, Izv. Vyssh. Uchebn. Zaved., Mat., 1995, no. 8, 63–73. (Russian); English transl., Russian Math. (Izv. VUZ), **39** (1995), no. 8, 60–70.

[143] *On approximation by linear superpositions*, Theory of functions and applications. Collection of works dedicated to the memory of M.M. Djrbashian, Lonys Publishing House, Yerevan, 1995.

[144] *The annihilator of linear superpositions*, Algebra i Analiz, **7** (1995), no. 3, 1–42. (Russian); English transl., St. Petersburg Math. J., **7** (1996), no. 3, 307–341.

[145] (with A.L. Garkavi and V.A. Medvedev) *On the existence of a best untform approximation of a function in two variables by sums $\varphi(x) + \psi(y)$*, Sibirsk. Mat. Zh., **36** (1995), no. 4, 819–827. (Russian); English transl., Siberian Math. J., **36** (1995), no. 4, 707–713.

[146] (with A.L. Garkavi and V.A. Medvedev) *On the existence of a uniform approximation of a function of several variables by the sums of functions of fewer variables*, Mat. Sb., **187** (1996), no. 5, 3–14. (Russian); English transl., Sb. Math., **187** (1996), no. 5, 623–634.

[147] *On representation and approximation by linear superpositions*, Semestr of approximation theory, Inst. of advanced Studies in Math. at the Technion, Haifa, 1994.

[148] *Best approximation by linear superpositions. (Approximate nomography)*, Transl. of Mathem. Monographs, **159**, Amer. Math. Soc., Providence, RI, 1997.

[149] (with V.A. Medvedev) *On extreme points of the space of measures, which are orthogonal to functions $\varphi(x) + \psi(y)$*, Voprosy mat., mekh. sploshnyh sred i primeneniya matem. metodov v stroitel'stve, Moscow State Civil Engin. Univ., Moscow, 1997, pp. 61–67. (Russian)

[150] *Modifications of the concept of analytic capacity*, Mathematical modelling and complex analysis. Tekhnika, Vilnius, 1996, pp. 32–35. (Russian)

[151] *Extremal problems for Golubev sums*, Mat. Zametki, **65** (1999), no. 5, 738–745. (Russian); English transl., Math. Notes, **65** (1999), no. 5–6, 620–626.

[152] *Golubev sums: a theory of extremal problems like the analytic capacity problem and of related approximation processes*, Uspekhi Mat. Nauk, **54** (1999), no. 4, 75–142. (Russian); English transl., Russian Math. Surveys, **54** (1999), no. 4, 753–818.

[153] *Some remarks to problems of approximation with prescribed rate*, Complex analysis, operators and related topics, Oper. Theory Adv. Appl., **113**, Birkhäuser, Basel, 2000, pp. 127–133.

[154] *10 lectures on elementary mathematics*, Moscow State Civil Engin. Univ., Moscow, 2000, pp. 3–114. (Russian)

[155] *Mathematical analysis. Differential calculus of functions of one and several variables*, Sovremennyi Gumanit. Univ., Moscow, 2000. (Russian)

[156] (with T.S. Kuzina) *On the structure of extremal functions on Riemann surfaces*, Geometric theory of functions, boundary value problems and their applications, Trudy Mat. Tsentra im. N.I. Lobachevskogo, **14**, Kazan. Mat. Obs., Kazan, 2002, pp. 95–97. (Russian)

[157] *New modifications of the concept of analytic capacity*, Geometric theory of functions, boundary value problems and their applications, Trudy Mat. Tsentra im. N.I. Lobachevskogo, **14**, Kazan. Mat. Obs., Kazan, 2002, pp. 298–302. (Russian)

[158] *Approximations by wedge elements taking into account the values of the approximating elements*, Izv. Vyssh. Uchebn. Zaved., Mat. 2002, no. 10, 71–84. (Russian); English transl., Russian Math. (Iz. VUZ), **46** (2002), no. 10, 69–82.

[159] *New relations for analytic capacity of sets and its modifications*, Voprosy matem., mekhaniki sploshnyh sred i primen. matem. metodov v stroit., Sbornik nauchn. trudov, **10**, Moscow Civil Engin. Univ., Moscow, 2003, pp. 6–10. (Russian)

[160] *On Lipschitz conditions for the metric projection*, Voprosy matem., mekhaniki sploshnyh sred i primen. matem. metodov v stroit., Sbornik nauchn. trudov, **10**, Moscow Civil Engin. Univ., Moscow, 2003, pp. 11–14. (Russian)

[161] *Duality relations in the theory of analytic capacity*, Algebra i Analiz, **15** (2003), no. 1, 3–62. (Russian); English transl., St. Petersburg Math. J., **15** (2004), no. 1, 1–40.

[162] *Approximation properties of certain sets in spaces of continuous functios*, Anal. Math., **29** (2003), no. 2, 87–105. (Russian)

[163*] (with T.S. Kuzina) *The structural formulae for extremal functions in Hardy classes on finite Riemann surfaces*, the present volume, pp. 37–57.

---

*When S.Ya. Khavinson passed away, the manuscript of this paper was left on his desk...

Operator Theory:
Advances and Applications, Vol. 158, 33–36

S.Ya. Khavinson: Photos

S.Ya. Khavinson, 1949 – the year of graduation
from Moscow State University

S.Ya. Khavinson with his parents, Kislovodsk, 1950

S.I. Zukhovitskii, G.Ts. Tumarkin and S.Ya. Khavinson, 1976.

S.Ya. Khavinson with some of his Ph.D. students:
V.Ya. Eiderman, T.S. Kuzina, M.V. Samokhin and M.P. Ovchintsev.
Conference on Function Theory, Chernogolovka, 1983.

S.Ya. Khavinson, 1987

S.Ya. Khavinson and A.G. O'Farrell.
St.-Petersburg, May 2001. The last conference.

Operator Theory:
Advances and Applications, Vol. 158, 37–57
© 2005 Birkhäuser Verlag Basel/Switzerland

The Structural Formulae for Extremal Functions in Hardy Classes on Finite Riemann Surfaces

S.Ya. Khavinson and T.S. Kuzina

Let $\mathfrak{M}$ be a finite (open) Riemann surface with n handles and boundary Γ consisting of closed analytic curves $\gamma_1, \ldots, \gamma_m$; $\mathfrak{M} \cup \Gamma = \overline{\mathfrak{M}}$ is a compact surface. In this article we study some "natural" extremal problems in the Hardy classes on $\mathfrak{M}$ and give a qualitative description of extremal functions in such problems. Extensions of classical theory of Hardy spaces and other related classes of analytic functions in the unit disk or simply connected domains to Riemann surfaces began with an important article by Parreau [1] and continued in several directions in a great many papers. We mention here the monograph by M. Heins [2] and the paper by W. Rudin [3] (who however considered only arbitrary planar domains). A number of topics were developed in a series of papers by S.Ya. Khavinson and G.Ts. Tumarkin. For planar domains the relevant results are contained in [4], where one could also find further references. In [5,6], the author developed factorization theory for various classes of analytic functions on finite open Riemann surfaces extending the classical results of R. Nevanlinna, V.I. Smirnov, F. & M. Riesz, G. Szegö in the unit disk. The purpose of the present paper is – by using factorization formulae from [5,6] – to obtain structural description of extremal functions in a wide class of extremal problems. Extremal problems in the classes of analytic functions in the disk were intensively studied in the last century, cf. [7–10]. The most detailed description of the structure of extremal functions in Hardy spaces in the disk is contained in [11–12]. Structural formulae for extremal functions in multiply connected domains in the plane are obtained in [13]. These and other results and further references are contained in the monographs [14–15]. An extremal problem for bounded analytic functions on a finite Riemann surface was first studied in L. Ahlfors' paper [16]. The variational method used there, extended in a

This research was supported by the Ministry of Education of Russia (Grant E 00-1.0-199). The first author was also supported by the Russian Foundation of Basic Research (Grant no 01-01-00608).

natural way the method used by the first author in an earlier paper [17]. The problem studied in [16–17] was rather special, similar to that in the Schwarz lemma. In the paper [17] the author also pointed out connections of that extremal problem to Painlevé's problem of characterizing removable singular sets for bounded analytic functions. Later on, it led to the development of important theory of analytic capacity of sets (cf., e.g., [18–21]). A. Read [22–23] and H. Royden [24] applied the duality method based on the Hahn-Banach theorem to study similar problems on finite Riemann surfaces not merely for bounded functions but also for general Hardy spaces functions. Applications of this method to problems involving analytic functions was initiated in the papers of S.Ya. Khavinson [25–26] and W.W. Rogosinski and H.S. Shapiro [12]. (This method provides the starting point for study of extremal problems in the monographs [7–10], [14–15] cited above.) Yet, the relevant study of dual extremal problems on finite Riemann surfaces, was limited to establishing "smooth" properties of the extremal functions near the boundary. Here, our goal is to describe the structure of extremal function in relation to the basic potential-theoretic functions on such surfaces: the Green and Neumann functions.

1. Preliminaries and notations

We shall denote the points on the surface $\mathfrak{M}$ or its boundary Γ by p, q, etc. Recall, that $\mathfrak{M}$ is a finite Riemann surface with h handles and the boundary Γ consisting of m closed analytic contours $\gamma_1, \ldots, \gamma_m$. The homological basis on $\mathfrak{M}$ consists of $L = 2h + m - 1$ cycles: choose any $m - 1$ boundary contours and meridians and parallels on each handle. The number L, the first Betti number for $\mathfrak{M}$, also defines genus of the closed surface $\widehat{\mathfrak{M}}$, the Shottky double of $\mathfrak{M}$ (cf. [27]). Denote by $G(p, q)$ the Green function of $\mathfrak{M}$ with pole at q. Fixing from now on a point $p_0 \in \mathfrak{M}$, define the Green kernel

$$P(p, q) = \frac{1}{2\pi} \frac{\dfrac{\partial}{\partial n_q} G(p, q)}{\dfrac{\partial}{\partial n_q} G(p_0, q)}, \quad q \in \Gamma, \; p \in \mathfrak{M}. \tag{1.1}$$

Here, $\dfrac{\partial}{\partial n_q}$ is the derivative in the inward normal direction to Γ at the point $q \in \Gamma$. The Neumann function (cf. [27]) $N(p, q_1, q_2)$ is a harmonic function of p on $\mathfrak{M}$ except for the logarithmic poles at q_1, and q_2: a positive pole at q_1, negative at q_2. Its defining property is

$$\frac{\partial}{\partial n_p} N(p, q_1, q_2) = 0, \quad p \in \Gamma. \tag{1.2}$$

The poles q_1, q_2 also may lie on Γ (one, or both of them), but then their order is doubled. This is easily seen if one extends $N(p, q_1, q_2)$ to the double $\widehat{\mathfrak{M}}$ by the Schwarz symmetry principle.

We denote by $*u(p)$ the conjugate harmonic function of a harmonic function $u(p)$ on $\mathfrak{M}$. The function $*N(p, q_1, q_2)$ does not have periods around boundary contours. Moreover, in view of the Cauchy-Riemann equations,

$$\frac{\partial}{\partial s} * N(p, q_1, q_2) = 0, \quad p \in \Gamma \tag{1.3}$$

(s is a parameter on Γ). Yet, $*N(p, q_1, q_2)$ is not single-valued. In view of (1.3), every branch of $*N$ is constant on each boundary contour. The function

$$\Psi(p, q_1, q_2) = \exp[N(p, q_1, q_2) + i * N(p, q_1, q_2)] \tag{1.4}$$

is meromorphic on $\mathfrak{M}$, has a single-valued modulus and every one of its single-valued branches has a constant argument on every boundary contour. Ψ also has a pole at q_1 (simple if $q_1 \in \mathfrak{M}$, and double when $q_1 \in \Gamma$) and a zero at q_2 (once again, a simple zero when $q_2 \in \mathfrak{M}$ and a double when $q_2 \in \Gamma$). Let D be a closed domain on $\overline{\mathfrak{M}}$, whose boundary ∂D consists of Jordan curves (in particular, D may be $\overline{\mathfrak{M}}$ itself). By $d\omega(p)$ we shall denote the harmonic measure (with respect to the domain D) on ∂D evaluated at $p \in D$. The harmonic measure at p_0 will simply be denoted by $d\omega$.

2. Hardy classes and certain other classes of analytic functions on the surface $\mathfrak{M}$

A single-valued analytic function $f(p)$ on $\mathfrak{M}$ belongs to the Hardy class $H_\delta(\mathfrak{M})$, $\delta > 0$, if the subharmonic function $|f(p)|^\delta$ has a harmonic majorant $u(p)$ on $\mathfrak{M}$, i.e.,

$$|f(p)|^\delta \le u(p). \tag{2.1}$$

This is equivalent to existence of $M > 0$ such that for any Jordan domain $D \subset \mathfrak{M}$

$$\int_{\partial D} |f(p)|^\delta \, d\omega \le M. \tag{2.2}$$

A function $f \in H_\delta(\mathfrak{M})$ has almost everywhere (with respect to $d\omega$) nontangential boundary values $f(q)$ on Γ. Moreover, a function $f(p) \in H_1(\mathfrak{M})$ can be represented by its boundary values $f(q)$ by Green's formula

$$f(p) = \int_\Gamma f(q) P(p, q) \, d\omega. \tag{2.3}$$

(The equality (2.3) generalizes a well-known G.M. Fikhtengoltz's theorem concerning representation H_1-functions by the Poisson integral.) Let D^ρ be a domain in $\mathfrak{M}$ whose boundary is defined by the level set Γ^ρ of the Green function:

$$\Gamma^\rho = \{p \in \mathfrak{M} : \quad G(p, p_0) = \rho\}.$$

For all sufficiently small ρ, Γ^ρ consists of contours $\gamma_1^\rho, \ldots, \gamma_m^\rho$ homotopic to $\gamma_1, \ldots, \gamma_m$ respectively. From point $q \in \Gamma$ drop the normal to Γ until it meets Γ^ρ at q_ρ.

Then, the following generalization of the classical F. Riesz theorem in the disk holds:

$$\int_\Gamma |f(q) - f(q_\rho)|^\delta \, d\omega \to 0, \quad \text{when } \rho \to 0. \tag{2.4}$$

In particular,

$$\int_\Gamma |f(q)|^\delta \, d\omega < +\infty. \tag{2.5}$$

The integral (2.5) gives the value of the least harmonic majorant of the function $|f(p)|^\delta$ evaluated at p_0. Property (2.4) is equivalent to uniform integrability with respect to the measure $d\omega$ of the integrals

$$\int_E |f(p)|^\delta \, d\omega, \quad E \subset \Gamma^\rho. \tag{2.6}$$

In view of the Luzin-Privalov uniqueness theorem, the class $H_\delta(\mathfrak{M})$ can be identified with the class $H_\delta(\Gamma)$ which consists of boundary values of functions in $H_\delta(\mathfrak{M})$. Inequality (2.5) then implies that $H_\delta(\Gamma)$ is a subspace of the Lebesgue space $L_\delta(\Gamma, d\omega)$ of functions summable to the power δ with respect to the measure $d\omega$ on Γ. For $\delta \geq 1$, the norm

$$\|f\|_\delta = \left\{ \int_\Gamma |f(q)|^\delta \, d\omega \right\}^{1/\delta} \tag{2.7}$$

turns $H_\delta(\Gamma)$ into a subspace of the Banach space $L_\delta(\Gamma, d\omega)$. One can also show that $H_\delta(\Gamma)$ is closed in L_δ. For $\delta = \infty$, the class $H_\infty(\mathfrak{M})$ consists of bounded and analytic functions on $\mathfrak{M}$. $H_\infty(\Gamma)$ is a subspace of $L_\infty(\Gamma, d\omega)$ of bounded measurable functions on Γ equipped with the usual Vrai sup norm. The following lemma of S.Ya. Khavinson is useful when one tries to justify taking the limits under the integral sign for functions on $\mathfrak{M}$. Let $H_\delta^1(\Gamma)$ denote the unit ball in the space $H_\delta(\Gamma)$.

Lemma 2.1. *Let a sequence of single-valued analytic on $\mathfrak{M}$ functions $\{f_n(p)\}$ converge uniformly on compact subsets $F \subset \mathfrak{M}$ to the limit function $f(p)$. Then the following hold.*

1. *If $\{f_n(p)\} \subset H_\infty^1$, then $f(p) \in H_\infty^1(\mathfrak{M})$ and for each $\Theta(q) \in L_1(\Gamma, d\omega)$*

$$\lim_{n \to \infty} \int_\Gamma f_n(q)\Theta(q) \, d\omega = \int_\Gamma f(q)\Theta(q) \, d\omega. \tag{2.8}$$

2. *If $\{f_n(p)\} \subset H_1^1(\mathfrak{M})$, then $f(p) \in H_1^1(\mathfrak{M})$ and (2.8) holds for every <u>continuous</u> function $\Theta(q)$ on Γ.*

3. *If $\{f_n(p)\} \subset H_\delta^1(\mathfrak{M})$, $1 < \delta < \infty$, then $f(p) \in H_\delta^1(\mathfrak{M})$ and (2.8) holds for all $\Theta(q) \in L_\eta(\Gamma, d\omega)$, where $\delta^{-1} + \eta^{-1} = 1$. In other words, for $\{f_n(p)\} \subset H_\infty^1$ weak $(*)$ convergence in $L^\infty(\Gamma, d\omega)$ holds. For $\{f_n(p)\} \subset H_\delta^1$, $1 < \delta < \infty$, weak $(*)$ convergence in $L_\delta(\Gamma, d\omega)$ holds, while for $\{f_n(p)\} \subset H_1^1$ convergence*

in the weak $(*)$ *topology of measures on* Γ *(weaker, than the weak topology in* L^1*) holds.*

We shall also need the following classes of single-valued analytic functions on $\mathfrak{M}$.

$A(\Gamma)$: this class consists of boundary values on Γ of functions analytic on $\mathfrak{M}$ and continuous on $\overline{\mathfrak{M}}$. $A(\Gamma)$ is viewed as the subspace of $C(\Gamma)$ of continuous functions on Γ with its usual max-norm. $A_1^1(\Gamma)$ is the unit ball in $A(\Gamma)$.

$$N^+(\mathfrak{M}) : \quad f(p) \in N^+ \quad \text{if} \quad \left\{ \int_E \ln^+ |f(p)| \, d\omega \right\}, \qquad E \subset \Gamma^\rho, \tag{2.9}$$

are uniformly integrable with respect to the harmonic measure.

$$N^{++}(\mathfrak{M}) : \quad f(p) \in N^{++} \quad \text{if} \quad \left\{ \int_E |\ln|f(p)|| \, d\omega \right\}, \qquad E \subset \Gamma^\rho, \tag{2.10}$$

are uniformly integrable with respect to the harmonic measure.

For the disk, the class N^+ was introduced by V.I. Smirnov [28] in connection with an important theorem of Polubarinova-Kochina. The class N^{++} was introduced and studied by I.I. Privalov [29]. (Often, in Russian literature, N^+ is denoted by D, and N^{++} by C.) It is easily verified that $H_\delta(\mathfrak{M}) \subset N^+(\mathfrak{M})$ for all $\delta > 0$. The above-mentioned Polubarinova-Kochina theorem yields in our context the following: if $f(p) \in N^+(\mathfrak{M})$ and (2.5) holds, then $f \in H_\delta(\mathfrak{M})$.

For more details concerning the topics discussed in this section the reader is referred to books [2,4] and articles [5–6]. The exposition in [4] is reduced to multiply connected domains, however boundary behavior of functions on $\mathfrak{M}$ can be easily reduced to that in multiply connected domains. Namely, take a contour Γ_1 on $\mathfrak{M}$, homologous to the boundary Γ and consider the "strip" domain $D \subset \mathfrak{M}$ whose boundary is $\Gamma \cup \Gamma_1$. This strip can be chosen so narrow that it is conformally equivalent to a schlicht domain. Then, map D conformally onto a multiply-connected domain $Q \subset \mathbb{C}$. This reduces study of boundary behavior of a function $f(p)$ near Γ to that of a transmitted function in Q near ∂Q. (Remark in passing that the classical version of the theory outlined in this section has been treated in numerous monographs, e.g., [7–10], [29–30], to name a few.)

3. Annihilators of Hardy classes

The following theorem distinguishes $H_\delta(\Gamma)$ functions among $L_\delta(\Gamma, d\omega)$.

Theorem 3.1 (Direct and converse Cauchy's theorems). *Let* $1 \leq \delta \leq \infty$. *If* $f \in H_\delta(\Gamma)$, *then*

$$\int_\Gamma f\alpha = 0 \tag{3.1}$$

holds for any abelian differential α *holomorphic on* $\overline{\mathfrak{M}}$. *Conversely, if* $f \in L_\delta(\Gamma, d\omega)$ *and* (3.1) *holds for all abelian differentials analytic on* $\overline{\mathfrak{M}}$ *then* $f \in H_\delta(\Gamma)$. *If, further,* $f \in C(\Gamma)$, *then* $f \in A(\Gamma)$.

42   S.Ya. Khavinson and T.S. Kuzina

Analytic on $\overline{\mathfrak{M}}$ here means analytic in an open domain S, such that

$$\overline{\mathfrak{M}} \subset S \subset \widehat{\mathfrak{M}}. \tag{3.2}$$

Theorem 3.1 was proved by Read [22–23] and was used to study some extremal problems. A simpler proof was given by Royden [24]. This theorem (with appropriate references) is contained in [2, p. 75, Theorem 6]. A more general case of a surface with a "bad" boundary was treated in [31]. We extend Theorem 3.1 somewhat further. For that we shall need the following general fact.

Proposition 3.2. *There exists an analytic differential α_0 on $\overline{\mathfrak{M}}$ which does not vanish anywhere on $\overline{\mathfrak{M}}$.*

Indeed, existence of such differential α_0 on an open Riemann surface S is a well-known fact in the theory of Riemann surfaces ([32, p. 205]). Suffices to take for S a domain in $\widehat{\mathfrak{M}}$ satisfying (3.2). The differential α_0 shall play an important role in all of the following constructions.

Recall that if X is a Banach space, $Y \subset X$ is a subspace of X and X^* is the dual of X, then the annihilator $Y^\perp$ of Y is the set of all functionals in X^* vanishing on Y. If Z is a subspace of X^*, the annihilator $_\perp Z$ is the subspace in X that consists of all common zeros of functionals in Z.

Introduce the following classes of (abelian) differentials of the first kind on $\mathfrak{M}$:

$$\widetilde{H}_\delta(\mathfrak{M}) = \{\beta = f\alpha_0, \quad f \in H_\delta(\mathfrak{M}), \quad 1 \le \delta \le \infty\}. \tag{3.3}$$

$\widetilde{H}_\delta(\Gamma)$ then denotes boundary values of differentials (3.3) on Γ. Note, that $\widetilde{H}_1(\Gamma)$ is a subspace in $C(\Gamma)^*$. Indeed, the latter is the space of all finite Borel measures on Γ, where the norm of a measure in $C(\Gamma)^*$ is its total variation. At the same time, a differential $\beta \in \widetilde{H}_1(\Gamma)$ obviously produces a measure on Γ and its total variation

$$\int |\beta| = \int |f\alpha_0|, \quad f \in H_1(\Gamma) \tag{3.4}$$

is finite. To see this, consider the complex Green function on $\mathfrak{M}$:

$$T(p, p_0) = G(p, p_0) + i * G(p, p_0) \tag{3.5}$$

($*$ stands for "conjugate" in the p-variable, of course.) Its differential $dT(p, p_0)$ is a meromorphic differential on $\mathfrak{M}$ and at points $q \in \Gamma$ its values along Γ (with respect to any parameterization) are:

$$dT = \left(\frac{\partial G(q, p_0)}{\partial s} + i\, \frac{\partial * G(q, p_0)}{\partial s} \right) ds = -i\, \frac{\partial G(q, p_0)}{\partial n_q}\, ds = -i\, d\omega \tag{3.6}$$

We have:

$$\beta = f\alpha_0 = \frac{f\alpha_0}{id\,T}\, d\omega. \tag{3.7}$$

The ratio $\frac{\alpha_0}{id\,T}$ is a meromorphic function on $\mathfrak{M}$, analytic on Γ and, in view of properties of G, α_0 and Γ (i.e., analyticity of the boundary contours), satisfying

$$0 < m \le \frac{\alpha_0}{id\,T} \le M < +\infty \tag{3.8}$$

with some constants m, M. Hence

$$m \int_\Gamma |f|\, d\omega \le \int_\Gamma |\beta| \le M \int_\Gamma |f|\, d\omega. \tag{3.9}$$

This gives a two-sided estimate of (3.4) in terms of $H_1(\Gamma)$-norm of f. In particular, (3.9) implies that total variation of the measure associated with the differential β is finite.

Proposition 3.3.

$$\perp \widetilde{H}_1(\Gamma) = A(\Gamma); \tag{3.10}$$

$$A(\Gamma)^\perp = \widetilde{H}_1(\Gamma); \tag{3.11}$$

$$\perp \widetilde{H}^\infty(\Gamma) = H_1(\Gamma); \tag{3.12}$$

$$H_1(\Gamma)^\perp = \widetilde{H}^\infty(\Gamma); \tag{3.13}$$

$$\perp \widetilde{H}_\eta(\Gamma) = H_\delta(\Gamma); \tag{3.14}$$

$$H_\delta(\Gamma)^\perp = \widetilde{H}_\eta(\Gamma), \tag{3.15}$$

where $1 < \delta, \eta < \infty$, $\quad \delta^{-1} + \eta^{-1} = 1$.

Proof. Let us prove (3.10)–(3.11).

1) Let $\varphi \in A(\Gamma)$ and $\beta = f\alpha_0 \in \widetilde{H}_1(\Gamma)$. Then, $\int_{\Gamma_\rho} \varphi\beta = \int_{\Gamma_\rho} \varphi f \alpha_0 = 0$, $\forall \rho > 0$ since the differential $\varphi f \alpha_0$ is analytic in $\overline{D^\rho}$. From this, by the F. Riesz theorem applied to the function f in β, it follows

$$\int_\Gamma \varphi\beta = \lim_{\rho \to 0} \int_{\Gamma_\rho} \varphi\beta = 0.$$

Hence, $\perp \widetilde{H}_1(\Gamma) \supset A(\Gamma)$. Now, let $\varphi \in C(\Gamma) \cap {}_\perp\widetilde{H}_1(\Gamma)$ and β be an analytic differential on $\overline{\mathfrak{M}}$. Then $f = \dfrac{\beta}{\alpha_0}$ is an analytic function on $\overline{\mathfrak{M}}$, so $f \in H_1(\Gamma)$. Hence analytic differentials β on $\overline{\mathfrak{M}}$ belong to $\widetilde{H}_1(\Gamma)$. Since $\varphi \in {}_\perp\widetilde{H}_1(\Gamma)$, φ annihilates all differentials analytic on $\overline{\mathfrak{M}}$. Then, the converse part of Theorem 3.1 of Cauchy-Read yields $\varphi \in A(\Gamma)$ and (3.10) is proved.

2) To prove (3.11), i.e.,

$$\left({}_\perp\widetilde{H}_1(\Gamma) \right)^\perp = \widetilde{H}_1(\Gamma),$$

it suffices, by the general theorem of functional analysis (cf. [33, Theorem 4.7]), to show that $\widetilde{H}_1(\Gamma)$ is weak $(*)$ closed in the space $C(\Gamma)^*$. Let a sequence of differentials $\{\beta_n = f_n\alpha_0\} \subset \widetilde{H}_1(\Gamma)$ converges weak $(*)$ in $C(\Gamma)^*$ to the measure μ on Γ. Then, total variations of measures associated with the differentials β_n are uniformly bounded. Then (3.9) implies, that there exists a constant $M > 0$ such that

$$\|f_n\| = \int_\Gamma |f_n|\, d\omega \le M < +\infty. \tag{3.16}$$

Show that the measures $\{f_n d\omega\}$ also converge weak $(*)$. Take an arbitrary $\psi \in C(\Gamma)$. We have:

$$\int_\Gamma f_n \psi\, d\omega = \int_\Gamma f_n \psi \alpha_0 \frac{id\,T}{\alpha_0} = \int_\Gamma \beta_n \psi \frac{id\,T}{\alpha_0}$$

and, hence, there exists a finite limit

$$\lim_{n\to\infty} \int_\Gamma f_n \psi\, d\omega.$$

In particular, $\forall p \in \mathfrak{M}$ there exists a finite limit

$$\lim_{n\to\infty} f_n(p) = \lim_{n\to\infty} \int_\Gamma f_n(q) P(p,q)\, d\omega.$$

$\left(P(p,q) = \dfrac{\frac{\partial}{\partial n_q} G(p,q)}{\frac{\partial}{\partial n_q} G(p_0,q)} \right.$ is a continuous function on Γ.) Thus, the sequence $\{f_n(p)\} \subset H_1(\mathfrak{M})$ converges in $\mathfrak{M}$ (on compact subsets) to an analytic function $f(p)$. Estimate (3.16) implies that $f(p) \in H_1(\mathfrak{M})$. Therefore, S.Ya. Khavinson's lemma from §2 implies that

$$\lim_{n\to\infty} \int_\Gamma f_n \Theta\, d\omega = \int_\Gamma f\Theta\, d\omega$$

for all functions $\Theta(q)$ continuous on Γ. Whence, we have

$$\int_\Gamma \Theta\, d\mu = \lim_{n\to\infty} \int_\Gamma \Theta\, \beta_n = \lim_{n\to\infty} \int_\Gamma f_n \alpha_0\, \Theta$$

$$= \lim_{n\to\infty} \int_\Gamma f_n \frac{\alpha_0}{id\,T} \Theta\, d\omega = \int_\Gamma f \frac{\alpha_0}{id\,T} \Theta\, d\omega = \int_\Gamma f\alpha_0 \Theta.$$

Thus, the measure μ, the limit of measures generated by differentials β_n, coincides with the measure associated with the differential $\beta = f\alpha_0 \in \widetilde{H}_1(\Gamma)$ since $f \in H_1(\mathfrak{M})$. We have proved that $\widetilde{H}_1(\Gamma)$ is weak $(*)$ closed, and hence (3.11).

Proofs of (3.12)–(3.13) are similar and we omit them. Proofs of (3.14)–(3.15) are even simpler, because we can exploit reflexivity of the corresponding Lebesgue spaces. $\qquad\square$

4. Extremal problems, duality, relations between extremal functions

Below, we shall write max (min) instead of sup (inf) in those cases when the fact that a corresponding extremum is attained is a part of the assertion.

Theorem 4.1.

1. *Let μ be a Baire measure on Γ. Then*

$$\sup_{f \in A^1(\Gamma)} \left| \int_\Gamma f \, d\mu \right| = \min_{\beta \in \tilde{H}_1(\Gamma)} \int_\Gamma |d\mu - \beta|. \tag{4.1}$$

2. *Let $Q(q)$ be a summable (wrt $d\omega$) function on Γ. Then*

$$\sup_{f \in A^1(\Gamma)} \left| \int_\Gamma f Q \, d\omega \right| = \max_{f \in H_\infty^1(\Gamma)} \left| \int_\Gamma f Q \, d\omega \right| = \min_{\varphi \in H_1(\Gamma)} \int_\Gamma \left| Q - \frac{\varphi \, \alpha_0}{id \, T} \right| d\omega. \tag{4.2}$$

3. *Let $Q(q)$ be a bounded measurable $(d\omega)$ function on Γ. Then*

$$\sup_{f \in H_1^1(\Gamma)} \left| \int_\Gamma f Q \, d\omega \right| = \min_{\varphi \in H_\infty(\Gamma)} \operatorname*{Vrai\,max}_{q \in \Gamma} \left| Q - \frac{\varphi \alpha_0}{id \, T} \right|. \tag{4.3}$$

(Vrai max is taken with respect to $d\omega$.) If $Q(q)$ is a continuous function on Γ, then the supremum in (4.3) can be replaced by max.

4. *Let $Q(q) \in L_\eta(\Gamma)$, $1 < \eta < \infty$. Then*

$$\max_{f \in H_\delta^1(\Gamma)} \left| \int_\Gamma f Q \, d\omega \right| = \min_{\varphi \in H_\eta(\Gamma)} \left\{ \int_\Gamma \left| Q - \frac{\varphi \, \alpha_0}{id \, T} \right|^\eta d\omega \right\}^{1/\eta}, \quad \eta^{-1} + \delta^{-1} = 1. \tag{4.4}$$

Proof. (4.1)–(4.4) are corollaries of the Hahn-Banach theorem and are justified in a straightforward manner by making use of Proposition 3.3. In order to show existence of the extremal functions in the left-hand sides of (4.2)–(4.4), one also needs to use the lemma on the weak convergence of boundary values from §2 (cf. also [14], where these statements are proved for the case of multiply connected domains in great detail). $\qquad \square$

Theorem 4.2. *The extremal functions f^* and φ^* in the left and right sides of (4.2)–(4.4) satisfy the following relations (θ denotes a real constant).*

1. *For (4.2):*

$$f^* \left[Q - \frac{\varphi^* \alpha_0}{id \, T} \right] = \left| Q - \frac{\varphi^* \alpha_0}{id \, T} \right| e^{i\theta}. \tag{4.5}$$

2. *For (4.3) (the case of a continuous function Q):*

$$f^* \left[Q - \frac{\varphi^* \alpha_0}{id \, T} \right] = \lambda \, |f^*| \, e^{i\theta}, \tag{4.6}$$

where λ is the value of both extrema in (4.3).

3. *For* (4.4):

$$f^* \left[Q - \frac{\varphi^* \alpha_0}{id\,T} \right] = \frac{e^{i\theta}}{\lambda^{\eta-1}} \left| Q - \frac{\varphi^* \alpha_0}{id\,T} \right|^{\eta} = \lambda\, e^{i\theta} |f^*|^{\delta}, \tag{4.7}$$

where θ is a real constant, λ is the value of both extrema in the left and right sides of (4.3) and (4.4) respectively. (4.5)–(4.7) hold at almost all $(d\omega)$ points of Γ and represent necessary and sufficient conditions for the functions f^* and φ^* to be extremal in respective equalities of Theorem 4.1.

Let R be a meromorphic differential on $\mathfrak{M}$, analytic on the boundary Γ. Consider the extremal problem

$$\max_{f \in H^1_\delta(\Gamma)} \left| \int_\Gamma fR \right|, \quad 1 \le \delta \le \infty. \tag{4.8}$$

Then, for Q in (4.5)–(4.7), we take the function

$$Q = \frac{R}{id\,T}. \tag{4.9}$$

This function is meromorphic on $\mathfrak{M}$ and analytic on the boundary Γ. Moreover, we have the following

Theorem 4.3. *The function*

$$P = f^* \left[\frac{R}{id\,T} - \frac{\varphi^* \alpha_0}{id\,T} \right], \tag{4.10}$$

where f^*, φ^* are extremals in (4.2)–(4.4) extends (by the Schwarz reflection principle) to a meromorphic function on the double $\widehat{\mathfrak{M}}$ of the surface $\mathfrak{M}$. The number of zeros on $\widehat{\mathfrak{M}}$ of the function P equals to the number of its poles inside $\mathfrak{M}$ (as usual, zeros on Γ are counted with half multiplicity.) If n is the number of poles of the differential R, then the (same) number of zeros or poles of the function P (on $\widehat{\mathfrak{M}}$ and $\mathfrak{M}$ respectively) equals

$$N = n + L = n + 2h + m - 1. \tag{4.11}$$

Proof. Extendibility of the function P to $\widehat{\mathfrak{M}}$ follows since by (4.5)–(4.7) this function has a constant argument θ on Γ. The latter, and the argument principle, also implies equality of the numbers of zeros and poles of P (P does not have poles on Γ and zeros on Γ are all of even orders.) The poles of P are at the poles of the differential R and zeros of the (meromorphic) differential $id\,T$. Since by (3.6) $id\,T = d\omega > 0$ on Γ, dT also extends by symmetry to the meromorphic differential on the double $\widehat{\mathfrak{M}}$ which has genus L. The Riemann-Roch theorem implies that the degree of any meromorphic differential on a compact surface of genus L is $2L - 2$ ([34, Theorem 10.11]). Whence, by symmetry, the degree of the divisor of $id\,T$ on $\mathfrak{M}$ equals $L - 1$. Since $id\,T$ has only one simple pole at p_0, the number of zeros $id\,T$ on $\mathfrak{M}$ is equal to L. This proves (4.11). $\qquad\square$

The main result, that we shall use to study the structure of extremals further, is contained in the following theorem.

Theorem 4.4. *Let $q_1, \ldots, q_N$ be the zeros of function P, and $\tilde{q}_1, \ldots, \tilde{q}_N$ be the poles. Then,*

$$P = c \prod_1^N \psi(p, \tilde{q}_j, q_j), \tag{4.12}$$

where c is a constant.

Note once more that the set of poles $\{\tilde{q}_1, \ldots, \tilde{q}_N\}$ of the function P consists of n poles of the differential R which defines the extremal problem and L zeros of the differential dT (i.e., the critical points of the Green function $G(p, p_0)$). Also, the point p_0, the pole of the differential dT belongs to the set $\{q_1, \ldots, q_N\}$ of zeros of the function P.

Proof. Denote by S the product in (4.12) and let

$$\sigma = \frac{P}{S}. \tag{4.13}$$

The function σ is analytic in $\overline{\mathfrak{M}}$, does not have zeros or poles, but, a priori, could turn out to be multi-valued. Of course, it does not have periods around boundary cycles, but could have nonzero periods around some of the basis cycles on the handles of our surface. This periods may occur because the function

$$\sum_j {}^* N(p, \tilde{q}_j, q_j) \tag{4.14}$$

could, in general, be multi-valued. If a period of (4.14) around the basis cycle K_i, $i = 1, \ldots, 2h$ equals a_i, then all periods of the sum (4.14) are given by the formula

$$\beta = a_1 n_1 + a_2 n_2 + \cdots + a_{2h} n_{2h}, \tag{4.15}$$

where $(n_1, n_2, \ldots, n_{2h})$ is an arbitrary set of integers (positive, or negative). The set of periods $\{\beta\}$ is countable and we shall enumerate all the periods. If for $p \in \mathfrak{M}$, $\sigma(p)$ is one value of a multivalued function σ at p, then all other values are given by $\sigma(p) e^{-i\beta_k}$, $k = 1, \ldots$, where $\{\beta_k\}$ are all the periods (4.15). On every boundary contour γ_j the arguments of all branches of σ are constant, hence boundary contours $\gamma_1, \ldots, \gamma_m$ are mapped by the function $w = \sigma(p)$ onto a countable set of segments that lie on the rays in the w-plane emanating from the origin $w = 0$. Denote by A the set of all images of the boundary Γ under mappings by all the branches of σ. The image of the surface $\mathfrak{M}$ under the mapping $w = \sigma(p)$ is an open connected set. This set is also bounded since $|\sigma(p)|$ is a single-valued continuous function on $\overline{\mathfrak{M}}$. Hence, the set A cannot exhaust the boundary of the image of $\mathfrak{M}$. Let w_0 be a boundary point of the image of $\mathfrak{M}$ that does not belong to A. Then, $\exists \{w_1 = \sigma(p_1), \ldots, w_k = \sigma(p_k), \ldots\} \subset \sigma(\mathfrak{M})$, $w_k \to w_0$, $k \to \infty$. Since $\overline{\mathfrak{M}}$ is compact we can assume without loss of generality that $p_k \to \tilde{p}$, $w_0 \in \sigma\{\tilde{p}\}$, $\tilde{p} \notin \Gamma$, otherwise in view of continuity and single-valuedness of all branches of σ near the boundary Γ, $w_0 \in A$. Let $\sigma(\tilde{p}) = w_0^0$ be a value of σ at $\tilde{p}$. By analyticity of σ, we can find a disk V centered at w_0^0 such that $V \subset \sigma(\mathfrak{M})$. Let Z be a preimage of V. Then $\{V_k = Ve^{-i\beta_k}\}$, $k = 1, \ldots$, is the set of all images of Z. Let w_k^0 be a "center" of V_k. (It is an image of $\tilde{p}$ under the mapping by the corresponding branch of σ.) We can assume that $w_k \in V_k$ since $\{p_k\} \to \tilde{p}$, $|w_k - w_k^0| \to 0$. Therefore, because $w_k \to w_0$, $w_k^0 \to w^0$ as well. But then w_0 is an interior point of V_k for all $k \geq k_0$. Hence, w_0 is an interior point of $\sigma(\mathfrak{M})$. But it was a boundary point. We have reached a contradiction, hence $\sigma(p) \equiv const = C$. $\qquad\square$

Let us yet give another, purely "analytic" proof. The function

$$u = \ln|\sigma(p)| \tag{4.16}$$

is a single-valued harmonic function on $\overline{\mathfrak{M}}$. The function $W = *u(p)$ is multi-valued and its values on different branches differ by additive constants. Hence dW is a well-defined harmonic differential. The differential

$$\beta = u\, dW$$

is then well defined on $\overline{\mathfrak{M}}$. Moreover, since every branch of $\sigma(p)$ has a constant argument on every boundary contour,

$$\int_\Gamma \beta = \int_\Gamma u\, \frac{\partial W}{\partial s}\, ds = 0.$$

But then

$$0 = \int_\Gamma u\, \frac{\partial W}{\partial s}\, ds = \int_\Gamma u\, \frac{\partial u}{\partial n}\, ds = \iint_{\mathfrak{M}} \left[\left(\frac{\partial u}{\partial x}\right)^2 + \left(\frac{\partial u}{\partial y}\right)^2 \right] dx\, dy,$$

the Dirichlet integral calculated with respect to some coordinate atlas on $\mathfrak{M}$. Hence, $\operatorname{grad} u = 0$ on $\mathfrak{M}$, i.e., $u \equiv const$ and $W \equiv const$, so $\sigma(p) \equiv const = C$.

5. Representation of extremal functions via Green's function and Neumann's function

Let $T(p, \tilde{p})$ be the complex Green function of $\mathfrak{M}$ with pole at $\tilde{p} \in \mathfrak{M}$,

$$T(p, \tilde{p}) = G(p, \tilde{p}) + i * G(p, \tilde{p}) \tag{5.1}$$

($*$ as above denotes conjugate harmonic function with respect to the variable p.) For $\tilde{p} = p_0$, (5.1) coincides with (3.5) and its differential dT has been a part of many preceding formulas.

Theorem 5.1. *Let* $q_1, \ldots, q_k$, $k \leq N = n + 2h + m - 1$, *be the zeros of the extremal function* f^* *in the problem* (4.8) *that lie inside* $\mathfrak{M}$. *Then the following hold.*

1) *For $\delta = \infty$,*

$$f^*(p) = C \exp\left[-\sum_1^k T(p, q_i)\right],\tag{5.2}$$

where $C : |C| = 1$, is a constant.

2) *For $1 \leq \delta < \infty$,*

$$f^*(p) = C_1 \exp\left[-\sum_1^k T(p, q_i)\right]\left\{\prod_1^N \psi(p, \tilde{q}_i, q_i)\right\}^{1/\delta}$$

$$\times \exp\left(-\sum_1^N T(p, \tilde{q}_i) + \sum_1^N T(p, q_i)\right)^{1/\delta},\tag{5.3}$$

where $\tilde{q}_i$ are the poles of the function P as in Theorem 4.4 and C_1 is a constant such that $\|f^*\|_\delta = 1$.

Proof. Consider on $\mathfrak{M}$ the harmonic function

$$u(p) = \ln|f^*(p)| + \sum_1^k G(p, q_i).\tag{5.4}$$

First show that the family of integrals

$$\int_E |u(p)|\, d\omega, \quad E \subset \partial D_\rho,\tag{5.5}$$

is uniformly absolutely continuous with respect to harmonic measure $d\omega$ for all sufficiently small $\rho > 0$. Indeed, $f^* \subset H_\delta(\mathfrak{M}) \subset N^+(\mathfrak{M})$, so the family

$$\int_E \ln^+ |f|\, d\omega\tag{5.6}$$

is uniformly absolutely continuous with respect to $d\omega$. Moreover, (4.10) yields

$$\frac{1}{f^*} = \left[\frac{R}{id\,T} - \frac{\varphi^* \alpha_o}{id\,T}\right] \cdot \frac{1}{P}.\tag{5.7}$$

The function P is analytic on Γ and then as is easily seen the function $1/P$ belongs to N^+ in a sufficiently narrow strips adjacent to the boundary Γ. The functions $\frac{R}{id\,T}$ and $\frac{\alpha_o}{id\,T}$ are also analytic on Γ while $\varphi^* \in H_\eta(\mathfrak{M}) \subset N^+(\mathfrak{M})$. Hence, the function

$$\left[\frac{R}{id\,T} - \frac{\varphi^* \alpha_0}{id\,T}\right]$$

belongs to the class H_η in such boundary "strips" and, accordingly to N^+. Therefore, each one the two factors in the representation (5.7) for $1/f^*$ belongs to N^+ in sufficiently narrow strips near the boundary and so does $1/f^*$. Hence, the family

$$\int_E \ln^+ \left|\frac{1}{f}\right|\, d\omega, \quad E \subset \partial D_\rho\tag{5.8}$$

is uniformly absolutely continuous with respect to $d\omega$ for all sufficiently small ρ. Combining (5.4), (5.5) and (5.8) with continuity of the Green function $G(p, q_i)$, $\delta = 1, \ldots, k$ near the boundary, we obtain that the family

$$\int_E |u(p)| \, d\omega, \quad E \subset \partial D_\rho \tag{5.9}$$

is uniformly absolutely continuous with respect to $d\omega$ for all small $\rho > 0$. This suffices to claim that $u(p)$ is representable by the Green integral of its boundary values (cf. [4], [6]):

$$u(p) = \int_E u(q) P(p, q) \, d\omega = \int_E \ln |f^*(q)| P(p, q) \, d\omega. \tag{5.10}$$

When $\delta = \infty$, $|f^*(q)| \equiv 1$ on Γ as is seen from (4.5), hence the last integral in (5.10) is equal to zero identically. Then, from (5.4) we obtain

$$\ln |f^*(p)| = -\sum_1^k G(p, q_i). \tag{5.11}$$

But this is equivalent to (5.2).

For $1 \le \delta < \infty$, (4.6), (4.7), (4.10), (4.12) and (5.4) imply that

$$u(q) = \ln |f^*(q)| = \frac{1}{\delta} \sum_1^k N(q, \widetilde{q}_i, q_i) + \ln C_1 \tag{5.12}$$

on Γ, where C_1 is a constant. Functions

$$v_j(p) = N(p, \widetilde{q}_j, q_j) - G(p, \widetilde{q}_j) + G(p, q_j) \tag{5.13}$$

are harmonic on $\mathfrak{M}$ and representable by the Green integral of their boundary values $N(q, \widetilde{q}_j, q_j)$. (In the case when the zero q_j is on Γ, the corresponding term $G(p, q_j) \equiv 0$. The poles $\widetilde{q}_j$ all lie inside $\mathfrak{M}$.) We obtain now from (5.4) and (5.10)–(5.13):

$$\ln |f^*(p)| = -\sum_1^k G(p, q_i) + \int_\Gamma u(q) P(p, q) \, d\omega \tag{5.14}$$

$$= -\sum_1^k G(p, q_i) + \int_\Gamma \left(\frac{1}{\delta} \sum_1^N N(q, \widetilde{q}_i, q_i) + \ln C_1 \right) P(p, q) \, d\omega$$

$$= -\sum_1^k G(p, q_i) + \frac{1}{\delta} \left\{ \sum_1^N N(p, \widetilde{q}_j, q_j) - \sum_1^N G(p, \widetilde{q}_j) + \sum_1^k G(p, q_j) \right\} + \ln C_1.$$

The latter is equivalent to (5.3). $\qquad\square$

Theorem 5.1 extends to finite Riemann surfaces the representation formulae for extremal functions in planar multiply connected domains derived in [13, 14]. In these papers it was also shown that the above formulae imply well-known earlier representations of extremals in the disk and annulus (the latter were obtained in [35]).

6. Representations of extremal functions via Schwarz kernels and Blaschke products

Formulas (5.2)–(5.3) represent a single-valued function f^* as product of functions, that are not in general single-valued. In this section we derive a different representation f^*, this time as product of single-valued functions. We shall make use of the ideas in [5, 6], where the general theory of factorization of single-valued functions on finite Riemann surfaces was developed.

Let $*P(p,q)$ be the conjugate harmonic function to the Green kernel $P(p,q)$ (with respect to the variable $p \in \mathfrak{M}$), $q \in \Gamma$ is fixed. Denote by $y_j(q)$, $j = 1, \ldots, L$ its period around the basis cycle k_j. [5, §1] and [6, §§3,5] contain explicit expressions for $y_j(q)$ in terms of differentials of a particular basis in the space of abelian differential of the first kind on $\mathfrak{M}$ and the Green function (this basis is constructed in [27, Ch. 4, §3]). In a nutshell, all formulas for $y_j(q)$ are given in terms of the Green function ([5, formula (1.3)], [6, (3.5)]). We need not give here the precise expressions. Choose on Γ open arcs $\Delta_1, \ldots, \Delta_L$ so that their closures $\overline{\Delta}_1, \ldots, \overline{\Delta}_L$ do not intersect and let $\omega_j(p)$ denote the harmonic measures of the arcs Δ_j, $j = 1, \ldots, L$. Denote by a_{ij} the period of $*\omega_j(p)$ around the basis cycle K_i, $i = 1, \ldots, L$. Arcs $\Delta_1, \ldots, \Delta_L$ are chosen so that the period matrix

$$A = \|a_{ij}\|, \quad i = 1, \ldots, L, \quad j = 1, \ldots, L \tag{6.1}$$

is nondegenerate. The existence of such choice of arcs is proved in [5, Lemma 2.1] and [6, Lemma 4.7]. Consider vector-functions $y(q) = (y_1(q), \ldots, y_L(q))$, $q \in \Gamma$ and $\omega(p) = (\omega_1(p), \ldots, \omega_L(p))$, $p \in \mathfrak{M}$ and define the vector $\lambda(q) = (\lambda_1(q), \cdots, \lambda_L(q))$ from the equation

$$\lambda(q)^T = A^{-1} \cdot y(q)^T. \tag{6.2}$$

(Here, T denotes the transpose.) The function

$$k(p,q) = P(p,q) - \omega(p)A^{-1}y(q)^T = P(p,q) - \sum_1^L \lambda_j(q)\omega_j(p) \tag{6.3}$$

has a single-valued conjugate $*k(p,q)$ (with respect to p) for each $q \in \Gamma$. The function

$$K(p,q) = k(p,q) + i*k(p,q) \tag{6.4}$$

defined on $\mathfrak{M} \times \Gamma$ possesses a number of properties similar to those of the Schwarz kernel in the disk ([6, §5], [5, §3]). As in [5,6], we shall call it the Schwarz kernel for $\mathfrak{M}$. We shall need the following identities

$$\int_{\Delta_j} K(p,q)\, d\omega \equiv 0, \quad p \in \mathfrak{M}, \quad j = 1, \ldots, L. \tag{6.5}$$

If we set

$$w_j(p) = \omega_j(p) + i * \omega_j(p), \quad j = 1, \ldots, L, \tag{6.6}$$

then $K(p, q)$ can be written as follows:

$$K(p, q) = P(p, q) + i * P(p, q) - \sum_1^L \lambda_j(q) w_j(p). \tag{6.7}$$

Similarly, we can find real numbers $d_1, \ldots, d_L$ such that the function

$$* \sum_1^L d_j \omega_j(p) \tag{6.8}$$

has the same periods around the basis cycles as the function $*T(p, q)$ with a fixed pole $q \in \mathfrak{M}$. Then, the function

$$B(p, q) = \exp\left[-T(p, q) + \sum_1^L d_j w_j(p) \right] \tag{6.9}$$

is a single-valued analytic function of p on $\mathfrak{M}$ that has one zero at $p = q$, is bounded on $\mathfrak{M}$ and continuous near the boundary Γ except for the end points of arcs Δ_j, $j = 1, \ldots, L$. The boundary values of $|B(p, q)|$ equal

$$|B(p, q)| = \begin{cases} 1, & p \in \Gamma \setminus \overline{\Delta}, \quad \Delta = \bigcup_{j=1}^L \Delta_j, \\ \exp d_j, & p \in \Delta_j, \quad j = 1, \ldots, L. \end{cases} \tag{6.10}$$

Following [5, §4 and 6, §6], we shall call function $B(p, q)$ the Blaschke factor in $\mathfrak{M}$. Convergence of infinite Blaschke products composed from such Blaschke factors has been investigated in [5, 6]. But our products will only contain a finite number of factors.

Theorem 6.1. *Let $q_1, \ldots, q_k$ be interior zeros of the function f^*, the extremal in the problem (4.8). Then, the following hold.*

   1. *For $\delta = \infty$,*

$$f^*(p) = cB(p)Q(p), \tag{6.11}$$

    where c, $|c| = 1$, is a constant, and the single-valued functions $B(p)$ and $Q(p)$ have the following form

$$B(p) = \prod_1^k B(p, q_i); \tag{6.12}$$

$$Q(p) = \exp\left(\sum_1^L \lambda_j w_j(p) \right), \tag{6.13}$$

and also,

$$\lambda_j = -\sum_1^k d_j^i, \tag{6.14}$$

where d_j^i are coefficients at $w_j(p)$ in the formula (6.9) for $B(p, q_i)$, $i = 1, \ldots, k$, $j = 1, \ldots, L$.

2. *For $1 \leq \delta < +\infty$,*

$$f^*(p) = cB(p)Q(p)\prod_1^L F_i(p), \tag{6.15}$$

where c is a constant and $B(p), Q(p), F_i(p)$, $i = 1, \ldots, L$ are single-valued analytic functions. Moreover, $B(p)$ has form (6.12), $Q(p)$ has form (6.13) with λ_j defined by (cf. (6.7)–(6.9))

$$\lambda_j = \int_\Gamma \lambda_j(q) \ln \left| \frac{f^*(q)}{B(q)} \right| d\omega, \tag{6.16}$$

while $F_i(p)$ has representation (6.25) below.

Proof. Consider the function

$$F(p) = \frac{f^*(p)}{B(p)}. \tag{6.17}$$

In the proof of Theorem 5.1 we have verified that $f^* \in N^{++}(\mathfrak{M})$. Dividing it by $B(p)$, a function that is "nice" near Γ, does not violate its inclusion in N^{++}, so $F \in N^{++}(\mathfrak{M})$. We have then for its boundary values:

$$|F(q)| = \frac{|f^*(q)|}{|B(q)|} = \begin{cases} |f^*(q)|, & q \in \Gamma \setminus \overline{\Delta} \\ |f^*(q)| \exp(-d_j), & q \in \Delta_j, \ j = 1, \ldots, L. \end{cases} \tag{6.18}$$

Here, the constants d_j are defined by

$$d_j = \sum_{i=1}^k d_j^i, \ j = 1, \ldots, L, \tag{6.19}$$

where d_j^i are the coefficients at $w_j(p)$ in the formula (6.9) for the Blaschke factor $B(p, q_i)$. The inclusion $F \in N^{++}(\mathfrak{M})$ means that the harmonic function $\ln |F(p)|$ is representable by Green's integral of its boundary values:

$$\ln |F(p)| = \int_\Gamma P(p, q) \ln |F(q)| d\omega. \tag{6.20}$$

54 S.Ya. Khavinson and T.S. Kuzina

Using (6.3)–(6.4) first and then (6.18) and (6.5), we obtain

$$\ln|F(p)| = \int_{\Gamma}\left[k(p,q) + \sum_1^L \lambda_j(q)\omega_j(p)\right]\ln|F(q)|\,d\omega$$

$$= \operatorname{Re}\int_{\Gamma} K(p,q)\ln|F(q)|\,d\omega + \sum_1^L \lambda_j\omega_j(p)$$

$$= \operatorname{Re}\left\{\int_{\Gamma\setminus\overline{\Delta}} K(p,q)\ln|f^*(q)|\,d\omega + \sum_1^L \int_{\Delta_j} K(p,q)[\ln|f^*(q)| - d_j]\right\} + \sum_1^L \lambda_j\omega_j(p)$$

$$= \operatorname{Re}\int_{\Gamma} K(p,q)\ln|f^*(q)|\,d\omega + \sum_1^L \lambda_j\omega_j(p), \tag{6.21}$$

where

$$\lambda_j = \int_{\Gamma}\lambda_j(q)\ln|F(q)|\,d\omega,$$

as was claimed in (6.16).

For $\delta = \infty$, $\ln|f^*(q)| \equiv 0$ and we obtain

$$f^*(p) = F(p)B(p) = \prod_1^k B(p,q_k)Q(p),$$

where $Q(p)$ has the form (6.13), and also

$$\lambda_j = -\sum_1^L d_j\int_{\Delta_j}\lambda_j(q)\,d\omega.$$

(On the other hand, comparing (6.11) and (5.2) we see that

$$\lambda_j = -d_j = -\sum_1^k d_j^i,$$

as was stated in (6.14)).

For $1 \leq \delta < +\infty$, the above argument yields

$$f^*(p) = B(p)F(p) = B(p)Q(p)\exp\int_{\Gamma} K(p,q)\ln|f^*(q)|\,d\omega. \tag{6.22}$$

So far we have not been keeping track of possible constants appearing when we switch from $\ln|F(p)|$ to $F(p)$. We shall continue this in the sequel and will only write down possible multiplicative constants in the final formula (6.15). Starting out from the expression (5.12) for $\ln|f^*|$, consider

$$\operatorname{Re}\frac{1}{\delta}\int\limits_{\Gamma} K(p,q)N(q,\widetilde{q}_i,q_i)\,d\omega$$

$$= \frac{1}{\delta}\left\{\int\limits_{\Gamma} P(p,q)N(q,\widetilde{q}_i,q_i)\,d\omega - \sum_1^L \mu_j^i\omega_j(p)\right\} \qquad (6.23)$$

$$= \frac{1}{\delta}\left\{N(p,\widetilde{q}_i,q_i) - G(p,\widetilde{q}_i) + G(p,q_i) - \sum_1^L \mu_j^i\omega_j(p)\right\},$$

where $i,j = 1,\ldots,L$,

$$\mu_j^i = \int\limits_{\Gamma} \lambda_j(q)\,N(q,\widetilde{q}_i,q_i)\,d\omega, \quad i,j=1,\ldots,L \qquad (6.24)$$

and $G(p,q_i) \equiv 0$ for $k+1 \leq i \leq N$ (i.e., when $q_i \in \Gamma$). Then (6.24) leads to single-valued analytic functions on $\mathfrak{M}$ defined by

$$F_i(p) = \exp\left\{\frac{1}{\delta}\int\limits_{\Gamma} K(p,q)\,N(q,\widetilde{q}_i,q_i)\,d\omega\right\}$$

$$= \left\{\Psi(p,\widetilde{q}_i,q_i)\exp\left[-T(p,\widetilde{q}_i) + T(p,q_i) - \sum_{j=1}^L \mu_j^i w_j(p)\right]\right\}^{1/\delta}. \qquad (6.25)$$

Thus, we have arrived at the representation (6.15) and Theorem 6.1 is proved. $\qquad\square$

References

[1] M. Parreau, *Sur les moyennes des fonctions harmoniques et analytiques et la classification des surfaces de Riemann*, Ann. Inst. Fourier, **3** (1951), 103–197.

[2] M. Heins, *Hardy classes on Riemann surfaces*, Lecture Notes in Mathematics, **98**, Springer-Verlag, Berlin–Heidelberg–New York, 1969.

[3] W. Rudin, *Analytic functions of class H_p*, Trans. Amer. Math. Soc., **78** (1955), no. 1, 46–66.

[4] S.Ya. Khavinson, *Factorization of analytic functions in finitely connected domains*, Moskov. Inzh.-Stroit. Inst. (Moscow Institute of Civil Engineering), Moscow, 1981. (Russian)

[5] S.Ya. Khavinson, *Factorization of univalent analytic functions on compact bordered Riemann surfaces*, Zapiski Nauchnych Seminarov LOMI, **170** (1989), no. 27, 285–313. (Russian); English transl., J. Soviet Math. **63** (1993), no. 22, 275–290.

[6] S.Ya. Khavinson, *Theory of factorization of single-valued analytic functions on compact Riemann surfaces with a boundary*, Uspekhi Mat. Nauk, **44** (1989), no. 4(268), 155–189. (Russian); English transl., Russian Math. Surveys **44** (1989), no. 4, 113–156.

[7] K. Hoffman, *Banach Spaces of Analytic Functions*, Prentice-Hall, Englewood Cliffs, New Jersey, 1962.

[8] P. Duren, *Theory of H^p Spaces*, Academic Press, New York, 1970.

[9] J. Garnett, *Bounded Analytic Functions*, Pure and Applied Math., vol. 96, Academic Press, New York, 1981.

[10] P. Koosis, *Introduction to H^p Spaces*, Cambridge Univ. Press, 1980.

[11] A.I. Macintyre, W.W. Rogosinsky, *Extremum problems in the theory of analytic functions*, Acta Math., **82** (1952), 275–325.

[12] W.W. Rogosinsky, H. Shapiro, *On certain extremum problems for analytic functions*, Acta Math., **90** (1954), no. 3, 287–318.

[13] S.Ya. Khavinson, *The representation of extremal functions in the classes E_p by the functions of Green and Neumann*, Matem. Zametki, **16** (1974), no. 5, 707–716. (Russian)

[14] S.Ya. Khavinson, *Foundations of the theory of extremal problems for bounded analytic functions and various generalizations of them*, Moskov. Inzh.-Stroit. Inst. (Moscow Institute of Civil Engineering), Moscow, 1981. (Russian); English transl., Amer. Math. Soc. Transl., Ser. 2, **129** (1986), 1–61.

[15] S.Ya. Khavinson, *Foundations of the theory of extremal problems for bounded analytic functions with additional conditions*, Moskov. Inzh.-Stroit. Inst. (Moscow Institute of Civil Engineering), Moscow, 1981. (Russian); English transl., Amer. Math. Soc. Transl., Ser. 2, **129**, (1986), 63–114.

[16] L.V. Ahlfors, *Open Riemann surfaces and extremal problems on compact subregions*, Commentarii Mathematical Helvetici, **24** (1950), 100–134.

[17] L.V. Ahlfors, *Bounded analytic functions*, Duke Math. J., **14** (1947), no. 1, 1–11.

[18] A.G. Vitushkin, *Analytic capacity of sets and problems in approximation theory*, Uspekhi Mat. Nauk, **22** (1967), no. 6, 141–199. (Russian); English transl., Russian Math. Surveys, **22** (1967), no. 6, 139–200.

[19] J. Garnett, *Analytic Capacity and Measure*, Lecture Notes in Math, **297**, Springer-Verlag, Berlin–New York, 1972.

[20] S.Ya. Khavinson, *Additional problems of the theory of removable sets of analytic functions*, Moskov. Inzh.-Stroit. Inst. (Moscow Institute of Civil Engineering), Moscow, 1982. (Russian)

[21] S.Ya. Khavinson, *Golubev sums: a theory of extremal problems like the analytic capacity problem and of related approximation processes*, Uspekhi Mat. Nauk, **54** (1999), no. 4, 75–142. (Russian); English transl., Russian Math. Surveys, **54** (1999), no. 4, 753–818.

[22] A. Read, *A converse of Cauchy's theorem and applications of extremal problems*, Acta Math., **100** (1958), no. 1–2, 1–22.

[23] A. Read, *Conjugate extremal problems of class H_p, $p > 1$*, Suomalais tiedeakat toimikuks. Sar. **AI**, 1958, no. 250/28, 1–8.

[24] H. Royden, *Algebras of bounded analytic functions on Riemann surfaces*, Acta Math., **114** (1965), 113–142.

[25] S.Ya. Khavinson, *On an extremal problem in the theory of analytic functions*, Uspekhi Mat. Nauk, **4** (1949), no. 4(32), 158–159. (Russian)

[26] S.Ya. Khavinson, *On certain extremal problems in the theory of analytic functions*, Uchenye Zapiski Moskov. Gos. Univ., Ser. Matematica, **148** (1951), no. 4, 133–143. (Russian)

[27] M. Shiffer and D.C. Spencer, *Functionals on Finite Riemann Surfaces*, Princeton Univ. Press, Princeton, 1954.

[28] E.D. Solomentsev, *On certain classes of subharmonic functions*, Izv. Acad. Nauk SSSR, Ser. Matem., **2** (1938), 571–582. (Russian)

[29] I.I. Privalov, *Boundary Properties of Analytic Functions*, Moskov. Gos. Univ., Moscow, 1941. (Russian)

[30] I.I. Privalov, *Boundary Properties of Analytic Functions*, Gostekhizdat, Moscow–Leningrad, 1950. (Russian); German transl., I. Privaloff, *Randeigenschaften analytischer Funktionen*, Deutscher Verlag der Wissenschaften, Berlin, 1956.

[31] M. Hasumi, *Hardy classes on plane domains*, Ark. Math., **16** (1978), no. 2, 213–227.

[32] O. Forster, *Lectures on Riemann Surfaces*, Graduate texts in Math., Springer–Verlag, 1981.

[33] G. Springer, *Introduction to Riemann Surfaces*, Addison–Wesley, 1957.

[34] W. Rudin, *Functional Analysis*, McGraw-Hill, 1973.

[35] O.B. Voskanyan, *On an extremal problem in the theory of Banach spaces of functions analytic in an annulus*, Izv. Arm. Acad. Nauk SSR, **VI** (1971), no. 5, 412–418. (Russian)

S.Ya. Khavinson and T.S. Kuzina
Moscow State Civil Engineering University
Yaroslavskoe Shosse 26
Moscow 129337
Russia

Operator Theory:
Advances and Applications, Vol. 158, 59–86

Extremal Problems for Nonvanishing Functions in Bergman Spaces

Dov Aharonov, Catherine Bénéteau, Dmitry Khavinson
and Harold Shapiro

Dedicated to the memory of Semeon Yakovlevich Khavinson.

Abstract. In this paper, we study general extremal problems for non-vanishing functions in Bergman spaces. We show the existence and uniqueness of solutions to a wide class of such problems. In addition, we prove certain regularity results: the extremal functions in the problems considered must be in a Hardy space, and in fact must be bounded. We conjecture what the exact form of the extremal function is. Finally, we discuss the specific problem of minimizing the norm of non-vanishing Bergman functions whose first two Taylor coefficients are given.

Mathematics Subject Classification (2000). Primary: 30A38, Secondary: 30A98.

1. Introduction

For $0 < p < \infty$, let

$$A^p = \{f \text{ analytic in } \mathbb{D} : (\int_{\mathbb{D}} |f(z)|^p dA(z))^{\frac{1}{p}} := \|f\|_{A^p} < \infty\}$$

denote the Bergman spaces of analytic functions in the unit disk $\mathbb{D}$. Here dA stands for normalized area measure $\frac{1}{\pi}dxdy$ in $\mathbb{D}$, $z = x + iy$. For $1 \le p < \infty$, A^p is a Banach space with norm $\| \ \|_{A^p}$. A^p spaces extend the well-studied scale of Hardy spaces

$$H^p := \{f \text{ analytic in } \mathbb{D} : (\sup_{0<r<1} \int_0^{2\pi} |f(re^{i\theta})|^p \frac{d\theta}{2\pi})^{\frac{1}{p}} := \|f\|_{H^p} < \infty\}.$$

The second author's research was partially supported by the AWM-NSF Mentoring Travel Grant and by a University Research Council Grant at Seton Hall University. The third author gratefully acknowledges support from the National Science Foundation.

60   D. Aharonov, C. Bénéteau, D. Khavinson and H.S. Shapiro

For basic accounts of Hardy spaces, the reader should consult the well-known monographs [Du, Ga, Ho, Ko, Pr]. In recent years, tremendous progress has been achieved in the study of Bergman spaces following the footprints of the Hardy spaces theory. This progress is recorded in two recent monographs [HKZ, DS] on the subject.

In H^p spaces, the theory of general extremal problems has achieved a state of finesse and elegance since the seminal works of S.Ya. Khavinson, and Rogosinski and Shapiro (see [Kh1, RS]) introduced methods of functional analysis. A more or less current account of the state of the theory is contained in the monograph [Kh2]. However, the theory of extremal problems in Bergman spaces is still at a very beginning. The main difficulty lies in the fact that the Hahn-Banach duality that worked such magic for Hardy spaces faces tremendous technical difficulty in the context of Bergman spaces because of the subtlety of the annihilator of the A^p space ($p \geq 1$) inside $L^p(dA)$. [KS] contains the first more or less systematic study of general linear extremal problems based on duality and powerful methods from the theory of nonlinear degenerate elliptic PDEs. One has to acknowledge, however, the pioneering work of V. Ryabych [Ry1, Ry2] in the 60s in which the first regularity results for solutions of extremal problems were obtained. Vukotić's survey ([Vu]) is a nice introduction to the basics of linear extremal problems in Bergman space. In [KS], the authors considered the problem of finding, for $1 < p < \infty$,

$$\sup\{|\int_{\mathbb{D}} \bar{w} f dA| : \|f\|_{A^p} \leq 1\}, \tag{1.1}$$

where w is a given rational function with poles outside of $\mathbb{D}$. They obtained a structural formula for the solution (which is easily seen to be unique) similar to that of the Hardy space counterpart of problem (1.1). Note here that by more or less standard functional analysis, problem (1.1) is equivalent to

$$\inf\{\|f\|_{A^p} : f \in A^p, l_i(f) = c_i, i = 1, \ldots, n\}, \tag{1.2}$$

where the $l_i \in (A^p)^*$ are given bounded linear functionals on A^p, $p > 1$. Normally, for l_i one takes point evaluations at fixed points of $\mathbb{D}$, evaluations of derivatives, etc... More details on the general relationship between problems (1.1) and (1.2) can be found in [Kh2, pp. 69-74]. For a related discussion in the Bergman spaces context, we refer to [KS, p. 960]. In this paper, we focus our study on problem (1.2) for nonvanishing functions. The latter condition makes the problem highly nonlinear and, accordingly, the duality approach does not work. Yet, in the Hardy spaces context, in view of the parametric representation of functions via their boundary values, one has the advantage of reducing the nonlinear problem for nonvanishing functions to the linear problem for their logarithms. This allows one to obtain the general structural formulas for the solutions to problems (1.1) or (1.2) for nonvanishing functions in Hardy spaces as well. We refer the reader to the corresponding sections in [Kh2] and the references cited there. Also, some of the specific simpler problems for nonvanishing H^p functions have recently been solved in [BK]. However, all the above-mentioned methods fail miserably in the context of Bergman spaces for the simple reason that there are no non-trivial Bergman functions that, acting as multiplication operators on Bergman spaces, are isometric.

Let us briefly discuss the contents of the paper. In Section 2, we study problem (1.2) for nonvanishing Bergman functions: we show the existence and uniqueness of the solutions to a wide class of such problems. Our main results are presented in Sections 2 and 3 and concern the regularity of the solutions: we show that although posed initially in A^p, the solution must belong to the Hardy space H^p, and hence, as in the corresponding problems in Hardy spaces in [Kh2], must be a product of an outer function and a singular inner function. Further, we show that that the solutions to such problems are in fact bounded. Moreover, led by an analogy with the Hardy space case, we conjecture that the extremal functions have the form

$$f^*(z) = \exp(\sum_{j=1}^{k} \lambda_j \frac{e^{i\theta_j} + z}{e^{i\theta_j} - z}) \prod_{j=1}^{2n-2} (1 - \bar{\alpha}_j z)^{\frac{2}{p}} \prod_{j=1}^{n}(1 - \bar{\beta}_j z)^{-\frac{4}{p}}, \qquad (1.3)$$

where $|\alpha_j| \leq 1$, $|\beta_j| < 1$, $\lambda_j < 0$, $n \geq 1$, $k \leq 2n - 2$. In Section 4, we sketch how, if one knew some additional regularity of the solutions, it would be possible to derive the form (1.3) for the solutions. In the context of linear problems, i.e., with the nonvanishing restriction removed, duality can be applied and then, incorporating PDE machinery to establish the regularity of the solutions to the dual problem, the structural formulas for the solutions of (1.1) and (1.2) are obtained (see [KS]). We must stress again that due to the nonlinear nature of extremal problems for nonvanishing functions, new techniques are needed to establish the regularity of solutions up to the boundary beyond membership in an appropriate Hardy class. In the last section, we discuss a specific case of Problem 1.2 with $l_1(f) = f(0)$ and $l_2(f) = f'(0)$. The study of this simple problem was initiated by D. Aharonov and H.S. Shapiro in unpublished reports [AhSh1, AhSh2], and B. Korenblum has drawn attention to this question on numerous occasions.

2. Existence and regularity of solutions

Consider the following general problem.

Problem 2.1. *Given n continuous linearly independent linear functionals $l_1, l_2, \ldots, l_n$ on A^p and given n points $c_1, c_2, \ldots, c_n$ in $\mathbb{C} - \{0\}$, find*

$$\lambda = \inf\{\|f\|_{A^p} : f \text{ is zero-free}, l_i(f) = c_i, 1 \leq i \leq n\}.$$

The set of zero-free functions satisfying the above interpolation conditions can in general be empty, so we will assume in what follows that this set is non void. Concerning existence of extremals, we have:

Theorem 2.2. *The infimum in Problem 2.1 is attained.*

Proof. (The following argument is well known and is included for completeness.) Pick a sequence f_k of zero-free functions in A^p such that $l_i(f_k) = c_i$ for every $1 \le i \le n$ and every $k = 1, 2, \ldots$, and such that $\|f_k\|_{A^p} \to \lambda$ as $k \to \infty$. Since these norms are bounded, there exists a subsequence $\{f_{k_j}\}$ and an analytic function f such that $f_{k_j} \to f$ as $j \to \infty$. By Hurwitz' theorem, f is zero-free. Moreover, $l_i(f) = c_i$ for every $1 \le i \le n$. By Fatou's lemma,

$$\left(\int_{\mathbb{D}} |f|^p dA \right)^{\frac{1}{p}} \le \lambda,$$

but by minimality of λ, we must actually have equality. Therefore f is extremal for Problem 2.1. $\qquad\square$

Let us now consider the special case of point evaluation. More specifically, let $\beta_1, \ldots, \beta_n \in \mathbb{D}$ be distinct points and let $l_i(f) = f(\beta_i)$, for $1 \le i \le n$. We will assume that none of the c_i is zero.

The following result shows that we need only solve the extremal problem in A^2 in order to get a solution in every A^p space $(p > 0.)$

Theorem 2.3. *If g is minimal for the problem*

$$\inf\{\|g\|_{A^2} : g \text{ is zero-free}, l_i(g) = b_i, 1 \le i \le n\},$$

where the b_i are elements of D, then $g^{\frac{2}{p}}$ is minimal for the problem

$$(*) \inf\{\|f\|_{A^p} : f \text{ is zero-free}, l_i(f) = c_i, 1 \le i \le n\},$$

where $c_i = l_i(g^{\frac{2}{p}})$.

Proof. The function $g^{\frac{2}{p}}$ is zero-free and

$$\int_{D} (|g(z)|^{\frac{2}{p}})^p dA(z) = \int_{D} |g(z)|^2 dA(z) < \infty,$$

so $g^{\frac{2}{p}}$ is in A^p. Moreover by definition, $g^{\frac{2}{p}}$ satisfies the interpolation conditions $c_i = l_i(g^{\frac{2}{p}})$.

Now suppose that $g^{\frac{2}{p}}$ is not minimal for the problem **(*)**. Then there exists $h \in A^p$ zero-free such that $c_i = l_i(h)$ and

$$\int_{D} |h(z)|^p dA(z) < \int_{D} |g(z)|^2 dA(z).$$

The function $h^{\frac{p}{2}}$ is a zero-free A^2 function such that

$$\|h^{\frac{p}{2}}\|_2 < \|g\|_2.$$

Moreover

$$l_i(h^{\frac{p}{2}}) = h^{\frac{p}{2}}(\beta_i) = c_i^{\frac{p}{2}} = (g^{\frac{2}{p}}(\beta_i))^{\frac{p}{2}} = g(\beta_i) = b_i.$$

This contradicts the minimality of g for the A^2 problem. $\qquad\square$

Notice that by the same argument, the converse also holds; in other words, if we can solve the extremal problem in A^p for some $p > 0$, then we can also solve the extremal problem in A^2. Therefore for the remainder of the paper, we will consider only the case $p = 2$. Notice that if we consider Problem (1.2) without the restriction that f must be zero-free, the solution is very simple and well known. Considering for simplicity the case of distinct β_j, the unique solution is the unique linear combination of the reproducing kernels $k(., \beta_j)$ satisfying the interpolating conditions, where

$$k(z, w) := 1/(1 - \bar{w}z)^2.$$

Since our functions are zero-free, we will rewrite a function f as $f(z) = \exp(\varphi(z))$, and solve the problem (relabelling the c_i)

$$\lambda = \inf\{\| \exp(\varphi(z))\|_{A^2} : \varphi(\beta_i) = c_i, 1 \leq i \leq n\}. \tag{2.1}$$

Theorem 2.4. *The extremal solution to Problem* (2.1) *is unique.*

Proof. Suppose φ_1 and φ_2 are two extremal solutions to (2.1), that is

$$\lambda = \|e^{\varphi_1}\|_{A^2} = \|e^{\varphi_2}\|_{A^2}$$

and

$$\varphi_1(\beta_i) = \varphi_2(\beta_i) = c_i$$

for every $1 \leq i \leq n$. Consider

$$\varphi(z) = \frac{\varphi_1(z) + \varphi_2(z)}{2}.$$

This new function satisfies $\varphi(\beta_i) = c_i$ for every $1 \leq i \leq n$, and therefore

$$\begin{aligned}
\lambda^2 &\leq \int_{\mathbb{D}} |e^{\varphi(z)}|^2 dA(z) \\
&= \int_{\mathbb{D}} |e^{\varphi_1(z)}||e^{\varphi_2(z)}| dA(z) \\
&\leq \|e^{\varphi_1}\|_{A^2}\|e^{\varphi_2}\|_{A^2} \text{ (by the Cauchy-Schwarz inequality)} \\
&= \lambda^2.
\end{aligned}$$

This implies that

$$|e^{\varphi_1(z)}| = C|e^{\varphi_2(z)}|$$

for some constant C. Since the function $e^{\varphi_1}/e^{\varphi_2}$ has constant modulus, it is a constant, which must equal 1 because of the normalization. The extremal solution to (2.1) is therefore unique. $\square$

Remark. We can generalize this theorem to some other linear functionals l_i. For instance, one may wish to consider linear functionals l_{ij}, $i = 1, \ldots, n$, $j = 0, \ldots, k_i$, that give the jth Taylor coefficients of f at β_i.

The next three lemmas are the technical tools needed to address the issue of the regularity of the extremal function: we want to show that the extremal function is actually a Hardy space function.

For integers $m \geq n$, consider the class P_m of polynomials p of degree at most m such that $p(\beta_i) = c_i$ for every $1 \leq i \leq n$. Let

$$\lambda_m = \inf\{\|e^{p(z)}\|_{A^2} : p \in P_m\}. \tag{2.2}$$

Lemma 2.5. $\qquad\qquad\qquad \lim_{m\to\infty} \lambda_m = \lambda.$

Proof. Notice that λ_m is a decreasing sequence of positive numbers bounded below by λ, so

$$\lim_{m\to\infty} \lambda_m \geq \lambda.$$

On the other hand, let φ^* be the extremal function for (2.1). Write

$$\varphi^*(z) = L(z) + h(z)g(z),$$

where L is the Lagrange polynomial taking value c_i at β_i, namely

$$L(z) = \sum_{i=1}^{n} c_i \frac{\prod_{k=1, k\neq i}^{n}(z - \beta_k)}{\prod_{k=1, k\neq i}^{n}(\beta_i - \beta_k)},$$

$h(z) = \prod_{i=1}^{n}(z - \beta_i)$, and g is analytic in $\mathbb{D}$. For each $0 < r < 1$, define

$$\varphi_r(z) := \varphi^*(rz).$$

Let $\varepsilon > 0$. Notice that there exists $\delta > 0$ such that if $\tilde{c}_i$ are complex numbers satisfying $|c_i - \tilde{c}_i| < \delta$ for $i = 1, \ldots, n$, then $|L(z) - \tilde{L}(z)| < \varepsilon$ (for every $z \in \mathbb{D}$), where $\tilde{L}$ is the Lagrange polynomial with values $\tilde{c}_i$ at β_i. We now pick r close enough to 1 so that

$$\|e^{\varphi^*} - e^{\varphi_r}\|_{A^2} < \varepsilon$$

and

$$|\varphi_r(\beta_i) - \varphi^*(\beta_i)| < \frac{\delta}{2} \quad \text{for } i = 1, \ldots, n.$$

Define $p_{m,r}$ to be the mth partial sum of the Taylor series of φ_r. Given any integer $N \geq n$, pick $m \geq N$ such that

$$\|e^{p_{m,r}(z)} - e^{\varphi_r(z)}\|_{A^2} < \varepsilon$$

and

$$|p_{m,r}(\beta_i) - \varphi_r(\beta_i)| < \frac{\delta}{2} \quad \text{for } i = 1, \ldots, n.$$

Let $\tilde{c}_i = p_{m,r}(\beta_i)$ for $i = 1, \ldots, n$ and let $\tilde{L}$ be the Lagrange polynomial taking values $\tilde{c}_i$ at β_i. Then we can write

$$p_{m,r}(z) = \tilde{L}(z) + h(z)q_{m-n,r}(z),$$

where $q_{m-n,r}$ is a polynomial of degree at most $m-n$. Notice that since $|p_{m,r}(\beta_i) - \varphi(\beta_i)| < \delta$ (for every $i = 1, \ldots, n$),

$$|L(z) - \tilde{L}(z)| < \varepsilon \quad \text{for every } z \in \mathbb{D}.$$

Define
$$p_m(z) = L(z) + h(z)q_{m-n,r}(z).$$
Then $p_m \in P_m$, and
$$
\begin{aligned}
|e^{p_m(z)} - e^{p_{m,r}(z)}|^2 &\leq |e^{p_{m,r}(z)}|^2(e^{|p_{m,r}(z)-p_m(z)|} - 1)^2 \\
&= |e^{p_{m,r}(z)}|^2(e^{|\tilde{L}(z)-L(z)|} - 1)^2 \\
&\leq |e^{p_{m,r}(z)}|^2(e^{\varepsilon} - 1)^2
\end{aligned}
$$
Therefore
$$
\begin{aligned}
\|e^{p_m} - e^{p_{m,r}}\|_{A^2} &\leq \|e^{p_{m,r}}\|_{A^2}(e^{\varepsilon} - 1) \\
&\leq C(e^{\varepsilon} - 1),
\end{aligned}
$$
where C is a constant depending only on $\|e^{\varphi^*}\|_{A^2}$. Therefore
$$\|e^{p_m(z)} - e^{\varphi^*(z)}\|_{A^2} \leq 2\varepsilon + C(e^{\varepsilon} - 1) = C_{\varepsilon},$$
which implies
$$\lambda_m \leq \|e^{p_m(z)}\|_{A^2} \leq C_{\varepsilon} + \lambda$$
for arbitrarily large m, where $C_{\varepsilon} \to 0$ as $\varepsilon \to 0$. Therefore
$$\lim_{m\to\infty} \lambda_m \leq \lambda.$$
Since we already have the reverse inequality, we can conclude that
$$\lim_{m\to\infty} \lambda_m = \lambda. \qquad \square$$

Lemma 2.6. *The extremal polynomial p_m^* in (2.2) exists, and for every polynomial ψ_{m-n} of degree at most $m - n$,*
$$\int_{\mathbb{D}} |e^{p_m^*(z)}|^2(z - \beta_1)\ldots(z - \beta_n)\psi_{m-n}(z)dA(z) = 0.$$

Proof. To prove the existence of the extremal polynomial p_m^*, consider the minimizing sequence p_m^k in (2.2). Without loss of generality, we can assume that the functions $e^{p_m^k}$ converge on compact subsets, and hence p_m^k converge pointwise in $\mathbb{D}$ to a polynomial $p_m^* \in P_m$. As above, applying Fatou's lemma, we see that p_m^* is in fact the extremal.

Define
$$F(\varepsilon) = \|\exp(p_m^*(z) + \varepsilon \prod_{i=1}^{n}(z - \beta_i)\psi_{m-n}(z))\|_{A^2}^2$$
where ψ_{m-n} is any polynomial of degree at most $m - n$. Then since p_m^* is extremal, $F'(0) = 0$.
$$
\begin{aligned}
F(\varepsilon) &= \int_{\mathbb{D}} |\exp(p_m^*(z) + \varepsilon \prod_{i=1}^{n}(z - \beta_i)\psi_{m-n}(z))|^2 dA(z) \\
&= \int_{\mathbb{D}} |\exp(p_m^*(z))|^2 \exp(2\varepsilon Re(\prod_{i=1}^{n}(z - \beta_i)\psi_{m-n}(z))dA(z)
\end{aligned}
$$

Therefore

$$F'(0) = \int_{\mathbb{D}} |\exp(p_m^*(z))|^2 2Re(\prod_{i=1}^{n}(z - \beta_i)\psi_{m-n}(z))dA(z) = 0.$$

Replacing ψ_{m-n} by $i\psi_{m-n}$ gives

$$\int_{\mathbb{D}} |\exp(p_m^*(z))|^2 2Re(\prod_{i=1}^{n}(z - \beta_i)i\psi_{m-n}(z))dA(z) = 0,$$

and therefore

$$\int_{\mathbb{D}} |\exp(p_m^*(z))|^2 \prod_{i=1}^{n}(z - \beta_i)\psi_{m-n}(z)dA(z) = 0$$

for every polynomial ψ_{m-n} of degree at most $m - n$. $\square$

Lemma 2.7. *For each $m \geq n$, $e^{p_m^*} \in H^2$, and these H^2 norms are bounded.*

Proof. Write

$$p_m^*(z) = L(z) + h(z)q_{m-n}(z),$$

where $L(z)$ is the Lagrange polynomial taking value c_i at β_i (for $i = 1, \ldots, n$), $h(z) = \prod_{i=1}^{n}(z - \beta_i)$, and q_{m-n} is a polynomial of degree at most $m - n$. We then have

$$\begin{aligned}
\int_{\mathbb{T}} |e^{p_m^*(e^{i\theta})}|^2 d\theta &= i\int_{\mathbb{T}} |e^{p_m^*(z)}|^2 z d\bar{z} \\
&= 2\int_{\mathbb{D}} \frac{\partial}{\partial z}(|e^{p_m^*(z)}|^2 z)dA(z) \quad \text{(by Green's formula)} \\
&= \int_{\mathbb{D}} |e^{p_m^*(z)}|^2 (p_m^{*'}(z)z + 1)dA(z).
\end{aligned}$$

We would like to show that this integral is bounded by $C\|e^{p_m^*(z)}\|_{A^2}^2$, where C is a constant independent of m. First notice that

$$zp_m^{*'}(z) = zL'(z) + zh'(z)q_{m-n}(z) + zh(z)q_{m-n}'(z).$$

Since $zq_{m-n}'(z)$ is a polynomial of degree at most $m - n$, Lemma 2.6 allows us to conclude that

$$\int_{\mathbb{D}} |e^{p_m^*(z)}|^2 zh(z)q_{m-n}'(z)dA(z) = 0.$$

On the other hand, $zL'(z)$ is bounded and independent of m, and therefore

$$|\int_{\mathbb{D}} |e^{p_m^*(z)}|^2 zL'(z)dA(z)| \leq C_1\|e^{p_m^*(z)}\|_{A^2}^2,$$

where C_1 is a constant independent of m. Therefore the crucial term is that involving $zh'(z)q_{m-n}(z)$. Write

$$q_{m-n}(z) = q_{m-n}(\beta_k) + (z - \beta_k)q_{m-n-1}(z),$$

where q_{m-n-1} is a polynomial of degree at most $m - n - 1$. Then

$$zh'(z)q_{m-n}(z) = z\{\sum_{k=1}^{n}[\prod_{i=1,i\neq k}^{n}(z - \beta_i)]\}\{q_{m-n}(\beta_k) + (z - \beta_k)q_{m-n-1}(z)\}$$

$$= \sum_{k=1}^{n}\{z\prod_{i=1,i\neq k}^{n}(z - \beta_i)\}q_{m-n}(\beta_k) + \sum_{k=1}^{n}\{\prod_{i=1}^{n}(z - \beta_i)\}zq_{m-n-1}(z).$$

Since $zq_{m-n-1}(z)$ is a polynomial of degree at most $m - n$, by Lemma 2.6, the contribution of the second big sum above, when integrated against $|e^{p_m^*(z)}|^2$, is zero. On the other hand, it is not hard to see that the polynomials q_{m-n} are (uniformly) bounded on the set $\{\beta_k : k = 1,\ldots,n\}$, and therefore their contribution is a bounded one, that is, there exists a constant C_2 such that

$$\int_{\mathbb{D}} |e^{p_m^*(z)}|^2 zh'(z)q_{m-n}(z)dA(z) \leq C_2\|e^{p_m^*(z)}\|_{A^2}^2.$$

We have therefore shown that there exist constants C and M, independent of m, such that

$$\int_{\mathbb{T}} |e^{p_m^*(e^{i\theta})}|^2 d\theta \leq C\|e^{p_m^*(z)}\|_A^2 = C\lambda_m \leq CM.$$

Thus the functions $e^{p_m^*}$ have uniformly bounded H^2 norms. $\qquad\square$

Theorem 2.8. $e^{\varphi^*} \in H^2$.

Proof. By an argument similar to that of Theorem 2.2 and by uniqueness of the extremal function for (2.1), there exists a subsequence $\{p_{m_k}^*\}$ of $\{p_m\}$ such that

$$e^{p_{m_k}^*} \to e^{\varphi^*}$$

pointwise as $k \to \infty$. For each fixed radius r, $0 < r < 1$, by Fatou's lemma,

$$\int_{\mathbb{T}} |\exp(\varphi^*(re^{i\theta}))|^2 d\theta \leq \liminf_{k\to\infty} \int_{\mathbb{T}} |\exp(p_{m_k}^*(re^{i\theta}))|^2 d\theta.$$

By Lemma 2.7, the right-hand side is bounded for all $0 < r < 1$, and therefore $e^{\varphi^*} \in H^2$. $\qquad\square$

The following corollary follows from Theorems 2.8 and 2.3.

Corollary 2.9. *Let $0 < p < \infty$, and let e^{φ^*} be the extremal function that minimizes the norm*

$$\lambda = \inf\{\|\exp(\varphi(z))\|_{A^p} : \varphi(\beta_i) = c_i, 1 \leq i \leq n\}.$$

Then e^{φ^*} is in H^p.

3. Another approach to regularity

In the following, we present a very different approach to showing the a priori regularity of the extremal function. It was developed by D. Aharonov and H.S. Shapiro in 1972 and 1978 in two unpublished preprints ([AhSh1, AhSh2]) in connection with their study of the minimal area problem for univalent and locally univalent functions. See also [ASS1, ASS2].

Given n points $\beta_1, \ldots, \beta_n$ of $\mathbb{D}$, and complex numbers $c_1, \ldots, c_n$ recall that L denotes the unique (Lagrange interpolating) polynomial of degree at most $n - 1$ satisfying

$$L(\beta_j) = c_j, j = 1, 2, \ldots, n. \tag{3.1}$$

As above, the polynomial h is defined by

$$h(z) := (z - \beta_1) \ldots (z - \beta_n).$$

We are considering, as before, Problem (1.2) when the functionals l_i are point evaluations at β_i, in A^2.

Recall that in order to get a nonvacuous problem, we assume that none of the c_j is zero. For a holomorphic function f in $\mathbb{D}$, let $L(f)$ denote the unique polynomial of degree at most $n - 1$ satisfying (3.1), with $c_j := f(\beta_j)$. Then, there is a unique function g analytic in $\mathbb{D}$ such that

$$f = hg + L(f).$$

Of course, $L(f)$ is bounded on $\mathbb{D}$ by $C \max |f(\beta_j)|$, where C is a constant depending on the $\{\beta_j\}$ and the $\{c_j\}$, but not on f.

Suppose now for each s in the interval $(0, s_0)$, a_s denotes a univalent function in $\mathbb{D}$ satisfying

$$a_s(0) = 0 \quad \text{and} \tag{3.2}$$

$$|a_s(z)| < 1 \quad \text{for } z \in \mathbb{D}. \tag{3.3}$$

(Thus, by Schwarz' lemma,

$$|a_s(z)| \leq |z| \quad \text{for } z \in \mathbb{D}.)$$

Let G_s denote the image of $\mathbb{D}$ under the map $z \to a_s(z)$.

Let now f be an extremal function for Problem (1.2), that is, it is a zero-free function in A^2 satisfying the interpolating conditions

$$f(\beta_j) = c_j, \quad j = 1, \ldots, n, \tag{3.4}$$

and having the least norm among such functions. Then, denoting

$$g_s(z) := f(a_s(z))a'_s(z), \tag{3.5}$$

we observe that the function f_s defined by

$$f_s(z) := g_s(z)L(f/g_s)(z) \tag{3.6}$$

is in A^2 and satisfies the interpolating conditions, since

$$f_s(\beta_j) = g_s(\beta_j)[f(\beta_j)/g_s(\beta_j)] = f(\beta_j).$$

Moreover, g_s is certainly zero-free, and hence so is f_s if we can verify that the polynomial $L(f/g_s)$ has no zeros in $\mathbb{D}$.

Now, we shall impose some further restrictions on the maps a_s. We assume that

$$|a_s(z) - z| \leq B(z)c(s) \text{ and} \tag{3.7}$$

$$|a_s'(z) - 1| \leq B(z)c(s) \tag{3.8}$$

where B is some positive continuous function on $\mathbb{D}$, and c is a continuous function on $(0, s_0]$ such that

$$c(s) \to 0 \text{ as } s \to 0. \tag{3.9}$$

With these assumptions, $a_s(z) \to z$ and $a_s'(z) \to 1$ for each z in $\mathbb{D}$, as $s \to 0$. Thus, $f(\beta_j)/g_s(\beta_j) \to 1$ as $s \to 0$, for each j. Thus, the polynomials

$$L_s := L(f/g_s)$$

of degree at most $n - 1$ tend to 1 on the set $\{\beta_1, \ldots, \beta_n\}$ as $s \to 0$, and hence they tend uniformly to 1 on $\mathbb{D}$. It follows that for s sufficiently near 0, L_s has no zeros in $\mathbb{D}$, and consequently f_s is zero-free.

Hence, for sufficiently small s, say $s < s_1$, f_s is a "competing function" in the extremal problem, and we have:

$$\|f\|_{A^2} \leq \|f_s\|_{A^2}. \tag{3.10}$$

Note that $L(f/g_s)$ differs from 1, uniformly for all z in $\mathbb{D}$, by a constant times the maximum of the numbers

$$\{|(f(\beta_j)/g_s(\beta_j)) - 1|, \; j = 1, 2, \ldots, n\}. \tag{3.11}$$

Now,

$$f(z)/g_s(z) - 1 = (f(z) - g_s(z))/g_s(z)$$

and since

$$|g_s(z)| = |f(a_s(z)||a_s'(z)| \to |f(z)| \text{ as } s \to 0,$$

by virtue of (3.7), (3.8), and (3.9) the numbers $|g_s(\beta_j)|$ remain greater than some positive constant as $s \to 0$. Consequently, the numbers (3.11) are, for small s, bounded by a constant times the maximum of the numbers

$$\{|f(\beta_j) - g_s(\beta_j)|, \; j = 1, 2, \ldots, n\}. \tag{3.12}$$

But,

$$\begin{aligned} |f(z) - g_s(z)| &= |f(z) - f(a_s(z))a_s'(z)| \\ &\leq |f(z) - f(a_s(z))| + |f(a_s(z))||1 - a_s'(z)| \end{aligned}$$

Using the estimates (3.7), (3.8) we find that the numbers (3.12) are bounded by a constant times $c(s)$, and therefore

$$L(f/g_s) = 1 + O(c(s)),$$

uniformly for z in $\mathbb{D}$, as $s \to 0$. Hence, from (3.10) and (3.6),

$$\|f\|_{A^2} \leq \|g_s\|_{A^2}(1 + Mc(s))$$

for some constant M, thus

$$\int_{\mathbb{D}} |f(z)|^2 dA \le \int_{\mathbb{D}} |f_s(z)|^2 dA + Nc(s)$$

for some new constant N. Since

$$\int_{\mathbb{D}} |f_s(z)|^2 dA \;=\; \int_{\mathbb{D}} |f(a_s(z)|^2 |a_s'(z)|^2 dA$$

$$=\; \int_{G_s} |f(z')|^2 dA(z') \text{ (changing variables by } z' = a_s(z))$$

and combining the two integrals yields:

(∗) *Under the assumptions made thus far, the area integral of $|f|^2$ over the domain D_s complementary to $G_s = a_s(\mathbb{D})$ in $\mathbb{D}$ does not exceed $Nc(s)$, where N is a constant and $c(s)$ is as in (3.7) and (3.8).*

To see the usefulness of (∗), let us first consider an almost trivial choice of a_s, namely

$$a_s(z) = (1 - s)z \text{ and } a_s'(z) = 1 - s.$$

Then, (3.7) and (3.8) hold with $c(s) = s$. Here G_s is the disk $\{|z| < 1 - s\}$, so (∗) asserts (denoting $t := 1 - s$): the integral of $|f|^2$ over the annulus $\{t < |z| < 1\}$, for all t sufficiently close to 1, is bounded by a constant times $1 - t$. Consequently, the mean value of $|f|^2$ over these annuli remains bounded. This, however, easily implies that f is in the Hardy class H^2 of the disk! So, we have given another proof of Theorem 2.8: extremals for the zero-free A^2 problem (1.2) always belong to H^2.

We can extract a bit more, namely that extremals are bounded in $\mathbb{D}$, with a more recondite choice of $a(s)$.

Let w denote a point of the unit circle $\mathbb{T}$, and s a small positive number. Let $G_{s,w}$ denote the crescent bounded by $\mathbb{T}$ and a circle of radius s internally tangent to $\mathbb{T}$ at w. (This circle is thus centered at $(1-s)w$.) Let $a_{s,w}$ be the unique conformal map of $\mathbb{D}$ onto $G_{s,w}$ mapping 0 to 0 and the boundary point w to $(1 - 2s)w$, and $b_{s,w}$ the z-derivative of $a_{s,w}$. We are going to show

Lemma 3.1. *With $a_{s,w}$ and $b_{s,w}$ in place of a_s, a_s' respectively, (3.7) and (3.8) hold, with $c(s) = s^2$, uniformly with respect to w.*

Assuming this for the moment, let us show how the boundedness of extremals follows. Applying (∗), we see that if f is extremal, the area integral of $|f|^2$ over the disk centered at $(1 - s)w$ of radius s does not exceed a constant (independent of w and s) times the area of this disk. Since $|f((1 - s)w)|^2$ does not exceed the areal mean value of $|f|^2$ over this disk, we conclude $|f((1 - s)w)|$ is bounded uniformly for all w in $\mathbb{T}$ and sufficiently small s, i.e., $|f|$ is bounded in some annulus $\{1 - s_0 < |z| < 1\}$, and hence in $\mathbb{D}$. We therefore have the following:

Theorem 3.2. *The extremal function f^* for Problem 2.1 is in H^∞.*

It only remains to prove Lemma 3.1.

Proof. First note that the arguments based on (3.7 and 3.8) leading to (*) only rely on the boundedness of the function B on compact subsets of $\mathbb{D}$, and more precisely on a compact subset containing all interpolation points β_j, $j = 1, \ldots, n$. Since clearly (3.8) follows (with a different choice of $B(.)$) from (3.7), we have only to verify (3.7). Also, by symmetry, it is enough to treat the case $w = 1$. We do so, and for simplicity denote $a_{s,1}$, $G_{s,1}$ by a_s, G_s respectively. Thus, a_s maps $\mathbb{D}$ onto the domain bounded by $\mathbb{T}$ and the circle of radius s centered at $1 - s$. Moreover $a_s(0) = 0$, and $a_s(1) = 1 - 2s$. Thus, we have a Taylor expansion

$$a_s(z) = c_{1,s}z + c_{2,s}z^2 + \cdots$$

convergent for $|z| < 1$. Moreover, it is easy to see from the symmetry of G_s that all the coefficients $c_{j,s}$ are real.

Under the map $Z = 1/(1-z)$, G_s is transformed to a vertical strip S in the Z plane bounded by the lines $\{Re\, Z = 1/2\}$ and $\{Re\, Z = 1/2s\}$. Thus, the function

$$h_s := 1/(1 - a_s)$$

maps $\mathbb{D}$ onto S and carries 0 into 1, and the boundary point 1 to ∞. Hence $u_s(e^{it}) := Re(h_s(e^{it}))$ satisfies

$$
\begin{aligned}
u_s(e^{it}) \;&=\; 1/2 \text{ for } |t| > t_0, \text{ and} \\
&=\; 1/2s \text{ for } |t| < t_0,
\end{aligned}
$$

where $t_0, 0 < t_0 < \pi$ is determined from

$$1 = \frac{1}{2\pi} \int_{\mathbb{T}} u_s(e^{it})dt = \frac{1}{2} + \frac{1-s}{2\pi s}t_0$$

hence

$$t_0 = (s/(1-s))\pi. \tag{3.13}$$

Now, we have a Taylor expansion

$$h_s(z) = 1 + b_{1,s}z + b_{2,s}z^2 + \cdots \tag{3.14}$$

where the $b_{j,s}$ are real, and so determined from

$$u_s(e^{it} = 1 + b_{1,s}\cos t + b_{2,s}\cos 2t + \cdots ,$$

i.e.,

$$b_{n,s} = \frac{2}{\pi} \int_0^{\pi} u_s(e^{it}) \cos(nt)dt$$

hence

$$b_{n,s} = \sin nt_0/nt_0 \qquad n = 1, 2, \ldots \tag{3.15}$$

where t_0 is given by (3.13).

We are now prepared to prove Lemma 3.1, i.e.,

$$|a_s(z) - z| \leq B(z)s^2. \tag{3.16}$$

We have

$$h_s(z) - \frac{1}{1-z} = \frac{1}{1 - a_s(z)} - \frac{1}{1-z} = \frac{a_s(z) - z}{(1-z)(1 - a_s(z))},$$

72   D. Aharonov, C. Bénéteau, D. Khavinson and H.S. Shapiro

so

$$|a_s(z) - z| \le 4|h_s(z) - \frac{1}{1-z}| \le 4 \sum_{n=1}^{\infty} |b_{n,s} - 1||z|^n. \tag{3.17}$$

But, from (3.15)

$$|b_{n,s} - 1| = |\frac{\sin nt_0}{nt_0} - 1|.$$

Since the function

$$\frac{(\sin x)/x - 1}{x^2}$$

is bounded for x real, we have for some constant N :

$$|\frac{\sin nt_0}{nt_0} - 1| \le N(nt_0)^2 \le N'n^2s^2$$

for small s, in view of (3.13), where N' is some new constant. Thus, finally, inserting this last estimate into (3.17),

$$|a_s(z) - z| \le N''s^2 B(z),$$

where

$$B(z) := \sum_{n=1}^{\infty} n^2|z|^n,$$

which is certainly bounded on compact subsets of $\mathbb{D}$, and the proof is finished. $\square$

Remark. This type of variation can be used to give another proof of the regularity and form of extremal functions in the non-vanishing H^p case, which were originally established in [Kh1, Kh2]. In what follows, we shall only discuss the case $p = 2$, since the case of other p follows at once via an analogue of Theorem 2.3 in the H^p setting.

For the sake of brevity, we only consider the following problem. Given complex constants $c_0, c_1, \dots, c_m$ with c_0 not zero (w.l.o.g. we could take $c_0 = 1$), let A be the subset of H^2 consisting of "admissible functions" f, i.e., those functions zero-free in $\mathbb{D}$ whose first $m+1$ Taylor coefficients are the c_j. We consider the extremal problem , to minimize $\|f\|_2 := \|f\|_{H^2}$ in the class A. The following argument is again an adaptation of a variational argument used by Aharonov and Shapiro in ([AhSh1, AhSh2]) for a different problem.

Proposition 3.3. *Every extremal is in the Dirichlet space, that is, satisfies*

$$\int_{\mathbb{D}} |f'(z)|^2 dA < \infty.$$

Proof. Let f be extremal, and $0 < t < 1$. Then,

$$f(z) = tf(tz)[f(z)/tf(tz)] = tf(tz)[S(z;t) + R(z;t)] \tag{3.18}$$

where S denotes the partial sum of order m of the Taylor expansion of $f(z)/tf(tz)$ $=: E(z;t)$ and R denotes the remainder $E - S$. Now,

$$|f(z) - f(tz)| \le C(1 - t),$$

uniformly for $|z| \leq 1/2$, where C is a constant depending on f, and this implies easily

$$|1 - E(z;t)| \leq C(1 - t)$$

for those z, and some (different) constant C. From this it follows easily that

$$S(z;t) = 1 + O(1 - t), \quad \text{uniformly for } z \in \mathbb{D}. \tag{3.19}$$

Moreover, from (3.18) we see that $tf(tz)S(z;t)$ has the same Taylor coefficients as f, through terms of order m. Also, (3.19) shows that S does not vanish in $\mathbb{D}$ for t near 1. We conclude that, for t sufficiently close to 1, $tf(tz)S(z;t)$ is admissible, and consequently its norm is greater than or equal to that of f, so we have

$$\int_{\mathbb{T}} |f(e^{is})|^2 ds \leq (\int_{\mathbb{T}} |tf(te^{is})|^2 ds)(1 + O(1 - t)),$$

or, in terms of the Taylor coefficients a_n of f,

$$\sum |a_n|^2 \leq [\sum |a_n|^2 t^{2n+2}](1 + O(1 - t))$$

so

$$\sum (1 - t^{2n+2})/(1 - t)|a_n|^2$$

remains bounded as $t \to 1$, which implies f has finite Dirichlet integral. $\qquad \square$

Corollary 3.4. *The extremal must be a polynomial of degree at most m times a singular function whose representing measure can only have atoms located at the zeros on $\mathbb{T}$ of this polynomial.*

Proof. As usual, for every $h \in H^\infty$, $(1 + wz^{m+1}h)f$ (where f is extremal, and w a complex number) is admissible for small $|w|$. Hence, as in the proof of Lemma 2.6, we obtain that f is orthogonal (in H^2!) to $z^{m+1}hf$. If $f = IF$, where I is a singular inner function and F is outer, since $|I| = 1$ a.e. on $\mathbb{T}$, it follows that F is orthogonal to $z^{m+1}FH^\infty$. Now, F is cyclic, so FH^∞ is dense in H^2, i.e., F is orthogonal to $z^{m+1}H^2$. Hence, F is a polynomial of degree at most m. For the product FI to have a finite Dirichlet norm, the singular measure for I must be supported on a subset of the zero set of F on $\mathbb{T}$ as claimed. Indeed, for any singular inner function I and any point $w \in \mathbb{T}$ where the singular measure for I has infinite Radon-Nikodym derivative with respect to Lebesgue measure,

$$\int_{\mathbb{D} \cap \{|z-w|<c\}} |I'|^2 dA = \infty,$$

because the closure of the image under I of any such neighborhood of w is the whole unit disk (cf. [CL, Theorem 5.4]). $\qquad \square$

4. A discussion of the conjectured form of extremal functions

In this section we provide certain evidence in support of our overall conjecture and draw out possible lines of attack that would hopefully lead to a rigorous proof in the future. Recall that the extremal function f^* in the problem (2.1):

$$\lambda = \inf\{\|\exp(\varphi(z))\|_{A^2} : \varphi(\beta_i) = c_i, 1 \leq i \leq n\}$$

is conjectured to have the form (1.3):

$$f^*(z) = C\frac{\prod_{j=1}^{2n-2}(1 - \bar{\alpha}_j z)^{\frac{2}{p}} \exp(\sum_{j=1}^{k} \lambda_j \frac{e^{i\theta_j}+z}{e^{i\theta_j}-z})}{\prod_{j=1}^{n}(1 - \bar{\beta}_j z)^{\frac{4}{p}}},$$

where C is a constant, $|\alpha_j| \leq 1$, $j = 1, \ldots, 2n - 2$, $|\beta_j| < 1$, $j = 1, \ldots, n$, $\lambda_j \leq 0$, $j = 1, \ldots k$, $k \leq 2n - 2$. As in the previous sections, we shall focus the discussion on the case $p = 2$, since the A^p extremals are simply the $2/p$th powers of those in A^2.

First, let us observe that if the solution to the problem for $p = 2$ in the whole space A^2, i.e.,

$$\lambda = \inf\{\|f(z)\|_{A^2} : f(\beta_j) = \exp(c_j), 1 \leq j \leq n\} \tag{4.1}$$

happens to be non-vanishing in $\mathbb{D}$, then it solves Problem (2.1.) The solution to Problem (4.1) is well known and is equal to a linear combination of the reproducing Bergman kernels at the interpolation points. That is,

$$f^*(z) = \sum_{j=1}^{n} \frac{a_j}{(1 - \bar{\beta}_j z)^2}, \tag{4.2}$$

where the a_j are constants, which does have the form (1.3) with singular inner factors being trivial.

Recall that a closed subset K of the unit circle $\mathbb{T}$ is called a Carleson set if

$$\int_{\mathbb{T}} \log \rho_K(e^{i\theta})d\theta > -\infty,$$

where $\rho_K(z) = \operatorname{dist}(z, K)$ (cf., e.g., [DS, p. 250].)

Now, if we could squeeze additional regularity out of the extremal function f^* in (2.1),the following argument would allow us to establish most of (1.3) right away. Namely

Theorem 4.1. *Assume that the support of the singular measure in the inner factor of the extremal function f^* in (2.1) is a Carleson set. Then the outer part of f^* is as claimed in (1.3).*

Remark. The regularity assumption for the singular factor of f^* is not unreasonable. In fact, some a priori regularity of extremals was the starting point in ([KS]) for the investigation of linear extremal problems in A^p, i.e., Problem 2.1 but without the non-vanishing restriction. There, the authors have been able to achieve the a priori regularity by considering a dual variational problem whose solution satisfied a nonlinear degenerate elliptic equation. Then, the a priori regularity results for solutions of such equations (although excruciatingly difficult) yielded the desired Lipschitz regularity of the extremal functions. Surprisingly, as we show at the end of the paper, even in the simplest examples of problems for non-vanishing functions in A^2, if the extremals have the form (1.3), they fail to be even continuous in the closed disk. This may be the first example of how some extremals in A^p and H^p differ qualitatively. Of course, the extremal functions for Problem 2.1 in the H^p context are all Lipschitz continuous (cf. Corollary 3.4). Unfortunately, in the context of highly nonlinear problems for non-vanishing functions (since the latter do not form a convex set) the direct duality approach fails at once. (Below, however, we will indicate another line of reasoning which may allow one to save at least some ideas from the duality approach.)

Proof. From the results of the previous sections, it follows that

$$f^* = FS, \tag{4.3}$$

where F is outer and S is a singular inner function whose associated measure $\mu \leq 0, \mu \perp d\theta$ is concentrated on the Carleson set K. Note that

$$S'(z) = S(z) \frac{1}{2\pi} \int_0^{2\pi} \frac{2d\mu(\theta)}{(e^{i\theta} - z)^2},$$

where

$$S(z) = \frac{1}{2\pi} \int_0^{2\pi} \frac{e^{i\theta} + z}{e^{i\theta} - z} d\mu(\theta).$$

So

$$|S'(z)| = O(\rho_K^{-2}(z)), \tag{4.4}$$

where ρ_K is the distance from z to the set K. By a theorem of Carleson (see [DS], p. 250), there exists an outer function $H \in C^2(\bar{\mathbb{D}})$ such that $K \subset \{\zeta \in \mathbb{T} : H^{(j)}(\zeta) = 0, j = 0, 1, 2\}$, and hence

$$H^{(j)}(z) = O(\rho_K(z)^{2-j}), j = 0, 1, 2 \tag{4.5}$$

when $z \to K$. (4.4) and (4.5) yield then that

$$(HS)' = H'S + HS' = hS, \tag{4.6}$$

where $h \in H^\infty(\mathbb{D})$. Recall from our discussions in Sections 2 and 3 that the extremal function f^* must satisfy the following orthogonality condition:

$$\int_{\mathbb{D}} |f^*|^2 \prod_{j=1}^n (z - \beta_j) g dA = 0 \tag{OC}$$

for all, say, bounded analytic functions g. Rewriting (OC) as

$$0 = \int_{\mathbb{D}} \bar{F}\bar{S}FS \prod_{j=1}^n (z - \beta_j) g dA \tag{4.7}$$

and noting that F is cyclic in A^2, so that we can find a sequence of polynomials p_n such that $Fp_n \to 1$ in A^2 (F is "weakly invertible" in A^2 in an older terminology), we conclude from (4.7) that $FS = f^*$ is orthogonal to all functions in the invariant subspace $[S]$ of A^2 generated by S that vanish at the points $\beta_1, \beta_2, \ldots, \beta_n$. In particular, by (4.6), f^* is orthogonal to all functions $\frac{\partial}{\partial z}(H \prod_{j=1}^{n}(z - \beta_j)^2 Sg)$ for all polynomials g, i.e.,

$$0 = \int_{\mathbb{D}} \bar{f}^* \frac{\partial}{\partial z}(H \prod_{j=1}^{n}(z - \beta_j)^2 Sg)dA. \tag{4.8}$$

Applying Green's formula to (4.8), we arrive at

$$0 = \int_{\mathbb{T}} \bar{f}^* H \prod_{j=1}^{n}(z - \beta_j)^2 Sgd\bar{z} = \int_{\mathbb{T}} \bar{F}H \prod_{j=1}^{n}(z - \beta_j)^2 g\frac{dz}{z^2}, \tag{4.9}$$

since $|S| = 1$ on $\mathbb{T}$. Finally, since H is outer and hence cyclic in H^2, there exists a sequence of polynomials q_n such that $Hq_n \to 1$ in H^2. Also there exists a sequence of polynomials p_n such that $p_n \to F$ in H^2, so replacing g by $p_n q_n g$, we obtain

$$0 = \int_{\mathbb{T}} |F|^2 \prod_{j=1}^{n}(z - \beta_j)^2 g\frac{dz}{z^2} \tag{4.10}$$

for all polynomials g. F.&M. Riesz' theorem (cf. [Du, Ga, Ho, Ko]) now implies that

$$|F|^2 = \frac{z^2 h}{\prod_{j=1}^{n}(z - \beta_j)^2} \quad \text{a.e. on } \mathbb{T} \tag{4.11}$$

for some $h \in H^1(\mathbb{D})$. The rest of the argument is standard (see for example [Du], Chapter 8.) Since

$$r(z) := \frac{z^2 h(z)}{\prod_{j=1}^{n}(z - \beta_j)^2} \geq 0$$

on $\mathbb{T}$, it extends as a rational function to all of $\hat{\mathbb{C}}$ and has the form

$$r(z) = C\frac{z^2 \prod_{j=1}^{2n-2}(z - \alpha_j)(1 - \bar{\alpha}_j z)}{\prod_{j=1}^{n}(z - \beta_j)^2(1 - \bar{\beta}_j z)^2} \tag{4.12}$$

where $|\alpha_j| \leq 1$, $j = 1, \ldots, 2n - 2$, are the zeros of r in $\bar{\mathbb{D}}$ (zeros on $\mathbb{T}$ have even multiplicity) and $C > 0$ is a constant. Thus, remembering that F is an outer function and so

$$\log F(z) = \frac{1}{4\pi} \int_{0}^{2\pi} \frac{e^{i\theta} + z}{e^{i\theta} - z} \log |F(e^{i\theta})|^2 d\theta,$$

we easily calculate from (4.11) and (4.12) that

$$F(z) = C\frac{\prod_{j=1}^{2n-2}(1 - \bar{\alpha}_j z)}{\prod_{j=1}^{n}(1 - \bar{\beta}_j z)^2}, |\alpha_j| \leq 1, \tag{4.13}$$

as claimed. $\qquad\qquad\qquad\qquad\qquad\qquad\qquad\qquad\qquad\qquad\qquad\qquad\qquad\qquad\square$

Several remarks are in order.

(i) If the inner part S of f^* is a cyclic vector in A^2, or, equivalently, by the Korenblum-Roberts theorem (see [DS], p. 249), its spectral measure puts no mass on any Carleson set $K \subset \mathbb{T}$, then (4.7) implies right away that f^* is orthogonal to all functions in A^2 vanishing at $\beta_1, \beta_2, \ldots, \beta_n$, and hence

$$f^* = \sum_{j=1}^{n} \frac{a_j}{(1 - \bar{\beta}_j z)^2}$$

is a linear combination of reproducing kernels. Thus, we have the corollary already observed in ([AhSh1, AhSh2]):

Corollary 4.2. *If f^* is cyclic in A^2, it must be a rational function of the form (4.2).*

(ii) On the other hand, if we could a priori conclude that the singular part S of f^* is atomic (with spectral measure consisting of at most $2n - 2$ atoms), then instead of using Carleson's theorem, we could simply take for the outer function H a polynomial $p \neq 0$ in $\mathbb{D}$ vanishing with multiplicity 2 at the atoms of S. Then following the above argument, once again we arrive at the conjectured form (1.3) for the extremal f^*.

Now, following S.Ya. Khavinson's approach to the problem (2.1) in the Hardy space context (see [Kh2, pp. 88 ff]), we will sketch an argument, which perhaps, after some refinement, would allow us to establish the atomic structure of the inner factor S, using only the a priori H^2 regularity.

For that, define subsets B_r of spheres of radius r in A^2 :

$$B_r := \{f = e^{\varphi} : \|f\|_{A^2} \leq r\},$$

where

$$\varphi(z) = \frac{1}{2\pi} \int_0^{2\pi} \frac{e^{i\theta} + z}{e^{i\theta} - z} d\nu(\theta), \tag{4.14}$$

$$d\nu = \log \rho(\theta) d\theta + d\mu, \tag{4.15}$$

and $\rho \geq 0$, $\rho, \log \rho \in L^1(\mathbb{T})$, $d\mu$ is singular and $d\mu \leq 0$. Consider the map Λ that maps the subsets B_r into $\mathbb{C}^n$, defined by

$$\Lambda(f) = (\varphi(\beta_j))_{j=1}^{n}.$$

More precisely, each φ is uniquely determined by the corresponding measure ν and vice versa. Hence, Λ maps the set of measures

$$\Sigma_r := \{\nu : \nu = s(\theta) d\theta + d\mu\}$$

satisfying the constraints

$$d\mu \leq 0 \text{ and } d\mu \text{ is singular} \tag{4.16}$$

$$\exp(s(\theta)), s(\theta) \in L^1(\mathbb{T}) \tag{4.17}$$

$$\|\exp(P(d\nu))\|_{L^2(\mathbb{D})} \leq r, \tag{4.18}$$

where

$$P(d\nu)(re^{i\alpha}) - \frac{1}{2\pi} \int_0^{2\pi} \frac{1 - r^2}{1 + r^2 - 2r\cos(\theta - \alpha)} d\nu(\theta)$$

78   D. Aharonov, C. Bénéteau, D. Khavinson and H.S. Shapiro

is the Poisson integral of ν, into $\mathbb{C}^n$ by

$$\Lambda(\nu) = (S(\nu)(\beta_j))_{j=1}^n.$$

Here

$$S(\nu)(z) = \frac{1}{2\pi} \int_0^{2\pi} \frac{e^{i\theta} + z}{e^{i\theta} - z} d\nu(\theta) \qquad (4.19)$$

stands for the Schwarz integral of the measure ν. Let us denote the image $\Lambda(\Sigma_r)$ in $\mathbb{C}^n$ by A_r. Repeating the argument in [Kh2] essentially word for word, we easily establish that for all $r > 0$, the sets A_r are open, convex, proper subsets of $\mathbb{C}^n$. (Convexity of A_r, for example, follows at once from the Cauchy-Schwarz inequality as in the proof of the uniqueness of f^* in Section 2.) If we denote by $\vec{c} = (c_1, \ldots, c_n)$ the vector of values we are interpolating in (2.1), then the infimum there is easily seen to be equal to

$$r_0 = \inf\{r > 0 : \vec{c} \in A_r\}.$$

Hence, our extremal function f^* (or equivalently $\varphi^* = \log f^*$) corresponds to a measure $d\nu^* \in \Sigma_{r_0}$ for which $\Lambda(\nu^*) \in \partial A_{r_0}$. So, to study the structure of extremal measures ν^* defining the extremals φ^* or $f^* = e^{\varphi^*}$, we need to characterize those $\nu^* \in \Sigma_r : \Lambda(\nu) \in \partial A_r$. From now on, without loss of generality, we assume that $r = 1$ and omit the index r altogether. Let $\vec{w} = (w_1, \ldots, w_n)$ be a finite boundary point of A. Then there exists a hyperplane H defined by $Re \sum_{j=1}^n a_j z_j = d$ such that for all $\vec{z} \in A$,

$$Re \sum_{j=1}^n a_j z_j \le d \quad \text{while} \quad Re \sum_{j=1}^n a_j w_j = d. \qquad (4.20)$$

Let ν^* denote a preimage $\Lambda^{-1}(\vec{w})$ in Σ. Using (4.14 and 4.15) we easily rephrase (4.20) in the following equivalent form:

$$\int_{\mathbb{T}} R(e^{i\theta}) d\nu(\theta) \le d \qquad (4.21)$$

for all ν satisfying (4.16), (4.17) and (4.18), $(r = 1)$ with equality holding for

$$\nu^* = s^* d\theta + d\mu^* \in \Lambda^{-1}(\vec{w}).$$

Then (4.14, 4.15 and 4.19) yield

$$R(e^{i\theta}) = \frac{1}{2\pi} Re(\sum_{j=1}^n a_j \frac{e^{i\theta} + \beta_j}{e^{i\theta} - \beta_j}), \qquad (4.22)$$

a rational function with $2n$ poles at $\beta_1, \ldots, \beta_n$ and $1/\bar{\beta}_1, \ldots, 1/\bar{\beta}_n$ that is real-valued on $\mathbb{T}$. Note the following (see [Kh2]):

Claim. *For d in (4.21) to be finite for all measures ν satisfying (4.16), (4.17) and (4.18), it is necessary that $R \ge 0$ on $\mathbb{T}$.*

Indeed, if $R(e^{i\theta})$ (which is continuous on $\mathbb{T}$) were strictly negative on a subarc $E \subset \mathbb{T}$, by choosing $d\nu = s\,d\theta$ with s negative and arbitrarily large in absolute value on E and fixed on $\mathbb{T} - E$, we would make the left-hand side of (4.21) go to $+\infty$ while still keeping the constraints (4.16), (4.17) and (4.18) intact, thus violating (4.21).

Now, if we knew that $R(e^{i\theta})$ had at least one zero at $e^{i\theta_0}$, we could easily conclude that the extremal measure ν^* in (4.21) can only have an atomic singular part with atoms located at the zeros of $R(e^{i\theta})$ on $\mathbb{T}$. Then, by the argument principle, since $R(e^{i\theta})$ cannot have more than n double zeros on $\mathbb{T}$, the argument sketched in Remark (ii) following Theorem 4.1 establishes the desired form of the extremal function f^*.

To see why a zero of R at $e^{i\theta_0}$ would yield the atomic structure of the singular part $d\mu^*$ of the extremal measure ν^* in (4.16), (4.17) and (4.18), simply note that if μ^* puts any mass on a closed set $E \subset \mathbb{T}$ where $R > 0$, we could replace μ^* by $\mu_1 = \mu^* - \mu^*|_E$ while compensating with a large negative weight at $e^{i\theta_0}$ not to violate (4.18). This will certainly make the integral in (4.21) larger, thus contradicting the extremality of ν^*. Unfortunately, however, we have no control over whether $R(e^{i\theta})$ vanishes on the circle or not, so this reasoning runs aground if we are dealing with (4.21) for $R > 0$ on $\mathbb{T}$. In order to establish the atomic structure of the singular part of the extremal measure ν^* in (4.21) for $R > 0$ on $\mathbb{T}$, we must come up with a variation of ν^* which would increase $\int_{\mathbb{T}} R(e^{i\theta})d\nu(\theta)$ without violating (4.18). This is precisely the turning point that makes problems in the Bergman space so much more difficult than in their Hardy space counterparts. For the latter, if we had simply gotten rid of the singular part μ^* in ν, i.e., divided our corresponding extremal function f^* by a singular inner function defined by μ^*, then we would not have changed the Hardy norm of f^* at all (while we would have dramatically increased the Bergman norm of f^*). This observation in addition to the elementary inequality $u \ln u - u > u \ln v - v$ for any $u, v > 0$, allowed S.Ya. Khavinson (see [Kh2]) to show that in the context of Hardy spaces, when (4.18) is replaced by a similar restriction on the Hardy norm of $\exp(S(\nu))$, if $R > 0$, the extremal measure ν^* is simply a constant times $\log R(e^{i\theta})d\theta$, and an easy qualitative description of extremals follows right away.

Now, in view of the above discussion, we cannot expect that for our problem, when $R > 0$ on $\mathbb{T}$, the extremal measure ν^* in (4.21) satisfying (4.16), (4.17) and (4.18) is absolutely continuous. But where should we expect the atoms of the singular part μ^* of the extremal ν^* to be located? We offer here the following conjecture.

Conjecture. *If $R > 0$ on $\mathbb{T}$, then the singular part μ^* of the extremal measure ν^* in (4.21) is supported on the set of local minimum points of R on $\mathbb{T}$.*

In other words, the singular inner part of the extremal function f^* for Problem 2.1 corresponding to the boundary point of A defined by the hyperplane (4.20) is atomic with atoms located at the local minima of R on $\mathbb{T}$.

The conjecture is intuitive in the sense that in order to maximize the integral in (4.21), we are best off if we concentrate all the negative contributions from the singular part of ν at the points where $R > 0$ is smallest. Note that this conjecture does correspond to the upper estimate of the number of atoms in the singular inner part of the extremal function f^* in (1.3). Indeed, R is a rational function of degree $2n$ and hence has $4n - 2$ critical points (i.e., where $R'(z) = 0$) in $\hat{\mathbb{C}}$. Since the number of local maxima and minima of R on $\mathbb{T}$ must be the same (consider $1/R$ instead), we easily deduce that R cannot have more than $2n - 2$ local minima (or maxima) on $\mathbb{T}$. (At least two critical points symmetric with respect to $\mathbb{T}$ must lie away from $\mathbb{T}$.)

One possible way to attempt to prove the conjecture using a variation of the extremal measure ν^* in (4.21) might be to divide the function f^* by a function G that would diminish the singular part μ^* of ν^*. Of course, a natural candidate for such a G would be the contractive divisor associated with the invariant subspace $[J]$ in A^2 generated by a singular inner function J built upon a part μ_0 of μ^* such that $\mu_0 \geq \mu^*$ (recall that $\mu^* \leq 0$), such that the support of μ_0 is a subset of the part of the circle that does not contain the local minima of R. Then (cf. [DuKS]) $G = hJ$, where h is a Nevanlinna function and $\|f^*/G\|_{A^2} \leq \|f^*\|$, so (4.18) is preserved. Unfortunately, $|h| > 1$ on $\mathbb{T} - supp(\mu_0)$, so the resulting measure ν defined by $\log(f^*/G) = S(\nu)$ may at least a priori actually diminish the integral in (4.21) instead of increasing it.

Finally, we remark that for the special case when the $\beta_j = 0$ and instead of Problem 2.1 we have the problem of finding

$$\inf\{\|f\|_{A^p} : f \neq 0, f^{(j)}(0) = c_j, j = 0, \ldots, n\}, \tag{4.23}$$

the conjectured general form of the extremal function f^* collapses to

$$f^*(z) = C \prod_{j=1}^{n} (1 - \bar{\alpha}_j z)^{\frac{2}{p}} \exp\left(\sum_{j=1}^{k} \lambda_j \frac{e^{i\theta_j} + z}{e^{i\theta_j} - z}\right), \tag{4.24}$$

where $|\alpha_j| \leq 1, j = 1, \ldots, n$, $k \leq n$, $\lambda_j \leq 0$. The difference in the degree of the outer part in (4.24) versus the rational function in (1.3) appears if one follows the proof of Theorem 4.1 word for word arriving at

$$|F|^2 = \frac{\prod_{j=1}^{n}(z - \alpha_j)(1 - \bar{\alpha}_j z)}{z^n}$$

instead of (4.12).

We shall discuss Problem 4.23 for $n = 2$ in great detail in the last section.

5. The minimal area problem for locally univalent functions

In this section we shall discuss a particular problem arising in geometric function theory and first studied by Aharonov and Shapiro in [AhSh1, AhSh2]. The problem is initially stated as that of finding

$$\inf\{\int_{\mathbb{D}}|F'(z)|^2 dA : F(0) = 0, F'(0) = 1, F''(0) = b, F'(z) \neq 0 \text{ in } \mathbb{D}\}. \qquad (5.1)$$

Problem (5.1) has the obvious geometric meaning of finding, among all locally univalent functions whose first three Taylor coefficients are fixed, the one that maps the unit disk onto a Riemann surface of minimal area. Setting $f = F'$ and $c = 2b$ immediately reduces the problem to a particular example of problems mentioned in (4.23), namely that of finding

$$\inf\{\int_{\mathbb{D}}|f|^2 dA : f \neq 0 \text{ in } \mathbb{D}, f(0) = 1, f'(0) = c\}. \qquad (5.2)$$

Assuming without loss of generality that c is real, we find that the conjectured form of the extremal function f in (5.2) is

$$f(z) = C(z - A)e^{\mu_0 \frac{z+1}{z-1}}, \qquad (5.3)$$

where $\mu_0 \geq 0$, and C, A, and μ_0 are uniquely determined by the interpolating conditions in (5.2). Of course, if $|c| \leq 1$ in (5.2), the obvious solution is

$$f^* = 1 + cz,$$

and hence, $F^* = z + \frac{c}{2}z^2$ solves (5.1), mapping $\mathbb{D}$ onto a cardioid. The nontrivial case is then when $|c| > 1$. All the results in the previous sections apply, so we know that the extremal for (5.2) has the form

$$f^* = hS,$$

where h is a bounded outer function and S is a singular inner function. As in Section 2, a simple variation gives us the orthogonality conditions (OC) as necessary conditions for extremality:

$$\int_{\mathbb{D}}|f^*|^2 z^{n+2} dA = 0, \; n = 0, 1, 2, \ldots. \qquad (5.4)$$

From now on, we will focus on the non trivial case of Problem 5.2 with $c > 1$. Thus, the singular inner factor of f^* is non trivial (cf. Corollary 4.2). In support of the conjectured extremal (5.3), we have the following proposition.

Proposition 5.1. *If the singular factor S of f^* has associated singular measure $d\mu$ that is atomic with a single atom, then*

$$f^*(z) = C(z - 1 - \mu_0)e^{\mu_0 \frac{z+1}{z-1}} \qquad (5.5)$$

where C and the weight μ_0 are uniquely determined by the interpolating conditions.

Remark. Although we have been unable to show that the singular inner factor for the extremal f^* is atomic, we offer some remarks after the proof that do support our hypothesis. If this is indeed the case, this would be, to the best of our knowledge, the first example of a "nice" extremal problem whose solution fails to be Lipschitz continuous or even continuous in the closed unit disk. All solutions to similar or even more general problems for non-vanishing H^p functions are Lipschitz continuous in $\bar{D}$ (cf. [Kh2] and the discussion in Section 4). Also, solutions to similar extremal problems in A^p without the non-vanishing restriction are all Lipschitz continuous in $\bar{\mathbb{D}}$ (cf. [KS]).

Proof. Our normalization ($c \in \mathbb{R}^+$) easily implies that the only atom of S is located at 1. So, $f^* = hS$, where S is a one atom singular inner function with mass μ_0 at 1, and h is outer. By Caughran's theorem ([Ca]), the antiderivative F^* of f^* has the same singular inner factor S and no other singular inner factors, i.e.,

$$F^* = HS, \tag{5.6}$$

where H is an outer function times perhaps a Blaschke product. Writing the orthogonality condition (5.4) in the form

$$\int_{\mathbb{D}} \bar{f^*} f^* z^2 p \, dA = 0$$

for any arbitrary polynomial p, and applying Green's formula, we obtain

$$\int_{\mathbb{T}} \bar{F^*} f^* z^2 p \, dz = 0 \tag{5.7}$$

for any arbitrary polynomial p. Using (5.6) and $S\bar{S} = 1$ a.e. on $\mathbb{T}$ yields

$$\int_{\mathbb{T}} \bar{H} h z^3 p \, d\theta = 0. \tag{5.8}$$

Since h is outer, hence cyclic in H^2, we can find a sequence of polynomials q_n such that $hq_n \to 1$ in H^2. Replacing p by $q_n p$ and taking a limit when $n \to \infty$ yields

$$\int_{\mathbb{T}} \bar{H} z^3 p \, d\theta = 0 \tag{5.9}$$

for all polynomials p. This last equation immediately implies that H is a quadratic polynomial. Now, $f^* = hS = (HS)' = H'S + HS'$, and $S' = \frac{2\mu_0}{(z-1)^2} S$. Since $f^* \in H^2$, H must have a double zero at 1 to cancel the pole of S'! Hence $H(z) = C(z-1)^2$, $F(z) = C(z-1)^2 S(z)$, and $f^*(z) = C(z - 1 - \mu_0) \exp(\mu_0 \frac{z-1}{z+1})$ as claimed. $\qquad\square$

We want to offer several additional remarks here.

(i) Obviously, the above calculations are reversible, so the function (5.5) does indeed satisfy the orthogonality condition (5.4) for the extremal.

(ii) The proof of Proposition 5.1 can be seen from a slightly different perspective. From Theorem 4.1, it already follows (assuming the hypothesis) that the outer part of f^* is a linear polynomial. Moreover, (5.7) implies that the antiderivative F^* of f^* is a noncyclic vector for the backward shift and hence has a meromorphic pseudocontinuation to $\hat{\mathbb{C}} - \mathbb{D}$ ([DSS]). Accordingly, F^* must be single-valued in a neighborhood of its only singular point $\{1\}$. This implies that $f^* = hS$ must have a zero residue at 1. (Otherwise F would have a logarithmic singularity there.) Calculating the residue of f^* at 1 for a linear polynomial h and an atomic singular factor S yields (5.5).

(iii) The only remaining obstacle in solving the extremal problem (5.2) is showing a priori that the singular inner factor of the extremal function is a one atom singular function. If one follows the outline given in Section 4, we easily find that for the problem (5.2), the function $R(e^{i\theta})$ in (4.22) becomes a rational function of degree 2, and since $R \geq 0$ on $\mathbb{T}$,

$$R(e^{i\theta}) = \text{const } \frac{(e^{i\theta} - a)(1 - \bar{a}e^{i\theta})}{e^{i\theta}} = \text{const } |e^{i\theta} - a|^2, \tag{5.10}$$

where $|a| \leq 1$. Thus, as we have seen in Section 4, we would be done if we could show that the one atom measure is the solution of the extremal problem

$$\max\{\int_{\mathbb{T}} R(e^{i\theta})d\mu(\theta) : \mu \leq 0, \mu \perp d\theta\} \tag{5.11}$$

where μ satisfies the constraint

$$\int_{\mathbb{D}} |h|^2 |S_\mu|^2 dA \leq 1 \tag{5.12}$$

for a given outer function h and R is given by (5.10). (Recall that S_μ is the singular inner function with associated singular measure μ.) Again, as noted previously, it is almost obvious when $|a| = 1$, since then we simply concentrate as much charge as needed at a to satisfy the constraint without changing the integral (5.11). Yet, in general, we have no control over where in $\mathbb{D}$ a appears.

(iv) Let $k(z)$ denote the orthogonal projection of $|f^*|^2$ onto the space of L^2 integrable harmonic functions in $\mathbb{D}$. The orthogonality condition (5.4) implies that $k(z)$ is a real harmonic polynomial of degree 1. Moreover, due to our normalization of the extremal problem (i.e., $c \in \mathbb{R}$), we can easily show that f^* in fact has real Taylor coefficients. Indeed, $f_1(z) := \overline{f^*(\bar{z})}$ satisfies the same interpolating conditions and has the same L^2-norm over $\mathbb{D}$, thus by the uniqueness of the extremal function, f_1 must be equal to f^*. Since f^* has real Taylor coefficients, the projection of $|f^*|^2$ is an even function of y, and thus

$$k(z) = A + Bx, \tag{5.13}$$

where $A = \int_{\mathbb{D}} |f^*|^2 dA$ and $B = 4 \int_{\mathbb{D}} z|f^*|^2 dA$. The orthogonality condition (5.4) now implies that the function $|f^*|^2 - k$ is orthogonal to all real-valued L^2 harmonic functions in $\mathbb{D}$. Using the integral formula in [DKSS2] (or [DS, Chapter 5, Section 5.3]), it follows that

$$\int_{\mathbb{D}} (|f^*(z)|^2 - k(z))s dA \geq 0 \tag{5.14}$$

for all functions s that are smooth in $\bar{\mathbb{D}}$ and subharmonic. The following corollary of (5.14) offers an unexpected application of the conjectured form of the extremal f^*.

Corollary 5.2. *Let $w \in \mathbb{T}$, and assume that $|f^*|^2/|z - w|^2 \in L^1(\mathbb{D})$. (Note that the conjectured extremal satisfies this condition at the point $w = 1$.) Then $k(w) \leq 0$. Thus, if f^* has the form (5.5), $B \leq -A$ in (5.13).*

84   D. Aharonov, C. Bénéteau, D. Khavinson and H.S. Shapiro

Proof. Choose $s(z) = 1/|rw - z|^2$ for $r > 1$. Applying (5.14) as $r \to 1^+$, we see that if $k(w) > 0$, the integral on the left must tend to $-\infty$, which violates (5.14). $\quad\square$

Calculating the classical balayage $U(e^{i\theta})$ of the density $|f^*|^2 dA$ to $\mathbb{T}$, i.e.,

$$U(e^{i\theta}) = Re \frac{1}{2\pi} \int_{\mathbb{D}} |f^*|^2 \frac{e^{i\theta} + z}{e^{i\theta} - z} dA, \qquad (5.15)$$

expanding the Schwarz kernel $\frac{e^{i\theta} + z}{e^{i\theta} - z}$ into the power series with respect to z and using the orthogonality condition (5.4) allows us to cancel all the terms containing powers of z of degree 2 and higher, so we arrive at

$$U(e^{i\theta}) = A + \frac{B}{2} \cos\theta,$$

where A and B are as in (5.13). Since $U > 0$ on $\mathbb{T}$ (it is a "sweep" of a positive measure!), it follows that

Corollary 5.3. $\qquad A > \frac{|B|}{2}, \quad i.e., \quad \int_{\mathbb{D}} |f^*|^2 dA > 2\left| \int_{\mathbb{D}} z|f^*|^2 dA \right|.$

A calculation confirms that for f^* as in (5.5), Corollary 5.3 does hold.

(v) If we denote the value of the minimal area in (5.1) by $A = A(b)$ and by a_3 the coefficient of z^3 in the Taylor expansion of the extremal function F^* (i.e, $F^*(z) = z + bz^2 + a_3 z^3 + \cdots$, where F^* is the anti-derivative of our extremal function f^*) then as was shown in ([AhSh2], Theorem 4, p. 21), the following equality must hold:

$$(3a_3 - 2b^2 - 1)A'(b) + 4bA(b) = 0. \qquad (5.16)$$

An involved calculation yields that the conjectured extremal function $F^* = \int f^*$, where f^* is as in (5.5), does indeed satisfy (5.16). This serves as yet one more justification of the conjectured form of the extremal. A number of other necessary properties of the extremal function are discussed in [AhSh1, AhSh2].

References

[AhSh1] D. Aharonov and H.S. Shapiro, *A minimal area problem in conformal mapping*, Part I, Preliminary Report, TRITA-MAT-1973-7, Royal Institute of Technology, Stockholm.

[AhSh2] D. Aharonov and H.S. Shapiro, *A minimal area problem in conformal mapping*, Part II, Preliminary Report, TRITA-MAT-1978-5, Royal Institute of Technology, Stockholm.

[ASS1] D. Aharonov, H.S. Shapiro, and A. Solynin, *A minimal area problem in conformal mapping*, J. Anal. Math. **78** (1999), 157–176.

[ASS2] D. Aharonov, H.S. Shapiro, and A. Solynin, *A minimal area problem in conformal mapping. II*, J. Anal. Math. **83** (2001), 259–288.

[BK]   C. Bénéteau and B. Korenblum, *Some coefficient estimates for H^p functions*, Complex Analysis and Dynamical Systems, Contemporary Mathematics **364** (2004), 5–14.

[Ca]   J. Caughran, *Factorization of analytic functions with H^p derivative*, Duke Math. J. **36** (1969), 153–158.

[CL]   E. Collingwood and A. Lowater, *Theory of cluster sets*, Cambridge University Press, 1966.

[DuKS]   P. Duren, D. Khavinson, H.S. Shapiro, *Extremal functions in invariant subspaces of Bergman spaces*, Illinois J. Math., **40** (1996), 202–210.

[DKSS1] P. Duren, D. Khavinson, H.S. Shapiro, and C. Sundberg, *Contractive zero-divisors in Bergman spaces*, Pacific J. Math., **157** (1993), no. 1, 37–56.

[DKSS2] P. Duren, D. Khavinson, H.S. Shapiro and C. Sundberg, *Invariant subspaces in Bergman spaces and the biharmonic equation*, Michigan Math. J., **41** (1994), no. 2, 247–259.

[DSS]   R. Douglas, H.S. Shapiro, A. Shields, *Cyclic vectors and invariant subspaces for the backward shift operator*, Ann. Inst. Fourier (Grenoble), **20** (1970), fasc. 1, 37–76.

[Du]   P. Duren, *Theory of H^p Spaces*, Pure and Appl. Math., Vol. 38, Academic Press, New York-London, 1970.

[DS]   P. Duren and A. Schuster, *Bergman Spaces of Analytic Functions*, Amer. Math. Soc., 2004.

[Ga]   J. Garnett, *Bounded Analytic Functions*, Pure and Applied Mathematics, 96. Academic Press, New York-London, 1981.

[GPS]   B. Gustafsson, M. Putinar, H.S. Shapiro, *Restriction operators, balayage and doubly orthogonal systems of analytic functions*, J. Funct. Anal., **199** (2003), no. 2, 332–378.

[HKZ]   H. Hedenmalm, B. Korenblum, K. Zhu, *Theory of Bergman Spaces*, Springer-Verlag, New York, 2000.

[Ho]   K. Hoffman, *Banach Spaces of Analytic Functions*, Prentice-Hall Inc., Englewood Cliffs, N. J. 1962.

[Kh1]   S.Ya. Khavinson, *On an extremal problem in the theory of analytic functions*, Uspekhi Math. Nauk, **4** (1949), no. 4 (32), 158–159. (Russian)

[Kh2]   S.Ya. Khavinson, *Two papers on extremal problems in complex analysis*, Amer. Math. Soc. Transl. (2) **129**. Amer. Math. Soc., Providence, RI, 1986.

[KS]   D. Khavinson and M. Stessin, *Certain linear extremal problems in Bergman spaces of analytic functions*, Indiana Univ. Math. J., **46** (1997), no. 3, 933–974.

[Ko]   P. Koosis, *Introduction to H^p Spaces*, London Math. Soc. Lecture Note Series, 40. Cambridge University Press, Cambridge–New York, 1980.

[Pr]   I.I. Privalov, *Boundary Properties of Analytic Functions*, Gostekhizdat, Moscow–Leningrad, 1950. (Russian)

[RS]   W.W. Rogosinski and H.S. Shapiro, *On certain extremum problems for analytic functions*, Acta Math., **90** (1953), 287–318.

[Ry1]   V.G. Ryabych, *Certain extremal problems*, Nauchnye soobscheniya R.G.U. (1965), 33–34. (Russian)

[Ry2]   V.G. Ryabych, *Extremal problems for summable analytic functions*, Siberian Math. J., **XXVIII** (1986), 212–217. (Russian)

[Vu]   D. Vukotić, *Linear extremal problems for Bergman spaces*, Expositiones Math., **14** (1996), 313–352.

Dov Aharonov
Department of Mathematics
Technion
32000 Haifa, Israel
e-mail: `dova@techunix.technion.ac.il`

Catherine Bénéteau
Department of Mathematics and Computer Science
Seton Hall University
South Orange
New Jersey 07079, USA
e-mail: `beneteca@shu.edu`

Dmitry Khavinson
Department of Mathematics
University of Arkansas
Fayetteville
Arkansas 72701, USA
e-mail: `dmitry@uark.edu`

Harold Shapiro
Department of Mathematics
Royal Institute of Technology
S-10044 Stockholm
Sweden
e-mail: `shapiro@math.kth.se`

Operator Theory:
Advances and Applications, Vol. 158, 87–94

Generalization of Carathéodory's Inequality and the Bohr Radius for Multidimensional Power Series

Lev Aizenberg

Abstract. In the present paper we generalize Carathéodory's inequality for functions holomorphic in Cartan domains in $\mathbf{C}^n$. In particular, in the case of functions holomorphic in the unit disk in $\mathbf{C}$, this generalization of Carathéodory's inequality implies the classical inequalities of Carahtéodory and Landau. As an application, new results on multidimensional analogues of Bohr's theorem on power series are obtained. Furthermore, the estimate from below of Bohr radius is improved for the domain $\mathcal{D} = \{z \in \mathbf{C}^2 : |z_1| + |z_2| < 1\}$.

Mathematics Subject Classification (2000). 32A05.

Keywords. Carathéodory's inequality, Bohr's theorem, Cartan domains, power series.

1. Carathéodory's inequality in several complex variables

Let $\mathcal{D} \subset \mathbf{C}^n$ be a circular domain, that is, a Cartan domain, characterized by the fact that if $z \in \mathcal{D}$ then $ze^{i\phi} \in \mathcal{D}$, where $z = (z_1, \ldots, z_n)$ and $0 \le \phi \le 2\pi$. Furthermore, we also assume that the domain $\mathcal{D}$ is a strongly starlike domain, that is, for every homothetic transformation $\lambda \overline{\mathcal{D}} \subset \mathcal{D}$, where $0 < \lambda < 1$.

It is a well-known fact that in domains of this type every holomorphic function can be expanded into a series of homogeneous polynomials. Now, we are ready to formulate our first lemma.

Lemma 1.1. *Let $\mathcal{D}$ be a Cartan, strongly starlike domain. Let f be a holomorphic function in $\mathcal{D}$, with expansion into homogeneous polynomials given by*

$$f(z) = \sum_{k=0}^{\infty} P_k(z), \ z \in \mathcal{D}, \qquad (1.1)$$

During the preparation of this work the author was supported by the Israel-Slovenia grant. Part of the work was completed in January-February 2001 during the author's stay at the Institute of Mathematics, Physics and Mechanincs, Ljubljiana, Slovenia.

where $P_k(z)$ is a homogeneous polynomial of degree k. If $\Re f(z) > 0$ for every $z \in \mathcal{D}$ then for every $k \geq 1$ the following inequality holds

$$|P_k(z)| \leq 2\Re P_0(z), \ \forall z \in \mathcal{D}. \tag{1.2}$$

Proof. It is known that for every $k = 1, 2, \ldots$, the polynomials $P_k(z)$ in the expansion (1.1) are given by

$$P_k(z) \quad = \quad \lim_{r \to 1-0} \frac{1}{2\pi r^k} \int_0^{2\pi} f(zre^{i\phi})e^{-ik\phi}d\phi$$

$$= \quad \lim_{r \to 1-0} \frac{1}{2\pi r^k} \int_0^{2\pi} \left(f(zre^{i\phi}) + \overline{f(zre^{i\phi})} \right) e^{-ik\phi}d\phi.$$

Therefore

$$|P_k(z)| \quad \leq \quad \lim_{r \to 1-0} \frac{1}{2\pi r^k} \int_0^{2\pi} 2\Re f(zre^{i\phi})d\phi = \lim_{r \to 1-0} \frac{1}{r^k} 2\Re f(0)$$

$$= \quad \lim_{r \to 1-0} \frac{1}{r^k} 2\Re P_0(z) = 2\Re P_0(z).$$

This completes the proof of the lemma. $\qquad\square$

Lemma 1.1 and the simple fact that every convex domain in the complex plane $\mathbf{C}$ is intersection of half-planes allow us to deduce the following theorem.

Theorem 1.1. *Let $\mathcal{D}$ be a Cartan domain and f be a function holomorphic in $\mathcal{D}$. If in the domain $\mathcal{D}$ the expansion (1.1) for the function f is valid and $f(\mathcal{D}) \subset G \subset \mathbf{C}$, then for every $z \in \mathcal{D}$ the following inequality holds for every $k \geq 1$*

$$|P_k(z)| \leq 2 \operatorname{dist}(P_0(z), \partial \widetilde{G}),$$

where $\widetilde{G}$ is the convex hull of G.

Remark 1.1. We point out here that in the above theorem $\widetilde{G}$ cannot be replaced by G.

Theorem 1.1 has a number of interesting corollaries.

Corollary 1.1. *Let $\mathcal{D}$ be a Cartan domain and f be a function holomorphic in $\mathcal{D}$. If $|f(z)| < 1$ for every $z \in \mathcal{D}$, then for every $k \geq 1$ the following holds*

$$|P_k(z)| \leq 2(1 - |P_0(z)|), \ \forall z \in \mathcal{D}. \tag{1.3}$$

Corollary 1.2. *Let $\mathcal{D}$ be a complete, bounded, Reinhardt domain. Let f be a function holomorphic in $\mathcal{D}$ with the corresponding multidimensional power series*

$$f(z) = \sum_{|\alpha| \geq 0} c_\alpha z^\alpha, \ z \in \mathcal{D}, \tag{1.4}$$

where $\alpha = (\alpha_1, \ldots, \alpha_n)$, $|\alpha| = \alpha_1 + \cdots + \alpha_n$, $z^\alpha = z_1^{\alpha_1} \ldots z_n^{\alpha_n}$, and all α_i are nonnegative integers. If $f(\mathcal{D}) \subset G \subset \mathbf{C}$, then for every α such that $|\alpha| \geq 1$ the following holds

$$|c_\alpha| \leq \frac{2 \operatorname{dist}(c_0, \partial \widetilde{G})}{d_\alpha(\mathcal{D})},$$

where $d_\alpha(\mathcal{D}) = \max_{\overline{\mathcal{D}}} |z^\alpha|$.

We point out here that the proof of Corollary 1.2 follows from Theorem 1.1 and the Cauchy estimates in the case of power series taken from [5].

In one complex variable Theorem 1.1 leads to

Corollary 1.3. *If in the unit disk $K = \{z_1 \in \mathbf{C} : |z_1| < 1\}$ the function f is a power series, that is,*

$$f(z_1) = \sum_{k=0}^{\infty} c_k z_1^k, \tag{1.5}$$

and $f(K) \subset G$, then for every $k \geq 1$ we have

$$|c_k| \leq 2 \operatorname{dist}(c_0, \partial \widetilde{G}).$$

The following result is the known Carathéodory's inequality [11]:

Corollary 1.4. *If in the unit disk $K = \{z_1 \in \mathbf{C} : |z_1| < 1\}$ the equality (1.5) holds and if $\Re f(z_1) > 0$ for every $z_1 \in K$ then for all $k \geq 1$*

$$|c_k| \leq 2\Re c_0.$$

Another classical inequality, known as Landau's inequality [15], is also deduced from Theorem 1.1:

Corollary 1.5. *If in the unit disk $K = \{z_1 \in \mathbf{C} : |z_1| < 1\}$ the equality (1.5) holds and if $|f(z_1)| < 1$ for every $z_1 \in K$ then for all $k \geq 1$*

$$|c_k| \leq 2(1 - |c_0|).$$

2. Bohr's theorem in several complex variables

Let us recall the theorem of H. Bohr [10], proven at the beginning of the 20th century.

Theorem 2.1. *If a power series (1.5) converges in the unit disk K and its sum has modulus less than 1, then*

$$\sum_{k=0}^{\infty} |c_k z_1^k| < 1$$

in the disk $\{z_1 \in \mathbf{C} : |z_1| < \frac{1}{3}\}$. Moreover the constant $\frac{1}{3}$ cannot be improved.

Formulations of Bohr's theorem in several complex variables appeared very recently. We recall some of them.

Given a complete Reinhardt domain $\mathcal{D}$, we denote by $R(\mathcal{D})$ the largest non-negative number r with the property that if the power series (1.4) converges in $\mathcal{D}$ and its sum has modulus less than 1, then

$$\sum_{|\alpha| \geq 0} |c_\alpha z^\alpha| < 1, \tag{2.1}$$

in the homothetic domain $r\mathcal{D}$. In [9] the following result is proved in the case when $\mathcal{D}$ is the unit polydisk

$$U^n = \{ z \in \mathbf{C}^n : |z_j| < 1, \ j = 1, \ldots, n \}.$$

Theorem 2.2. *For $n > 1$ one has*

$$\frac{1}{3\sqrt{n}} < R(U^n) < \frac{2\sqrt{\log n}}{\sqrt{n}}.$$

We see from Theorem 2.2 that $R(U^n) \to 0$ as $n \to \infty$. If $\mathcal{D}$ is the hypercone

$$\mathcal{D}_1^n = \{ z \in \mathbf{C}^n : |z_1| + \cdots + |z_n| < 1 \},$$

then the situation is quite different as the following theorem, taken from [1], shows.

Theorem 2.3. *For $n > 1$ one has*

$$\frac{1}{3e^{\frac{1}{3}}} < R(\mathcal{D}_1^n) \leq \frac{1}{3}.$$

For further estimates of $R(\mathcal{D})$ in the domains

$$\mathcal{D}_p^n = \{ z \in \mathbf{C}^n : |z_1|^p + \cdots + |z_n|^p < 1 \},$$

where $1 \leq p < \infty$, we refer the reader to [7]. For other generalizations of Bohr's theorem see [2], [3], [4], [6], [14], [8]. It is noteworthy that in [13] the lower bound from Theorem 2.2 is refined and is $\sqrt{\frac{\log n}{n \log \log n}}$ times a constant.

As it was pointed out already in [1], it seems more natural to consider not a single number in the Bohr problem in $\mathbf{C}^n$, but the largest subdomain $\mathcal{D}_B$ of $\mathcal{D}$, such that (2.1) holds. At this stage, we state the following new result in this direction.

Theorem 2.4. *If the power series (1.4) converges in the unit ball*

$$\mathcal{D}_2^n = \{ z \in \mathbf{C}^n : |z_1|^2 + \cdots + |z_n|^2 < 1 \}$$

and the modulus of its sum is less than 1, then (2.1) holds in the hypercone

$$\frac{1}{3}\mathcal{D}_1^n = \{ z \in \mathbf{C}^n : |z_1| + \cdots + |z_n| < 1 \}$$

and the constant $\frac{1}{3}$ cannot be improved.

The proof of Theorem 2.4 is based on the following lemma.

Lemma 2.1. *If $P_k(z)$ is the homogeneous polynomial*

$$P_k(z) = \sum_{|\alpha|=k} c_\alpha z^\alpha$$

and $|P_k(z)| < 1$ in the ball $\mathcal{D}_2^n$, then

$$\sum_{|\alpha|=k} |c_\alpha z^\alpha| < 1$$

for every z in the hypercone $\mathcal{D}_1^n$.

Proof. Using induction, one can prove the inequality

$$\sqrt{\frac{|\alpha|^{|\alpha|}}{\alpha_1^{\alpha_1}\ldots\alpha_n^{\alpha_n}}} \leq \frac{|\alpha|!}{\alpha_1!\ldots\alpha_n!}. \tag{2.2}$$

It follows from Corollary 1.2 that

$$|c_\alpha| \leq \frac{1}{d_\alpha(\mathcal{D}_2^n)} = \sqrt{\frac{|\alpha|^{|\alpha|}}{\alpha_1^{\alpha_1}\ldots\alpha_n^{\alpha_n}}}.$$

From this, the relation (2.2) implies that for every $z \in \mathcal{D}_1^n$ we have

$$
\begin{aligned}
\sum_{|\alpha|=k} |c_\alpha z^\alpha| &\leq \sum_{|\alpha|=k} \frac{|\alpha|!}{\alpha_1!\ldots\alpha_n!}|z_1|^{\alpha_1}\ldots|z_n|^{\alpha_n} \\
&= (|z_1| + \cdots + |z_n|)^k < 1.
\end{aligned}
$$

The proof of the lemma is now complete. $\qquad\square$

We remark here that the condition $|P_k(z)| < 1$ in the ball $\mathcal{D}_2^n$ cannot be replaced by the analogous inequality in the hypercone $\mathcal{D}_1^n$. To see that, we consider the second-order homogeneous polynomial

$$Q(z) = \frac{\sqrt{3}}{2}(z_1^2 + z_2^2) + 3iz_1z_2.$$

Then for $z \in \overline{\mathcal{D}}_1^2$, we have $|Q(z)| \leq 1$, but

$$\max_{\overline{\mathcal{D}}_1^2}\left[\frac{\sqrt{3}}{2}(|z_1|^2 + |z_2|^2) + 3|z_1z_2|\right] = \frac{3+\sqrt{3}}{4} > 1.$$

Now, we are ready to return to the proof of Theorem 2.4.

Proof of Theorem 2.4. From the assumption $|f(z)| < 1$ in the ball $\mathcal{D}_2^n$ and the Corollary 1.1 we get that for every $k \geq 1$ the estimate

$$\left|\sum_{|\alpha|=k} c_\alpha z^\alpha\right| \leq 2(1 - |c_0|)$$

holds for every $z \in \mathcal{D}_2^n$. From this inequality and Lemma 2.1 we deduce that

$$\sum_{|\alpha|=k} |c_\alpha z^\alpha| \leq 2(1 - |c_0|)$$

for every $z \in \mathcal{D}_1^n$. Now, if $z \in \frac{1}{3}\mathcal{D}_1^n$, then

$$\sum_{|\alpha|\geq 0} |c_\alpha z^\alpha| \leq |c_0| + 2(1 - |c_0|) \sum_{k=1}^{\infty} \frac{1}{3^k} = 1.$$

The fact that the constant $\frac{1}{3}$ is sharp can be seen if we consider a function of the type $f(z_1, 0, \ldots, 0)$. This completes the proof of the theorem. $\qquad\square$

In order to improve the estimate $R(\mathcal{D}_1^n) > \frac{1}{3e^{\frac{1}{3}}} = 0.238844$ taken from Theorem 2.3 in the case $n = 2$, we use the following lemma from [12], [16].

Lemma 2.2. *Let*

$$F(t) = \sum_{j=-n}^{n} a_j e^{ijt}$$

be a real trigonometric polynomial. Then for $k > \frac{n}{2}$ the following inequality holds

$$|a_o| + |a_{-k}| + |a_k| \leq \max_t |F(t)|.$$

The last theorem of the paper is the following one.

Theorem 2.5.

$$R(\mathcal{D}_1^2) > 0.304236.$$

Proof. If for every $z \in \overline{\mathcal{D}}_1^2$ one has that

$$|P_k(z)| = \left| \sum_{|\alpha|=k} c_\alpha z^\alpha \right| \leq 1,$$

then in particular this holds for the points of the form $(\frac{1}{2}, \frac{1}{2}e^{it})$ and we obtain

$$\left| \sum_{\alpha_2=0}^{k} c_{k-\alpha_2, \alpha_2} e^{it\alpha_2} \right| \leq 2^k.$$

From this we deduce that

$$2^{2k} \left| P_k\left(\frac{1}{2}, \frac{e^{it}}{2}\right) \right|^2 = \sum_{\alpha_1+\alpha_2=k} |c_{\alpha_1,\alpha_2}|^2 + c_{0,k}\bar{c}_{k,0}e^{ikt}$$
$$+ \ \bar{c}_{0,k}c_{k,0}e^{-ikt} + \ldots \leq 2^{2k}.$$

Lemma 2.2 implies

$$\sum_{\alpha_1+\alpha_2=k} |c_{\alpha_1,\alpha_2}|^2 + 2|c_{0,k}||c_{k,0}| \leq 2^{2k},$$

or

$$\sum_{\alpha_1=1}^{k-1} |c_{\alpha_1,k-\alpha_1}|^2 + (|c_{0,k}| + |c_{k,0}|)^2 \le 2^{2k}. \tag{2.3}$$

We now want to compute

$$\max_{\substack{x_1+x_2\le 1 \\ x_1,x_2\ge 0}} \sum_{\alpha_1+\alpha_2=k} |c_{\alpha_1,\alpha_2}| x_1^{\alpha_1} x_2^{\alpha_2} = A_k$$

under conditions (2.3) and $|c_{k,0}| \le 1$, $|c_{0,k}| \le 1$. It is not very difficult to show that

$$A_k = \frac{1 + \sqrt{(k-1)(2^{2(k-1)} - 1)}}{2^{k-1}}.$$

Furthermore, we now consider the equation

$$\sum_{k=1}^{\infty} A_k x^k = \sum_{k=1}^{\infty} \frac{1 + \sqrt{(k-1)(2^{2(k-1)} - 1)}}{2^{k-1}} x^k = \frac{1}{2}. \tag{2.4}$$

Using the program Mathematica 3.0 [17], we estimated that the equation (2.4) has a root greater than $x_0 = 0.304236$. In addition, if $z \in x_0 \mathcal{D}_1^2$ then

$$\sum_{|\alpha|=0}^{\infty} |c_\alpha| |z^\alpha| < |c_0| + \left(\sum_{k=1}^{\infty} A_k x^k\right) 2(1 - |c_0|) < 1.$$

This completes the proof of the theorem. $\qquad\square$

Remark 2.1. The paper by Boas and Khavinson [9] contains essentially the following result (Remark 4 in [1]): If the power series (1.4) converges in the unit polydisk U^n and the modulus of its sum is less than 1, then (2.1) holds in the ball $\frac{1}{3}\mathcal{D}_2^n$ and the constant $\frac{1}{3}$ cannot be improved.

Remark 2.2. The estimates of Bohr radius from the Theorem 2.3 hold also for every Reinhardt domain of the type

$$\mathcal{D} = \{z \in \mathbf{C}^n : \phi(|z_1|, \ldots, |z_n|) < 0\},$$

where ϕ is a convex function, that is, $\mathcal{D}$ is the union of hypercones

$$\{z \in \mathbf{C}^n : a_1|z_1| + \cdots + a_n|z_n| < 1\}.$$

The same holds about the estimates from Theorem 2.5.

Acknowledgement. The author thanks J. Globevnik, E. Liflyand, and A. Vidras for their help during the preparation of this paper.

References

[1] L. Aizenberg, *Multidimensional analogues of Bohr's theorem on power series*, Proc. Amer. Math. Soc. **128** (2000), 1147–1155.

[2] L. Aizenberg, *Bohr Theorem*, Encyclopedia of Mathematics, Supplement II (ed. M. Hazewinkel), Kluwer, Dordrecht, 2000, 76–78.

[3] L. Aizenberg, A. Aytuna, P. Djakov, *An abstract approach to Bohr's phenomenon*, Proc. Amer. Math. Soc. **128** (2000), 2611–2619.

[4] L. Aizenberg, A. Aytuna, P. Djakov, *Generalization of Bohr's theorem for bases in spaces of holomorphic functions of several complex variables* , J. Math. Anal. Appl. **258** (2001), 428–447.

[5] L. Aizenberg, B.S. Mityagin, *The spaces of functions analytic in multicircular domains*, Sibir. Math. J. **1** (1960), 1953–1970 (Russian).

[6] L. Aizenberg, N. Tarkhanov, *A Bohr phenomenon for elliptic equations*, Proc. London Math. Soc. **82** (2001), 385–401.

[7] H.P. Boas *Majorant series*, J. Korean Math. Soc. **37** (2000), 321–337.

[8] C. Bénéteau, A. Dahlner, D. Khavinson, *Remarks on the Bohr phenomenon*, Comput. Methods Func. Theory **4** (2004), 1–19.

[9] H.P. Boas, D. Khavinson, *Bohr's power series theorem in several variables*, Proc. Amer. Math. Soc. **125** (1997), 2975–2979.

[10] H. Bohr, *A theorem concerning power series*, Proc. London Math.Soc. **13** (1914), 1–5.

[11] C. Carathéodory, *Über den Variabilitätsbereich der Koeffizienten der Potenzreihen, die gegebene Werte nicht annehmen*, Math. Ann. **64** (1907), 95–115.

[12] J.G. van der Corput, C. Visser, *Inequalities concerning polynomials and trigonometric polynomials*, Neder. Acad. Wetensch. Proc. **49** (1946), 383–392.

[13] A. Defant, L. Frerick, *A logarithmic lower bound for the Bohr radii*, 2004, Preprint.

[14] P.B. Djakov, M.S. Ramanujan, *A remark on Bohr's theorem and its generalizations*, J. Analysis, **8** (2000), 65–77.

[15] E. Landau, D. Gaier, *Darstellung und Begründung einiger neuerer Ergebnisse der Funktionentheorie*, Springer-Verlag, 1986.

[16] D.S. Mitrinovic, J.E. Pecaric, A.M. Frank, *Classical and new inequalities in Analysis*, Kluwer, Dordrecht, 1993.

[17] Wolfram Research, *Mathematica 3.0*, 1996.

Lev Aizenberg
Department of Mathematics
Bar-Ilan University
52900 Ramat-Gan
Israel
e-mail: `aizenbrg@math.biu.ac.il`

Operator Theory:
Advances and Applications, Vol. 158, 95–99
© 2005 Birkhäuser Verlag Basel/Switzerland

Remarks on the Value Distribution of Meromorphic Functions

J.M. Anderson

In memory of S. Ya. Khavinson

1. Introduction

Let $f(z)$ be a meromorphic function in the complex plane $\mathbb{C}$. We assume acquaintance with the standard definitions of the Nevanlinna theory as given in [2]. Define the spherical derivative $\rho\left(f(z)\right)$ of $f(z)$ by

$$\rho\left(f(z)\right) = \frac{|f'(z)|}{1 + |f(z)|^2}$$

and set

$$\mu(r, f) = \sup\{\rho(f(z)) : |z| = r\}.$$

We also let $D(w, r)$ be the radius of the largest disk centred on w with $|w| = r$ in which $f(z)$ does not assume the values $0, 1, \infty$. We set

$$D(r) = \min\{D(w, r) : |w| = r\}.$$

The following lemma is due to Pommerenke [4].

Lemma 1. *Let $f(z)$ be analytic and $\neq 0, 1$ in $|z| < r$ where $r > 0$. Then*

$$r\rho(f(0)) \leq 4\sqrt{2}.$$

Applying this Lemma to the point on $|z| = r$ where $\mu(r)$ is attained we see that

$$D(r)\mu(r) \leq 4\sqrt{2}.$$

In [1] the following theorems have been proved (among others).

Theorem 1. *Suppose that $f(z)$ is a meromorphic function such that*

$$\liminf_{r \to \infty} \frac{T(r, f)}{r^\sigma} \leq K$$

for some $\sigma > 0$ and $0 < K < \infty$. Then

$$\limsup_{r \to \infty} \frac{r^{\sigma-1}}{\mu(r)} \geq \left(24\sqrt{2}K\sigma^2\right)^{-1}. \tag{1.1}$$

Theorem 2. *Suppose that $f(z)$ is a meromorphic function of lower order b for some $b > 0$. Then*

$$\limsup_{r \to \infty} \frac{T(r,f)}{r\mu(r)} \geq \left(126\sqrt{2}b^2\right)^{-1}. \tag{1.2}$$

Theorem 3. *Suppose that $f(z)$ is a meromorphic function such that*

$$\liminf_{r \to \infty} \frac{T(r,f)}{(\log r)^2} = \infty,$$

but, for some $b \geq 2$,

$$\liminf_{r \to \infty} \frac{\log T(r,f)}{\log \log r} = b.$$

Then

$$\limsup_{r \to \infty} \frac{T(r,f)}{r\mu(r)(\log r)^2} \geq \left(96\sqrt{2}b^2\right)^{-1}. \tag{1.3}$$

An examination of the proofs of these theorems shows that, in fact, (1.1), (1.2) and (1.3) can be replaced by the stronger conclusions

$$\limsup_{r \to \infty} r^{\sigma-1}D(r) \geq (6\sigma^2 K)^{-1}, \tag{1.4}$$

$$\limsup_{r \to \infty} \frac{T(r,f)D(r)}{r} \geq \left(\frac{63}{2}b^2\right)^{-1}, \tag{1.5}$$

$$\limsup_{r \to \infty} \frac{T(r,f)D(r)}{r(\log r)^2} \geq (24b^2)^{-1}, \tag{1.6}$$

respectively. We then apply Lemma 1 to points on $|z| = r$ at which $\mu(r)$ is attained and the theorems follow immediately.

2. Remarks

The purpose of this note, which is purely expository, is to point out the connection between the results and the much deeper-lying results of [3]. In that paper the functions which are considered are entire and the disks are centred on the points of maximum modulus. The disks, moreover, are disks where the value 0 alone is omitted. The results are expressed in terms of a proximate order function, rather than directly with $T(r)$, but have the advantage of being of a lim inf nature. Moreover the corresponding results to (1.4), for example, depends only on σ and not on σ^2.

3. Proof of (1.4)

The proofs of Theorems $1, 2, 3$ depend on an increasingly more complicated selection of sequences along which the $\limsup$ is attained. To prove (1.4) we set

$$n(r) = n(r, 0, f) + n(r, 1, f) + n(r, \infty, f),$$
$$N(r) = N(r, 0, f) + N(r, 1, f) + N(r, \infty, f),$$

so that by Nevanlinna's first fundamental theorem,

$$N(r) \le 3T(r, f) + O(1), \quad r \to \infty. \tag{3.1}$$

Suppose that for some a there is an integer p_0 with

$$n\left(a(p+1)^{\frac{1}{\sigma}}\right) - n(ap^{\frac{1}{\sigma}}) \ge 1$$

for all integers $p \ge p_0$. This implies that

$$n(r) \ge \left(\frac{r}{a}\right)^{\sigma} + O(1), \quad r \to \infty,$$

which in term implies that

$$T(r, f) \ge (3\sigma a^{\sigma})^{-1} r^{\sigma} + O(\log r), \quad r \to \infty,$$

by (3.1). This contradicts the assumption of Theorem 1 unless

$$a \le (3\sigma K)^{-\frac{1}{\sigma}}. \tag{3.2}$$

For such a choice of a there is a sequence $\{p_s\}$ of integers tending to ∞ as $s \to \infty$ such that

$$n\left(a(p_s + 1)^{\frac{1}{\sigma}}\right) - n\left(ap_s^{\frac{1}{\sigma}}\right) = 0.$$

Putting

$$t_s = \frac{1}{2}\left[a(p_s + 1)^{\frac{1}{\sigma}} + ap_s^{\frac{1}{\sigma}}\right], \quad d_s = \frac{1}{2}\left[a(p_s + 1)^{\frac{1}{\sigma}} - ap_s^{\frac{1}{\sigma}}\right],$$

we see that any circle centred on the circle $|z| = t_s$ of radius d_s can have no $0, 1$ or ∞ points of $f(z)$. Thus

$$\limsup_{r \to \infty} r^{\sigma-1} D(r) \ge \limsup_{s \to \infty} (t_s)^{\sigma-1} d_s.$$

But $t_s \sim ap_s^{\frac{1}{\sigma}}$, $d_s \sim \frac{a}{2\sigma} p_s^{\frac{1}{\sigma}-1}$ as $s \to \infty$, which yields that

$$\limsup_{s \to \infty} (t_s)^{\sigma-1} d_s = \frac{a^{\sigma}}{2\sigma}.$$

Since a is any number satisfying (3.2) we get

$$\limsup_{r \to \infty} r^{\sigma-1} D(r) \ge (6\sigma^2 K)^{-1}$$

which is (1.4). The proofs of (1.5) and (1.6) are similar but more complicated.

4. A lemma

In view of the remarks at the end of [1] it is perhaps desirable to give a proof of the following lemma.

Lemma 2. *Let $h(x)$ be a positive non-decreasing function for $x > 0$ and set*

$$H(x) = \int_0^x h(t)dt.$$

Suppose that for some $b > 1$

$$\liminf_{x \to \infty} \frac{\log H(x)}{\log x} = b. \tag{4.1}$$

Then, for $p \in N$

$$\liminf_{p \to \infty} p^2 \left\{ \frac{H(p+1) - H(p) - h(p)}{H(p)} \right\} \le b(b-1), \tag{4.2}$$

and the right-hand side is best possible.

Proof. If the $\liminf$ above is denoted by α then, arguing as in [1] p. 298 we see that

$$\liminf_{x \to \infty} \frac{\log H(x)}{\log x} \ge \frac{1}{2} \left\{ 1 + (1 + 4\alpha)^{\frac{1}{2}} \right\}.$$

This contradicts (4.1) if

$$\frac{1}{2} \left\{ 1 + (1 + 4\alpha)^{\frac{1}{2}} \right\} > b,$$

i.e., if $\alpha > b(b-1)$.

If we take $h(x) = x^{b-1}$ for $b > 1$ then the $\liminf$ in (4.2) is $\frac{b(b-1)}{2}$. The necessary counterexample is

$$h(x) = (p+1)^b - p^b, \quad p < x \le p+1, \quad H(x) = p^b, \quad p < x \le p+1.$$

Then the $\liminf$ becomes

$$\liminf_{p \to \infty} p^2 \left[\frac{(p+1)^b + (p-1)^b - 2p^b}{p^b} \right] = b(b-1)$$

as required. $\square$

5. Concluding Remarks

In view of Lemma 2 above we have

Theorem 3′. *Suppose that $f(z)$ is a meromophic function such that*

$$\liminf_{p \to \infty} \frac{T(r, f)}{(\log r)^2} = \infty,$$

but for some $b \ge 2$

$$\liminf_{r \to \infty} \frac{\log T(r, f)}{\log \log r} = b.$$

Then

$$\limsup_{r\to\infty} \frac{T(r)D(r)}{r(\log r)^2} \geq (24b(b-1))^{-1}.$$

Proof. The proof follows as in Theorem 6 of [1]. For any $p \in \mathbb{N}$ we set

$$d_p = \left[4\left(1 + n\left(e^{p+\frac{1}{2}}\right) - n(e^p)\right) \right]^{-1}$$

and choose t_p so that

$$p + d_p \leq \log t_p < p + \frac{1}{2} - d_p \quad \text{and} \quad n(t_p e^{-d_p}) = n(t_p e^{d_p}).$$

Then as before we conclude that

$$\limsup_{r\to\infty} \frac{T(r)D(r)}{r(\log r)^2} \geq \left[24 \liminf_{p\to\infty} p^2 \frac{N(e^{p+1}) - N(e^p) - n(e^p)}{N(e^p)} \right]^{-1}$$

and the result follows from Lemma 2. $\square$

It would be of interest to study whether there is an analogue of Theorem 1 of [3] for functions of zero order, with some suitable substitute for the notion of proximate order. In many cases the critical growth rate is $O(\log r)^2$ and this might also be the case here. Further investigation is required.

References

[1] J.M. Anderson and S. Toppila, *The growth of the spherical derivative of a meromorphic function of finite lower order*, J. London Math. Soc., **27** (1983), 289–305.

[2] W.K. Hayman, *Meromorphic Functions*, Clarendon Press, Oxford, 1964.

[3] I.V. Ostrovskii and A.E. Ureyen, *Distance between a maximum modulus point and zero set of an entire function*, Complex Variables, **48** (2003), 583–598.

[4] Ch. Pommerenke, *Estimates for normal meromorphic functions*, Ann. Acad. Sci. Fenn. Ser. A1 Math., **476** (1970), 1–10.

J.M. Anderson
Mathematics Department
University College London
WC1E 6BT
United Kingdom

Operator Theory:
Advances and Applications, Vol. 158, 101–108

Approximation Problems on the Unit Sphere in $\mathbf{C}^2$

John T. Anderson and John Wermer

Dedicated to the memory of S. Ya. Khavinson

1. Introduction

Let X be a compact subset of $\mathbf{C}^n$. We denote by $R_0(X)$ the algebra of all functions $\frac{P}{Q}$ where P and Q are polynomials on $\mathbf{C}^n$ and $Q \neq 0$ on X, and we denote by $R(X)$ the uniform closure of $R_0(X)$ in the space $C(X)$ of continuous functions on X. We are interested in finding conditions on X that imply that $R(X) = C(X)$, i.e., that each continuous function on X is the uniform limit of a sequence of rational functions holomorphic in a neighborhood of X.

We denote by $h_r(X)$ the rationally convex hull of X, defined as the set of points $y \in \mathbf{C}^n$ such that every polynomial Q with $Q(y) = 0$ vanishes at some point of X. The following is a necessary condition for the equality $R(X) = C(X)$:

$$h_r(X) = X. \tag{1}$$

If (1) holds, we say that X is *rationally convex*. Rational convexity is invisible when studying rational approximation on plane sets, since every compact plane set satisfies (1).

In this article we shall be concerned with the special case of this question when X is a closed subset of the unit sphere $\partial B := \{(z,w) : |z|^2 + |w|^2 = 1\}$ in $\mathbf{C}^2$.

The first result on this problem was obtained by Richard Basener [4] in 1972. Basener constructed a family of rationally convex sets $X_E \subset \partial B$ for which $R(X) \neq C(X)$. Let E be a compact subset of the open unit disk $D := \{z \in \mathbf{C} : |z| < 1\}$. For each $z \in D$ we put

$$\gamma_z = \{w \in \mathbf{C} : |z|^2 + |w|^2 = 1\}.$$

Definition 1. $X_E = \{(z,w) : z \in E \text{ and } w \in \gamma_z\}.$

Definition 2. A *Jensen measure* for a point $z \in E$, relative to the algebra $R(E)$, is a probability measure σ on E such that

$$\log |f(z)| \leq \int_E \log |f| \, d\sigma \quad \text{for all } f \in R(E).$$

For information on Jensen measures, see [6].

Definition 3. The set E is of *type (β)* if for all $z \in E$, the only Jensen measure for z relative to $R(E)$ is the point mass δ_z.

Theorem 1 (Basener, [4]). *Let E be a compact subset of the open unit disk. Assume that $R(E) \neq C(E)$, and that E is of type (β). Then X_E is rationally convex and $R(X_E) \neq C(X_E)$.*

In the converse direction, Basener showed the following (see Section 3 of [5]): if X_E is rationally convex, then E is of type (β). We note that if a compact plane set E is of type (β) and $R(E) \neq C(E)$, then E has empty interior and the complement of E is infinitely connected. Sets E with property (β) satisfying $R(E) \neq C(E)$ are known to exist; see the remarks on the "Swiss cheese" sets below.

Corollary. *Let E be of type (β). Then each closed subset Y of X_E is rationally convex.*

Proof. Fix a point $x \in \mathbf{C}^2 \setminus Y$. If x lies outside X_E, then there exists a polynomial P with $P(x) = 0$ and $P \neq 0$ on X_E, hence $P \neq 0$ on Y. If x lies in X_E, then x belongs to $\partial B \setminus Y$. It follows that there exists a linear function L with $L(x) = 1$ and $|L| < 1$ on $\partial B \setminus \{x\}$. Then $L - 1$ vanishes at x but not on Y. $\square$

This corollary provides us with a large collection of rationally convex subsets of ∂B on which to test the question: what is required of a subset Y of ∂B, beyond rational convexity, in order that the equality $R(Y) = C(Y)$ may hold? We make the following conjecture:

Conjecture. *Let E be a set of type (β) and let f be a continuous complex-valued function defined on E such that $|f(z)| = \sqrt{1 - |z|^2}$ for all $z \in E$. Denote by Γ_f the graph of f in $\mathbf{C}^2$:*

$$\Gamma_f = \{(z, f(z)) : z \in E\}$$

Then $R(\Gamma_f) = C(\Gamma_f)$.

In Section 2 we shall prove a special case of this conjecture in Theorem 2.

Swiss Cheeses. The classical example of a compact plane set E without interior such that $R(E) \neq C(E)$ is the so-called "Swiss cheese" of S. Mergelyan and A. Roth (see [6]). Fix a closed disk $\overline{D_0} \subset D$ and choose a countable family of disjoint open disks $D_j, j = 1, 2, \ldots$ contained in D_0, in such a way that

$$E \equiv \overline{D_0} \setminus \bigcup_{j=1}^{\infty} D_j$$

has empty interior. We assume that $\sum_1^\infty r_j < \infty$, where r_j is the radius of D_j. It follows that $R(E) \neq C(E)$ (see [6]). McKissick and others have constructed Swiss cheeses with property (β).

At the end of this article, in the Appendix, we show that X_E can be regarded as a three-dimensional "Swiss Cheese".

2. Rational approximation on graphs in ∂B

In our paper [3], entitled "Rational Approximation on the Unit Sphere in $\mathbf{C}^2$", we treated cases of the conjecture stated in the Introduction. To obtain the equality $R(X) = C(X)$ for certain subsets X of ∂B, we imposed on X a strengthening of the rational convexity condition which we called the "hull-neighborhood property" (see Theorem 2.5 of [3]).

It turns out that for a graph Γ_f in ∂B, where f is a function defined on a set of type (β) and satisfying a mild regularity condition, we can dispense with the assumption of the hull-neighborhood condition. We have the following:

Theorem 2. *Let E be a set of type (β) and let f be a continuous function defined on the open unit disk D and satisfying a Hölder condition*

$$|f(z) - f(z')| \leq M|z - z'|^\alpha, \quad \text{for all } z, z' \in D$$

where M and α are constants, $0 < \alpha < 1$, and assume $|f(z)| = \sqrt{1 - |z|^2}$ for all $z \in D$. Let Γ_f denote the graph of f over E. Then $R(\Gamma_f) = C(\Gamma_f)$.

Before beginning the proof, we give some preliminaries. Our proof will be based on a transform of measures on ∂B, given by G. Henkin in 1977 in [7], which generalizes the Cauchy transform of measures on plane sets. Let μ be a complex measure on ∂B. In [7], Henkin defined the kernel

$$H(\zeta, z) = \frac{\overline{\zeta_1} z_2 - \overline{\zeta_2} z_1}{|1 - \langle z, \zeta \rangle|^2} \tag{2}$$

on $\partial B \times \partial B \setminus \{z = \zeta\}$, where $\langle , \rangle$ denotes the standard Hermitian inner product in $\mathbf{C}^2$. Henkin's transform is the function

$$K_\mu(\zeta) = \int_{\partial B} H(\zeta, z) d\mu(z).$$

Then $K_\mu \in L^1(\partial B)$ and K_μ is smooth on $\partial B \setminus \text{supp}(\mu)$ (see also [10]).

If μ is orthogonal to polynomials, Henkin showed that

$$\int \phi \, d\mu = \frac{1}{4\pi^2} \int_{\partial B} K_\mu \, \overline{\partial} \phi \wedge \omega, \quad \text{where } \omega(z) = dz_1 \wedge dz_2 \tag{3}$$

for all $\phi \in C^1(\partial B)$. It follows that if X is a closed subset of ∂B and μ is a measure supported on X with μ orthogonal to $R(X)$, then (3) holds. In [9] H.P. Lee and J. Wermer proved that in this setting if X is rationally convex, then K_μ extends from $\partial B \setminus X$ to the interior of B as a holomorphic function, again denoted K_μ, by abuse of language.

For each $a \in \mathbf{C}$ we put as earlier

$$\gamma_a = \{w \in \mathbf{C} : |a|^2 + |w|^2 = 1\} \quad \text{and we put} \quad \Delta_a = \{(a, w) : |a|^2 + |w|^2 < 1\}.$$

So Δ_a is the disk on the complex line $\{z = a\}$ bounded by the circle $\{(a, w) : w \in \gamma_a\}$. For each a, K_μ restricted to Δ_a is analytic. Without loss of generality we shall assume $E \subset D_0 := \{z : |z| < 1 - \epsilon_0\}$ for some $\epsilon_0 > 0$.

In the proof of Theorem 2 we shall make use of the following four results in our paper [3] (proved as Lemma 2.3, Lemma 2.2, Lemma 2.6, and formula (14) of Section 4 of that paper, respectively):

Lemma 2.1. *Let μ be a measure on ∂B and put $X = supp(\mu)$. Then for all $a \in D$ and for all $w \in \gamma_a$, we have*

$$|K_\mu(a, w)| \leq \frac{4\|\mu\|}{dist^4((a, w), X)}. \tag{4}$$

Here $\|\mu\|$ denotes the total variation of the measure μ. In the next lemma, m_3 refers to three-dimensional Hausdorff measure.

Lemma 2.2. *Let X be a rationally convex subset of ∂B with $m_3(X) = 0$. Let μ be a measure on X with $\mu \perp R(X)$. If the holomorphic extension of K_μ to B belongs to the Hardy space $H^1(B)$, then $\mu \equiv 0$.*

Lemma 2.3. *With the notations of the preceding lemma, assume that for some $s > 0$, the restriction of K_μ to Δ_a lies in $H^s(\Delta_a)$ for almost all $a \in D_0$. Then $K_\mu \in H^1(B)$ and so, by Lemma 2.2, $\mu = 0$.*

Lemma 2.4. *Let f and Γ_f be as in Theorem 2. Fix a measure μ orthogonal to $R(\Gamma_f)$. There exists a constant c, depending on only on μ, such that for all $a \in D$ and for all $w \in \gamma_a$, we have*

$$|w - f(a)|^{2/\alpha} \leq c \cdot dist^2((a, w), \Gamma_f). \tag{5}$$

We are now ready to begin the proof of Theorem 2.

Lemma 2.5. *Let f be as in Theorem 2 and let μ be a measure on Γ_f orthogonal to $R(\Gamma_f)$. There exists a constant κ depending only on μ such that for all $a \in \overline{D_0}$, setting $r = \sqrt{1 - |a|^2}$, we have*

$$\int_0^{2\pi} |K_\mu(a, re^{i\phi})|^{\alpha/8} \, d\phi \leq \kappa. \tag{6}$$

Proof. Fix $a \in \overline{D_0}, \phi \in [0, 2\pi]$ and put $w = re^{i\phi}$. By (5),

$$\frac{1}{dist^4((a, w), \Gamma_f)} \leq \frac{c^2}{|w - f(a)|^{4/\alpha}}.$$

The estimate (4) then gives

$$|K_\mu(a, w)| \leq \frac{4\|\mu\|c^2}{|w - f(a)|^{4/\alpha}} \quad \text{and so} \quad |K_\mu(a, w)|^{\alpha/8} \leq \frac{c_2}{|w - f(a)|^{1/2}}$$

where c_2 is a constant depending only on μ. Thus

$$\int_0^{2\pi} |K_\mu(a, re^{i\phi})|^{\alpha/8} \, d\phi \le c_2 \int_0^{2\pi} \frac{d\phi}{|re^{i\phi} - f(a)|^{1/2}}.$$

We write $f(a) = re^{i\phi_0}$. Then the right-hand side equals

$$c_2 \int_0^{2\pi} \frac{d\phi}{r^{1/2}|e^{i\phi} - e^{i\phi_0}|^{1/2}} \le \frac{c_2}{r^{1/2}} \int_0^{2\pi} \frac{d\theta}{|e^{i\theta} - 1|^{1/2}}.$$

Note that the integral on the right-hand side of the last inequality is finite. Also, since $|a| < 1 - \epsilon_0$, there exists $r_0 > 0$ such that $r > r_0$ for every $a \in \overline{D_0}$. So

$$\int_0^{2\pi} |K_\mu(a, re^{i\phi})|^{\alpha/8} \, d\phi \le \frac{c_2}{r_0^{1/2}} \int_0^{2\pi} \frac{d\theta}{|e^{i\theta} - 1|^{1/2}}.$$

Denoting this last expression by κ, we get (6). $\square$

Lemma 2.6. *Fix $a \in \overline{D_0} \setminus E$. Put $r = \sqrt{1 - |a|^2}$. For $R < r$ we have*

$$\int_0^{2\pi} |K_\mu(a, Re^{i\theta})|^{\alpha/8} \, d\phi \le \kappa \tag{7}$$

where κ is the constant in (6).

Proof. Since a lies outside E, $\overline{\Delta_a}$ is disjoint from Γ_f, so the restriction of K_μ to Δ_a extends continuously to $\overline{\Delta_a}$. It is well known that the function

$$R \to \int_0^{2\pi} |K_\mu(a, Re^{i\phi})|^{\alpha/8} \, d\phi$$

is monotonic on $0 < R < r$ and continuous on $0 \le R \le r$. So (6) implies (7). $\square$

Lemma 2.7. *Fix a_0 in E. Fix $R < \sqrt{1 - |a_0|^2}$. Then*

$$\int_0^{2\pi} |K_\mu(a, Re^{i\theta})|^{\alpha/8} \, d\phi \le \kappa \tag{8}$$

where κ is the constant in (6).

Proof. Choose a sequence $\{a_n\}$ converging to a_0 such that $a_n \in D_0 \setminus E$ for each n. For n large, then, $R < \sqrt{1 - |a_n|^2}$. By Lemma 2.6,

$$\int_0^{2\pi} |K_\mu(a_n, Re^{i\theta})|^{\alpha/8} \, d\phi \le \kappa, \quad n \gg 1.$$

Also $K_\mu(a_n, Re^{i\theta}) \to K_\mu(a_0, Re^{i\theta})$ uniformly on $0 \le \phi \le 2\pi$ as $n \to \infty$ since K_μ is continuous on $\mathrm{int}(B)$. By continuity, then, we get (8). $\square$

Lemmas 2.6 and 2.7 say that for all $a_0 \in \overline{D_0}$, K_μ restricted to Δ_{a_0} lies in $H^{\alpha/8}(\Delta_{a_0})$. Lemma 2.3 then yields that $\mu \equiv 0$. Since this holds for each μ orthogonal to $R(\Gamma_f)$, we conclude that $R(\Gamma_f) = C(\Gamma_f)$, and so Theorem 2 is proved. $\square$

Appendix: Geometric interpretation of the sets X_E

We shall show that the sets X_E lying in ∂B can be seen as three-dimensional analogues of the Swiss cheese E in $\mathbf{C}$. We denote k-dimensional Hausdorff measure by m_k.

The Swiss cheese E is constructed by removing a countable family of disjoint open disks D_j from a closed disk $\overline{D_0}$. The following properties hold:

(i) $\sum_{j=1}^{\infty} m_1(\partial D_j) < \infty$

(ii) $m_2(E) > 0$

(iii) The measure dz restricted to the union of the circles ∂D_j (properly oriented) is finite on E and is orthogonal to $R(E)$.

It follows immediately from (iii) that $R(E) \neq C(E)$.

Let us now start with a family of disks D_j in $\mathbf{C}$ as above and let us replace each D_j by the open solid torus $T_j = \{(z,w) \in \partial B | z \in D_j\}, j = 1, 2, \ldots$, with $T_0 = \{(z,w) \in \partial B | z \in D_0\}$. Set

$$E^* = \overline{T_0} \setminus \bigcup_{j=1}^{\infty} T_j.$$

Then E^* is a compact subset of ∂B with the following properties:

(i′) $\sum_{j=1}^{\infty} m_2(\partial T_j) < \infty$

(ii′) $m_3(E^*) > 0$

(iii′) The measure $\mu = dz \wedge dw$ restricted to the union of the boundaries ∂T_j (properly oriented) is finite on E^* and is orthogonal to $R(E^*)$.

Properties (i′) and (ii′) follow immediately from Fubini's Theorem and properties (i) and (ii) of the Swiss Cheese. As for (iii′), the finiteness of $\mu = dz \wedge dw$ follows from assumption (i′) together with the following assertion.

Claim. *Let M be a smooth two (real-) dimensional submanifold of $\mathbf{C}^2$, S a Borel subset of M, and m_2 two-dimensional Hausdorff measure. Then*

$$\|\mu\|(S) \leq m_2(S).$$

Here $\|\mu\|$ denotes the total variation measure of μ.

Proof. Identify $\mathbf{C}^2$ with $\mathbf{R}^4$ using coordinates

$$z = x + iy, \quad w = u + iv.$$

We may assume that near S, M is given parametrically, i.e., is the image of a smooth map Φ from a neighborhood of the origin in $\mathbf{R}^2$ to M. Using coordinates (ξ, η) in $\mathbf{R}^2$, let E_1, E_2 be the images of the tangent vectors $\partial/\partial\xi$ and $\partial/\partial\eta$ under the differential of Φ, so

$$E_1 = (x_\xi, y_\xi, u_\xi, v_\xi), \quad E_2 = (x_\eta, y_\eta, u_\eta, v_\eta)$$

as vectors in $\mathbf{R}^4$, where subscripts denote partial derivatives. It is standard that the two-dimensional volume form on M is given by (the area of the parallelogram spanned by E_1, E_2):

$$dV = \sqrt{\det(g)}\, d\xi d\eta$$

where g is the 2×2 matrix with entries $g_{ij} = E_i \cdot E_j$, $i, j = 1, 2$ and $\cdot$ is the usual inner product in $\mathbf{R}^4$. It is also well known that

$$m_2(S) = \int_S dV.$$

On the other hand, writing $dx = x_\xi d\xi + x_\eta d\eta$, etc., we obtain

$$dz \wedge dw = (dx + idy) \wedge (du + idv) = (A + iB)\, d\xi \wedge d\eta$$

where

$$A = x_\xi u_\eta - x_\eta u_\xi - y_\xi v_\eta + y_\eta v_\xi$$

and

$$B = x_\xi v_\eta - x_\eta v_\xi + y_\xi u_\eta - y_\eta u_\xi.$$

To establish the claim it suffices to show that

$$\det(g) \geq A^2 + B^2. \tag{9}$$

A calculation gives

$$\det(g) - (A^2 + B^2) = (x_\xi y_\eta - x_\eta y_\xi + v_\eta u_\xi - v_\xi u_\eta)^2$$

which establishes (9) and completes the proof of the claim. $\qquad \square$

To prove the assertion of (iii$'$) that $dz \wedge dw$ is orthogonal to $R(E^*)$, we argue as follows: fix a rational function $f = P/Q$, where P, Q are polynomials with $Q \neq 0$ on E^*. The set $\{Q = 0\} \cap \overline{T_0}$ is contained in $\bigcup_{j=1}^\infty T_j$. By Heine-Borel, there exists an integer such that this set is contained in $\bigcup_{j=1}^n T_j$. We put $\Omega_n = T_0 \setminus \bigcup_{j=1}^n T_j$. Then f is holomorphic on $\overline{\Omega_n}$. By Stokes' Theorem applied to the form $f dz \wedge dw$ on Ω_n, we have

$$\int_{\partial \Omega_n} f dz \wedge dw = \int_{\Omega_n} \bar\partial f \wedge dz \wedge dw.$$

The right-hand side of this equation vanishes, since f is analytic on a neighborhood of $\overline{\Omega_0}$. The left-hand side approaches $\int_{E^*} f\, d\mu$ as $n \to \infty$. So $\int_{E^*} f\, d\mu = 0$. Thus μ is orthogonal to f. Since this holds for each $f \in R_0(E^*)$, we have μ orthogonal to $R(E^*)$.

Finally, we remark that $R(E^*) \neq C(E^*)$ clearly follows from (iii$'$).

It is clear that E^* coincides with X_E, by the definition of X_E in the Introduction. For an arbitrary Swiss cheese E, X_E will not be rationally convex.

There is a substantial literature related to the approximation questions treated in this article. The references below list some of the relevant papers, as well as those papers specifically cited in this article.

References

[1] J. Anderson and J. Cima, *Removable Singularities for L^p CR functions*, Mich. Math. J. **41** (1994), 111–119.

[2] J. Anderson and A. Izzo, *A Peak Point Theorem for Uniform Algebras Generated by Smooth Functions On a Two-Manifold*, Bull. London Math. Soc. **33** (2001), 187–195.

[3] J. Anderson, A. Izzo and J. Wermer, *Rational Approximation on the Unit Sphere in* $\mathbf{C}^2$, Mich. J. Math. **52** (2004), 105–117.

[4] R.F. Basener, *On Rationally Convex Hulls*, Trans. Amer. Math. Soc. **182** (1973), 353–381.

[5] R.F. Basener, *Rationally Convex Hulls and Potential Theory*, Proc. Royal Soc. Edinburgh, **74A** (1974/5), 81–89.

[6] T. Gamelin, *Uniform Algebras*, 2nd ed. Chelsea, New York, 1984.

[7] G.M. Henkin, *The Lewy equation and analysis on pseudoconvex manifolds*, Russian Math. Surveys, **32:3** (1977); Uspehi Mat. Nauk **32:3** (1977), 57–118.

[8] H.P. Lee, *Tame measures on certain compact sets*, Proc. AMS (1980), 61–67.

[9] H.P. Lee and J. Wermer, *Orthogonal Measures for Subsets of the Boundary of the Ball in* $\mathbf{C}^2$, in *Recent Developments in Several Complex Variables*, Princeton University Press, 1981, 277–289.

[10] N.Th. Varopoulos, *BMO functions and the $\overline{\partial}$-equation*, Pac. J. Math. **71**, no. 1 (1977), 221–273.

John T. Anderson
Department of Mathematics and Computer Science
College of the Holy Cross
Worcester
MA 01610-2395, USA
e-mail: `anderson@mathcs.holycross.edu`

John Wermer
Department of Mathematics
Brown University
Providence
RI 02912, USA
e-mail: `wermer@math.brown.edu`

Operator Theory:
Advances and Applications, Vol. 158, 109–130
© 2005 Birkhäuser Verlag Basel/Switzerland

Conformal Maps and Uniform Approximation by Polyanalytic Functions

Joan Josep Carmona and Konstantin Yu. Fedorovskiy

Abstract. In this paper we study some properties of Carathéodory domains and their conformal mappings onto the unit disk. Using these results we obtain some new conditions for uniform approximability of functions by polyanalytic polynomials and polyanalytic rational functions on compact subsets of the complex plane.

Mathematics Subject Classification (2000). 30E10, 30G30, 30C20.

1. Introduction

In this paper we deal with some problems of uniform approximation by polyanalytic polynomials and polyanalytic rational functions on compact subsets of the complex plane $\mathbb{C}$.

Let $U \subset \mathbb{C}$ be an open set. Denote by $\mathrm{Hol}(U)$ the space of all holomorphic functions in U. We recall that a function f is called *polyanalytic of order n* (or, shorter, n-analytic) in U if it is of the form

$$f(z) = \bar{z}^{n-1} f_{n-1}(z) + \cdots + \bar{z} f_1(z) + f_0(z), \tag{1.1}$$

where $n \in \mathbb{N}$ and $f_0, \ldots, f_{n-1} \in \mathrm{Hol}(U)$. Denote by $\mathrm{Hol}_n(U)$ the space of all n-analytic functions in U. Notice, that $\mathrm{Hol}_n(U)$ consists of all continuous functions f on U such that $\bar{\partial}^n f = 0$ in U in the distributional sense, where $\bar{\partial}$ is the standard Cauchy-Riemann operator in $\mathbb{C}$ (i.e., $\bar{\partial} = \partial/\partial\bar{z} = \frac{1}{2}(\partial/\partial x + i\partial/\partial y)$).

In what follows a *polynomial* and a *rational function* will mean a complex-valued polynomial and a rational function in the complex variable z respectively. By *n-analytic polynomials* and by *n-analytic rational functions* we mean the functions of the form (1.1) where $f_0, \ldots, f_{n-1}$ are polynomials and rational functions

The first author was partially supported by grants 2000SGR-00059, 2001SGR-00172 of Generalitat de Catalunya and BFM 2002-04072-C02-02 of Ministerio de Ciencia y Tecnologia.
The second author was partially supported by grants NSh-2040.2003.1 of the Program "Leading Scientific Schools", YSF 2001/2-16 Renewal of INTAS and 04-01-00720 of RFBR.

respectively. Denote by $\mathcal{P}_n$ the set of all n-analytic polynomials. Also for a compact set $X \subset \mathbb{C}$ we denote by $\mathcal{R}_n(X)$ the set of all n-analytic rational functions f such that the corresponding rational functions $f_0, \ldots, f_{n-1}$ in (1.1) have their poles outside X. Put $\mathcal{P} := \mathcal{P}_1$ and $\mathcal{R}(X) := \mathcal{R}_1(X)$.

In the most general form the problem we are interested in can be formulated as follows: *given a compact set $X \subset \mathbb{C}$, which conditions on X are necessary and sufficient in order that each function which is continuous on X and n-analytic on its interior can be uniformly on X approximated (with an arbitrary accuracy) by n-analytic polynomials or by n-analytic rational functions with poles outside some appropriately chosen compact set $Y \supseteq X$?*

The investigation of this problem was started in 1980-th (see [TW, Ca, Wa], where some sufficient approximability conditions were obtained). In 1990-th several results on approximability of functions by polyanalytic polynomials have been obtained using the concepts of *Nevanlinna* and *locally Nevanlinna domains* which were introduced in [Fe1] and [CFP] and turned out to be fairly useful. The recent progress in the themes under consideration is related with usage of some concepts analogous to the concept of a Nevanlinna domain and a "reductive" approach (see [BGP1, BGP2, Za]). The last approach allows us to conclude (under some suitable assumptions) that one approximability property takes place on a compact set X whenever one has certain similar properties on some appropriately chosen (and more simple) compact subsets of X. Some other bibliographical notes concerning the matter may be found in [Fe2]. Our investigation of the problem under consideration is based on studying of special properties of conformal mappings of Carathéodory domains onto the unit disk and on some special results concerning the structure of measures that are orthogonal to rational functions on certain type of compact sets in $\mathbb{C}$.

The paper is organized as follows. In Section 2 we explain some properties of Carathéodory domains and obtain one useful generalization of Carathéodory extension theorem (see Theorem 1 and Corollary 1). In Section 3 we study the structure of measures being orthogonal to rational functions with poles outside some kind of compact sets in $\mathbb{C}$ (see Theorem 2, Propositions 3 and 4). In Section 4 we obtain (see Theorems 3 and 4) new approximability conditions in the problem under consideration for some special unions X of compact sets. These conditions have a reductive nature and are substantially based on the concepts of a Nevanlinna and locally Nevanlinna domains. In Theorem 5 we present one interesting rigidity property of Nevanlinna domains.

We need to fix some notation. Denote by $\overline{\mathbb{C}}$ the standard one-point compactification of $\mathbb{C}$, that is $\overline{\mathbb{C}} = \mathbb{C} \cup \{\infty\}$. Put $\mathbb{D} := \{w \in \mathbb{C} : |w| < 1\}$ and $\mathbb{T} := \{\xi \in \mathbb{C} : |\xi| = 1\}$. We also denote by $[a, b]$ the closed line segment, joining two points a and b in $\mathbb{C}$ and by $B(a, r)$ the open disk with center at $a \in \mathbb{C}$ and radius $r > 0$. For a set E we denote by E° its interior, by $\overline{E}$ its closure, by ∂E its boundary and by $\mathfrak{S}(E)$ the set of all connected components of E°. By a *continuum* one means a non-empty connected compact set in $\mathbb{C}$ and by a *contour* a closed Jordan curve in $\mathbb{C}$. For a contour Γ we denote by $D(\Gamma)$ the Jordan domain bounded by Γ.

If X is a compact set, then $\widehat{X}$ means the union of X and all bounded connected components of its complementary. Usually, $\widehat{X}$ is called the *polynomial convex hull* of X, because $\widehat{X} = \{z \in \mathbb{C} : \text{for each } p \in \mathcal{P} \text{ one has } |p(z)| \leq \max_{x \in X} |p(x)|\}$.
We recall, that a compact set X is called a *Carathéodory compact set* if $\partial X = \partial \widehat{X}$.

As usual, $\mathrm{H}^1(\mathbb{D})$ and $\mathrm{H}^\infty(\mathbb{D})$ are the standard Hardy spaces in $\mathbb{D}$. By Fatou's theorem for each $f \in \mathrm{H}^1(\mathbb{D})$ and for almost all $\xi \in \mathbb{T}$ there exists the angular boundary value $f(\xi)$ of f at ξ. We denote by $F(f)$ the set of all such points $\xi \in \mathbb{T}$. This set is called the Fatou set of f. In view of [Po1, Proposition 6.5] $F(f)$ is a Borel set. Also, for an open set U we denote by $\mathrm{H}^\infty(U)$ the set of all bounded holomorphic functions in U.

2. Some properties of the Carathéodory domains

In what follows, if Ω is some bounded simply connected domain in $\mathbb{C}$, then we will always assume that φ is a fixed conformal mapping from $\mathbb{D}$ onto Ω and ψ is its inverse.

Let Ω be a bounded simply connected domain. We start with the following question which have appeared naturally in the theory of conformal mappings and which will be one of the central points in the further consideration: *what it is possible to say about extension of φ onto $\overline{\mathbb{D}}$ and about extension of ψ onto $\overline{\Omega}$?*

The well-known Carathéodory extension theorem [Po1, Theorem 2.6] gives an answer to both questions in the case of Jordan domains: *φ can be extended to a homeomorphism of $\overline{\mathbb{D}}$ onto $\overline{\Omega}$ if and only if Ω is a Jordan domain.* If we only ask about continuous (not necessary homeomorphic) extension of φ the answer is given by [Po1, Theorem 2.1] that says: *φ has a continuous extension to $\overline{\mathbb{D}}$ if and only if $\partial\Omega$ is locally connected.* We always will keep the notation φ for the extension to $\overline{\mathbb{D}}$ if it exists.

In the general case, the Carathéodory prime end theorem [Po1, pag. 18] guarantee the existence of the extension of φ to a homeomorphism of $\overline{\mathbb{D}}$ onto the union of Ω and the space of prime ends of Ω (see [Po1, pag. 30]), but this union differs from $\overline{\Omega}$ in any admissible geometrical and/or topological sense.

Put
$$\partial_a \Omega := \{\varphi(\xi) : \ \xi \in F(\varphi)\}.$$
This set is called an *accessible part* of $\partial\Omega$. By [Po1, Propositions 2.14 and 2.17] $\partial_a\Omega$ is the set of all points in $\partial\Omega$ which are accessible from Ω by some curve and therefore, $\partial_a\Omega$ depends only on Ω, but not on the choice of φ. Mazurkiewich [Ma, Theorem 2] proved the following result: let E be a compact subset of $\overline{\mathbb{C}}$ and let E_a be the set of all points of E which are accessible from $\overline{\mathbb{C}} \setminus E$ by some curve; then E_a is a Borel set. Applying this result to the set $E = \overline{\mathbb{C}} \setminus \Omega$ and taking into account that $\partial_a\Omega = (\overline{\mathbb{C}} \setminus \Omega)_a$ we conclude that $\partial_a\Omega$ is a Borel set.

Recall, that a bounded domain Ω is called a *Carathéodory domain* if $\partial\Omega = \partial\Omega_\infty$, where Ω_∞ is the unbounded connected component of the set $\overline{\mathbb{C}} \setminus \overline{\Omega}$. In particular, each Carathéodory domain Ω is simply connected and possesses the property $\Omega = (\overline{\Omega})^{\circ}$.

Proposition 1. *Let Ω be a Carathéodory domain and $\zeta \in \partial_a \Omega$. Then there exists a unique point $t = t(\zeta) \in F(\varphi) \subseteq \mathbb{T}$ such that $\zeta = \varphi(t(\zeta))$.*

Proof. Take a point $\zeta \in \partial_a \Omega$. There exists at least one point $t \in F(\varphi)$ such that $\varphi(t) = \zeta$. Assume that there exist $t_1, t_2 \in F(\varphi)$, $t_1 \neq t_2$ but $\varphi(t_1) = \varphi(t_2) = \zeta$. Let $\ell_j = [0, t_j]$, $j = 1, 2$. Then $\gamma = \varphi(\ell_1 \cup \ell_2)$ is a contour in $\Omega \cup \{\zeta\}$. The continuum $\ell_1 \cup \ell_2$ separates $\mathbb{D}$ onto two open circular sectors Δ_1 and Δ_2 (i.e., $\Delta_1 \cup \Delta_2 = \mathbb{D} \setminus (\ell_1 \cup \ell_2)$). Since γ is a contour in $\Omega \cup \{\zeta\}$, then one of these sectors, says Δ_1, possesses the property $\varphi(\Delta_1) \subset D(\gamma)$.

The existence of radial limits of φ almost everywhere on the nondegenerate arc $\overline{\Delta_1} \cap \mathbb{T}$, implies that $D(\gamma) \cap \partial\Omega \neq \varnothing$. Therefore, there exists at least one point $\zeta' \in D(\gamma) \cap \partial\Omega$. Since Ω is the Carathéodory domain, then $\zeta' \in D(\gamma) \cap \partial\Omega_{\infty}$. One has that there exists at least one point $z' \in D(\gamma) \cap \Omega_{\infty}$. Since Ω_{∞} is connected, then it is possible to find an infinite polygonal line $\ell_{\infty} \subset \Omega_{\infty}$ which goes from z' to ∞. It follows from the Jordan curve theorem, that $\ell_{\infty} \cap \gamma \neq \varnothing$. But this is a contradiction because $\gamma \subset \Omega \cup \{\zeta\}$. Hence $t_1 = t_2 = t(\zeta)$. $\qquad\square$

Let Ω be a Carathéodory domain in $\mathbb{C}$ and assume that φ has the following normalization: $\varphi(0) = z_0 \in \Omega$ and $\varphi'(0) > 0$.

Fix a sequence of a rectifiable contours $\{\Gamma_m\}_{m=1}^{\infty}$ such that $\Omega \subset D(\Gamma_m) \subset D(\Gamma_{m-1})$ so that Γ_m tends to $\partial\Omega$ as $m \to \infty$. The existence of such sequence is proving by consideration of a fixed conformal map h from $\mathbb{D}$ onto Ω_{∞} with $h(0) = \infty$ and setting $\Gamma_m := h(\{t : |t| = 1 - \frac{1}{m+1}\})$. Take a sequence of the conformal maps $\{\varphi_m\}_{m=1}^{\infty}$ from $\mathbb{D}$ onto $D(\Gamma_m)$ such that $\varphi_m(0) = z_0$ and $\varphi_m'(0) > 0$. Since each $D(\Gamma_m)$ is a Jordan domain, then each φ_m can be extended to a homeomorphism of $\overline{\mathbb{D}}$ onto $\overline{D(\Gamma_m)}$ (which we also denote by φ_m). The following convergence properties of $\{\varphi_m\}_m$ are consequences of the *Carathéodory kernel theorem* (see [Go, Chapter II, Section 5, Theorem 1]):

$$\varphi_m \rightrightarrows \varphi \text{ on compact subsets of } \mathbb{D} \text{ as } m \to \infty,$$
$$\varphi_m^{-1} \rightrightarrows \varphi^{-1} \text{ on compact subsets of } \Omega \text{ as } m \to \infty.$$

For each $\zeta \in \partial_a \Omega$ the point $t(\zeta)$ that appears in Proposition 1 will be denoted by $\varphi^{-1}(\zeta)$.

Theorem 1. *Let Ω be a Carathéodory domain and $\{\Gamma_m\}_m$ and $\{\varphi_m\}_m$ be as above.*

(1) *For each $\zeta \in \partial_a \Omega$ one has $\varphi_m^{-1}(\zeta) \to \varphi^{-1}(\zeta)$ as $m \to \infty$.*

(2) *For each bounded connected component G of $\mathbb{C} \setminus \overline{\Omega}$ one has $|\varphi_m^{-1}(z)| \to 1$ uniformly on $\overline{G}$.*

Proof. Let us prove the assertion (1). Take a point $\zeta_0 \in \partial_a \Omega$ and put $t_0 = t(\zeta_0) = \varphi^{-1}(\zeta_0)$. Denote by ℓ the radius $[0, t_0]$ and put $L = \varphi(\ell)$. Then L is a Jordan arc which goes, in Ω, from z_0 to ζ_0.

For each m we consider a point $\zeta_m \in \Gamma_m$ which is a nearest point to ζ_0. For each $m \geq 1$ we put $L_m := L \cup [\zeta_0, \zeta_m]$ and $\ell_m := \varphi_m^{-1}(L_m)$. Put $t_m = \varphi_m^{-1}(\zeta_m)$ and note, that each $\ell_m = \varphi_m^{-1}(L) \cup \varphi_m^{-1}([\zeta_0, \zeta_m])$ is the union of two consecutive Jordan arcs in $\mathbb{D} \cup \{t_m\}$. It is clear, that the sequence $\{\ell_m\}_m$ accumulates to a subset Λ of $\overline{\mathbb{D}}$. It means that Λ is the set of all points $w \in \overline{\mathbb{D}}$ such that there exists a sequence $\{w_{m_j}\}_j$ with the following properties: $w_{m_j} \in \ell_{m_j}$ and $w_{m_j} \to w$ as $j \to \infty$.

The set Λ possesses some special properties. Namely, one has:

(i) Λ *is a continuum*. Since Λ is a *compact set* let us prove that Λ is *connected*. Indeed, otherwise would exist two open sets U_1 and U_2 such that $\Lambda \subset U_1 \cup U_2$, $U_1 \cap U_2 = \varnothing$, $\Lambda \cap U_j \neq \varnothing$ for $j = 1, 2$ and suppose $0 \in U_1$. Consider the first point $w_m \in \ell_m$ such that $w_m \in \overline{U_1} \setminus U_1$ (such point w_m exists because ℓ_m is the union of two Jordan arcs and $\ell_m \not\subset U_1$ for big enough m). If w is an accumulation point of $\{w_m\}_m$, then $w \in \overline{U_1} \setminus U_1$ and $w \notin U_2$ because $\overline{U_1} \cap U_2 = \varnothing$, but $w \in \Lambda$ and this is a contradiction.

(ii) $\Lambda \subset \ell \cup \mathbb{T}$. This is consequence of the following arguments. If there exists $w \in \Lambda$, $w \notin \ell \cup \mathbb{T}$, then we can find $z_{m_j} \in L_{m_j}$ and $\varepsilon > 0$, such that

$$\overline{B(w, \varepsilon)} \cap (\ell \cup \mathbb{T}) = \emptyset, \quad w_j := \varphi_{m_j}^{-1}(z_{m_j}) \in \overline{B(w, \varepsilon)} \quad \text{and} \quad w = \lim_{j \to \infty} w_j.$$

Since $\varphi_{m_j} \to \varphi$ uniformly on $\overline{B(w, \varepsilon)}$ as $j \to \infty$, one has

$$\lim_{j \to \infty} z_{m_j} = \lim_{j \to \infty} \varphi_{m_j}(w_j) = \lim_{j \to \infty} \varphi(w_j) = \varphi(w).$$

This means that $\varphi(w)$ belongs to the closure of $L \cup \left(\bigcup_j [\zeta_0, \zeta_j] \right)$, but this is impossible since $\varphi(w) \in \Omega \setminus L$.

(iii) $\ell \subset \Lambda$. Really, for each $\varepsilon > 0$, we consider the segment $\ell_\varepsilon = [0, (1-\varepsilon)t_0]$. Since $\varphi_m^{-1} \rightrightarrows \varphi^{-1}$ uniformly on $\varphi(\ell_\varepsilon)$, then one has $\ell_\varepsilon \subset \Lambda$ for each $\varepsilon > 0$. Therefore $\ell \subset \Lambda$.

(iv) *The set* $\Lambda \cap \mathbb{T}$ *is connected*. Indeed, otherwise, $\Lambda \cap \mathbb{T} = E_1 \cup E_2$, where $E_1 \subset \mathbb{T}$ and $E_2 \subset \mathbb{T}$ are nonempty, closed and disjoint. Assume that $t_0 \in E_1$. Then, $\Lambda = (\ell \cup E_1) \cup E_2$ and hence Λ is not connected, which gives a contradiction. Therefore, $\Lambda = \ell \cup \alpha$, where α is some closed subarc of $\mathbb{T}$.

In order to prove (1) we need to show that $\Lambda = \ell$. If $\Lambda \neq \ell$, then $\alpha \neq \{t_0\}$. Let $w'_m \in \ell_m$ be a nearest point to t_0 and let ℓ'_m be the subcontinuum $\varphi_m^{-1}(L'_m)$ where L'_m is $[\varphi_m(w'_m), \zeta_m]$ if $\varphi_m(w'_m) \notin L$ and $L''_m \cup [\zeta_0, \zeta_m]$ otherwise, where L''_m is the subarc of L that joints $\varphi_m(w'_m)$ with ζ_0.

We claim that $\varphi_m(w'_m) \to \zeta_0$. If this would be not true, then there exists some partial sequence $\{w'_{m_j}\}_j$ such that $z'_{m_j} := \varphi_{m_j}(w'_{m_j}) \to z' \neq \zeta_0$. Since $z'_{m_j} \in L_{m_j} = L \cup [\zeta_0, \zeta_{m_j}]$ and $\zeta_{m_j} \to \zeta_0$ as $j \to \infty$, then $z' \in L \setminus \{\zeta_0\}$. Since

114 J.J. Carmona and K.Yu. Fedorovskiy

$\varphi_{m_j}^{-1} \rightrightarrows \varphi^{-1}$ in Ω, then one has

$$t_0 = \lim_{j\to\infty} w'_{m_j} = \lim_{j\to\infty} \varphi_{m_j}^{-1}(z'_{m_j}) = \lim_{j\to\infty} \varphi^{-1}(z'_{m_j}) = \varphi^{-1}(z') \in \mathbb{D},$$

but this is a contradiction. Since $\varphi_m(\ell'_m)$ is a subcontinuum of L_m and its extremes converge to ζ_0 we obtain, that

$$\operatorname{diam}\left(\varphi_m(\ell'_m)\right) \to 0 \text{ as } m \to \infty.$$

By [Po1, Corollary 1.4] we have, that

$$\sup_{z\in\overline{\mathbb{D}}}(1 - |z|^2)|\varphi'_m(z)| \le A < \infty$$

for all m, where A is an absolute constant. Notice that ℓ'_m is either a Jordan arc or the union of two consecutive Jordan arcs. Then, by [Po2, Theorem 9.2] applied to ℓ'_m or to each of the arcs that form ℓ'_m, we conclude, that

$$\operatorname{diam}(\ell'_m) \to 0 \text{ as } m \to \infty, \tag{2.1}$$

which means, that $t_m \to t_0$ as $m \to \infty$ and hence, $\alpha = \{t_0\}$.

Finally, taking into account (2.1), we have

$$\lim_{m\to\infty} \varphi_m^{-1}(\zeta_0) = \lim_{m\to\infty} \varphi_m^{-1}(\zeta_m) = t_0 = \varphi^{-1}(\zeta_0),$$

which ends the proof of (1).

We are going to prove the assertion (2). Let G be a bounded connected component of $\mathbb{C} \setminus \overline{\Omega}$. Assume that $|\varphi_m^{-1}|$ does not converge uniformly to 1 on $\overline{G}$. This implies the existence of a sequence $\{z_k\}_k$ in G and a subsequence $\{\varphi_{m_k}^{-1}\}_k$ of $\{\varphi_k^{-1}\}_k$ such that $|\varphi_{m_k}^{-1}(z_k)| \le r < 1$ for all k. Let $w_k := \varphi_{m_k}^{-1}(z_k)$. Considering a subsequence of $\{w_k\}_k$ if it is necessary we may assume that $w_k \to w_0, |w_0| \le r < 1$. Since φ_{m_k} converge uniformly on the compact set $\bigcup_{k=0}^{\infty}\{w_k\}$ we obtain

$$\varphi(w_0) = \lim_{k\to\infty} \varphi_{m_k}(w_k) \in \overline{G}.$$

But $\varphi(w_0) \in \Omega$ and $\overline{G} \cap \Omega = \varnothing$ and we arrive to a contradiction. $\square$

Remind, that ψ is the inverse mapping for φ. Proposition 1 and Theorem 1 allow us to state the following corollary which will be frequently and implicitly used later.

Corollary 1. *Let Ω be a Carathéodory domain. Then φ and ψ can be extended to Borel measurable functions (denoted also by φ and ψ) on $\mathbb{D} \cup F(\varphi)$ and $\Omega \cup \partial_a\Omega$ respectively and such that*

$$\begin{aligned}
\psi(\varphi(\xi)) &= \xi \quad \text{for all} \quad \xi \in F(\varphi) \\
\varphi(\psi(\zeta)) &= \zeta \quad \text{for all} \quad \zeta \in \partial_a\Omega.
\end{aligned} \tag{2.2}$$

Indeed, the function φ is a Borel function on $\mathbb{D} \cup F(\varphi)$ because of [Po2, pag. 331] and ψ is a Borel function on $\Omega \cup \partial_a\Omega$ in view of Theorem 1(1).

Example 1. Let us give an example of a Carathéodory domain Ω such that $\partial\Omega = \partial_a\Omega$ and Ω is not a Jordan domain.

$$\Omega := \mathbb{D} \setminus \bigcup_{n=1}^{\infty} \{z : \ (2n+1)^{-1} \leq \operatorname{Re} z \leq (2n)^{-1}, \ \operatorname{Im} z \geq 0\}.$$

Let f be some conformal mapping from $\mathbb{D}$ onto Ω. Observe that there exists a point $\xi \in \mathbb{T}$ such that $[0, i] = C(f, \xi)$ (where $C(g, a)$ means the total cluster set of a function g at a point a), and for each $\zeta \in [0, i]$ there exists some point $t \in \mathbb{T}$ such that $\zeta = f(t)$. So it is clear that $\partial\Omega = \partial_a\Omega$ and that Ω is not a Jordan domain.

Example 1 shows how Theorem 1 generalize the Carathéodory extension theorem for the sufficiently wide class of non Jordan domains. Notice, that if $\partial\Omega = \partial_a\Omega$, then $\varphi_m^{-1}(\zeta) \to \varphi^{-1}(\zeta)$ for *all* $\zeta \in \partial\Omega$ and therefore, φ^{-1} extends to the function belonging to the first Baire class in $\overline{\Omega}$. It is worth to compare this fact with [Go, Chapter II, Section 3, Theorem 1].

The Carathéodory domain possesses the following property, which seems fairly interesting and useful.

Proposition 2. *Let Ω be a Carathéodory domain in $\mathbb{C}$ and let G be a bounded connected component of $\mathbb{C} \setminus \overline{\Omega}$. Then the set $\partial_a\Omega \cap \partial G$ consists of at most 1 point.*

Proof. Assume the opposite, which means, that there exists at least two different points, says ζ_1 and ζ_2 in $\partial_a\Omega \cap \partial G$. Then there exist two different points t_1 and t_2 in $F(\varphi) \subset \mathbb{T}$ such that $\zeta_s = \varphi(t_s)$ for $s = 1, 2$. Denote by ℓ_1 and ℓ_2 the radii $[0, t_1]$ and $[0, t_2]$ respectively and put $L_s = \varphi(\ell_s)$ for $s = 1, 2$. Set $B_s(r) := \overline{B(\zeta_s, r)}$ for any $r > 0$ and for $s = 1, 2$.

Take some $\delta > 0$ such that $B_1(\delta) \cap B_2(\delta) = \varnothing$. Denote by $\zeta_s^*(\delta)$ for $s = 1, 2$ the first intersection point of L_s with $B_s(\delta)$ and by $L_s(\delta)$ the subarc of L_s from z_0 to $\zeta_s^*(\delta)$. Let $\lambda(\delta) \subset G$ be some polygonal line in G that connects $B_1(\delta)$ with $B_2(\delta)$ and such that $\lambda(\delta) \cap B_s(\delta) = \{\zeta_s^{**}(\delta)\}$ for $s = 1, 2$.

Denote by $M(\delta)$ the continuum $L_1(\delta) \cup L_2(\delta) \cup B_1(\delta) \cup B_2(\delta) \cup \lambda(\delta)$. It follows from the Jordan curve theorem, that $\mathbb{C} \setminus M(\delta)$ has only two connected components, says $U(\delta)$ the bounded one. Indeed, $U(\delta)$ is the bounded connected component of the complementary of the Jordan curve which is the union of $L_s(\delta)$ (for $s = 1, 2$), $\lambda(\delta)$ and appropriately chosen subarcs of ∂B_s, joining the points $\zeta_s^*(\delta)$ and $\zeta_s^{**}(\delta)$ for $s = 1, 2$ respectively.

The continuum $\ell_1 \cup \ell_2$ separates $\mathbb{D}$ onto two open sectors. The image by φ of one of them cuts $U(\delta)$. Therefore, there exists $\zeta \in \partial\Omega \cap U(\delta)$ and $r > 0$ such that $B(\zeta, r) \subset U(\delta)$. Since Ω is a Carathéodory domain, it is possible to find some point $z \in B(\zeta, r) \cap \Omega_\infty$. Then, there exists an infinite polygonal line $\Pi \subset \Omega_\infty$ joints the point z with ∞. Since Π is connected, then $\Pi \cap M(\delta) \subset B_1(\delta) \cup B_2(\delta)$.

Now let us choose $\delta_1 < \delta$ such that $B_s(\delta_1) \cap \Pi = \varnothing$ for $s = 1, 2$ and repeat the construction of $M(\cdot)$ and $U(\cdot)$ using δ_1 instead of δ. Observe, that $z \in B(\zeta, r) \subset U(\delta_1)$. Therefore we reach to a contradiction, because $\Pi \cap M(\delta_1) \neq \varnothing$, which is possible only if $\Pi \cap (B_1(\delta_1) \cup B_2(\delta_1)) \neq \varnothing$. But $\Pi \cap (B_1(\delta_1) \cup B_2(\delta_1)) = \varnothing$ by the election of δ_1. $\qquad\square$

Example 2. It is worth to note, that there exists a Carathéodory domain Ω having a bounded component G of $\mathbb{C} \setminus \overline{\Omega}$ such that $\partial_a\Omega \cap \partial G$ consists only of one point. Indeed, if

$$\Omega_1 := \left(\bigcup_{n=1}^{\infty} \left\{ z : \tfrac{1}{2n+1} < \operatorname{Re} z < \tfrac{1}{2n}, \ |\operatorname{Im} z| < \pi - \tfrac{1}{n} \right\} \right)$$
$$\cup \left\{ z : \ |\operatorname{Im} z| < \operatorname{Re} z, \ 0 < \operatorname{Re} z < \tfrac{1}{2} \right\},$$

then the desired example may be obtained as $\Omega := \exp\Omega_1$ and $G = \mathbb{D}$.

Corollary 2. *If Ω be a Carathéodory domain in $\mathbb{C}$ with $\partial\Omega = \partial_a\Omega$, then $\mathbb{C} \setminus \overline{\Omega}$ is connected.*

3. Structure of measures orthogonal to rational functions

Let $X \subset \mathbb{C}$ be a compact set. Denote by $\mathrm{C}(X)$ the space of all complex-valued continuous functions on X endowed with the uniform norm $\|f\|_X = \max_{z \in X} |f(z)|$. Put $\mathrm{A}(X) = \mathrm{C}(X) \cap \mathrm{Hol}(X^\circ)$. Define $\mathrm{P}(X)$ and $\mathrm{R}(X)$ to be the closures in $\mathrm{C}(X)$ of the subspaces $\{p_{|X} : \ p \in \mathcal{P}\}$ and $\{g_{|X} : \ g \in \mathcal{R}(X)\}$ respectively.

In what follows, a *measure* means a finite complex-valued Borel measure on $\mathbb{C}$. By $\mathrm{Supp}\,(\mu)$ one denotes the support of the measure μ, and by $|\mu|$ – the corresponding *variational measure*. If μ is a measure and E is a Borel set then $\mu_{|E}$ means the restriction of μ to the set E, i.e., $\mu_{|E}(E_1) = \mu(E \cap E_1)$ for each Borel set E_1. The notation $\mu \ll \nu$ (respectively, $\mu \perp \nu$) means, as usual, that the measure μ is absolutely continuous with respect to ν (respectively, that μ and ν are mutually singular).

If μ is some measure on X, then one says that μ is orthogonal to some subclass $V \subseteq \mathrm{C}(X)$ if $\int f \, d\mu = 0$ for each $f \in V$ and writes this fact as $\mu \perp V$.

For a rectifiable contour or arc γ one writes $d\zeta_{|\gamma}$, or $d\zeta$ if it is clear from the context what γ we deal with, for the measure on γ which acts (as a functional in the space $\mathrm{C}(\gamma)$) by the formula $d\zeta_{|\gamma}(f) = \int_\gamma f(\zeta) \, d\zeta$. In the case when γ is some subarc of $\mathbb{T}$ we use the complex parameters ξ or t (instead of ζ). We denote by $d\mathrm{A}$ the planar Lebesgue measure in $\mathbb{C}$.

Let Ω be a Carathéodory domain. Take a function $u \in \mathrm{L}^1(\mathbb{T})$ so that $\tau := u \, d\xi$ is a measure on $\mathbb{T}$. Define the measure $\varphi(\tau)$ on $\partial\Omega$ by the formula

$$\varphi(\tau)(E) := \tau(\psi(E \cap \partial_a\Omega))$$

for every Borel set $E \subset \partial\Omega$ or, equivalently,

$$\int g(\zeta) \, d\varphi(\tau)(\zeta) = \int_{F(\varphi)} g(\varphi(\xi))u(\xi) \, d\xi = \int_{\mathbb{T}} g(\varphi(\xi))u(\xi) \, d\xi \qquad (3.1)$$

for every $g \in \mathrm{C}(\overline{\Omega})$.

In the sequel ω will denote the measure $\omega := \varphi(d\xi)$. In fact, ω is a measure on $\partial_a \Omega$ and has no atoms. Furthermore, it follows from (2.2) and (3.1), that

$$\varphi(u\,d\xi) = (u \circ \psi)\,\omega. \tag{3.2}$$

Let U be a bounded not empty open set, $E \subset \partial U$ be a Borel set and $z \in U$ be a point. Then, as usual, $\omega(z, E, U)$ denotes the harmonic measure of E evaluated with respect to U and z. Remind, that $\omega(z_1, \cdot, U) \ll \omega(z_2, \cdot, U)$ for all z_1, z_2 that belong to the same connected component of U.

Observe, that $\omega(\varphi(0), \cdot, \Omega) = \varphi(|d\xi|/(2\pi))$. Using the definitions of $\omega(z, \cdot, \Omega)$ and $\omega(\cdot)$ and the fact that $|d\xi(E)| \geq c(\delta)|d\xi|(E)$ for each Borel set $E \subset \mathbb{T}$ such that $\operatorname{diam}(E) \leq \delta$, where $c(\delta) \to 1$ as $\delta \to 0$, it is possible to show, that

$$\omega(\varphi(0), \cdot, \Omega) = |\omega(\cdot)|/(2\pi).$$

E. Bishop had provided in [Bi1, Bi2] a fruitful investigation of the structure of certain measures orthogonal to rational functions on Carathéodory compact sets. The following results are essentially (but only implicitly) stated in [Bi1, Bi2].

Theorem 2.

(1) *Let Ω be a Carathéodory domain in $\mathbb{C}$ and μ be a measure on $\partial\Omega$ such that $\mu \perp \mathcal{R}(\overline{\Omega})$. Then there exists a function $h \in \mathrm{H}^1(\mathbb{D})$ such that*

$$\mu = (h \circ \psi)\,\omega. \tag{3.3}$$

(2) *Let X be a Carathéodory compact set in $\mathbb{C}$ such that $X^\circ \neq \varnothing$ and μ be the measure on ∂X such that $\mu \perp \mathcal{R}(X)$. Then*

$$\mu = \sum_{\Omega \in \mathfrak{S}(X)} \mu_\Omega, \tag{3.4}$$

where $\mu_\Omega = \mu_{|\partial\Omega} \perp \mathcal{R}(\overline{\Omega})$ and the series converges in norm in the space of measures on ∂X.

The proof of Theorem 2 may be obtained, passing throughout both papers [Bi1, Bi2] which are based on studies of the concept of an analytic differential that represents a measure. We recall, that an analytic differential in a domain Ω is the differential form $g(z)\,dz$ where $g \in \operatorname{Hol}(\Omega)$. The analytic differential $g(z)\,dz$ represents the measure μ on $\partial\Omega$ if the sequence of measures $\{g(\zeta)\,d\zeta_{|\gamma_j}\}_j$, where $\{\gamma_j\}_j$ is some sequence of rectifiable contours such that $D(\gamma_j) \subset D(\gamma_{j+1}) \subset \Omega$ and $D(\gamma_j) \uparrow \Omega$ as $j \to \infty$, converges in the weak-star topology of the space of measures on $\overline{\Omega}$ to μ. Observe, that the analytic differential $g(z)\,dz$ in Ω is defined even in the case when $\partial\Omega$ is not rectifiable. We consider that it is interesting to present a direct, and free from the concept of analytic differentials, proof of Theorem 2.

The Cauchy transform of a measure μ is the function defined as

$$\widehat{\mu}(z) = \frac{1}{2\pi i} \int \frac{d\mu(\zeta)}{\zeta - z}$$

for dA-almost all $z \in \mathbb{C}$. It is well known, that $\widehat{\mu}$ is holomorphic off $\mathrm{Supp}\,(\mu)$ and $\overline{\partial}\widehat{\mu} = \frac{i}{2}\mu$ in the distributional sense.

Proof of Theorem 2. Let us denote by Ω_j, where $j \in J$ and J is some finite or countable set of indexes, all elements of the set $\mathfrak{S}(X)$. It is clear, that each Ω_j with $j \in J$, is a Carathéodory domain.

Step 1. *There exists a Carathéodory continuum Y such that $X \subseteq Y$ and $X^\circ = Y^\circ$.*

Proof. In order to prove this assertion we consider for each integer $k \geq 1$ the family $\mathcal{D}_k$ of the dyadic squares of the generation k, i.e.,

$$\mathcal{D}_k = \left\{ Q = \left[\frac{j_1}{2^k}, \frac{j_1+1}{2^k}\right) \times \left[\frac{j_2}{2^k}, \frac{j_2+1}{2^k}\right) : \ j_1, j_2 \in \mathbb{Z} \right\}.$$

Define the subfamily $\mathcal{D}_k(X)$ that consists of all squares $Q \in \mathcal{D}_k$ such that $X \cap \overline{Q} \neq \varnothing$, put $F_k := \bigcup_{Q \in \mathcal{D}_k(X)} Q$ and suppose $F_{k,1}, \ldots, F_{k,r_k}$ to be the closures of the polynomial hulls of the connected components of F_k. In such a case one has that $X \subset F_{1,1}^\circ \cup \cdots \cup F_{1,r_1}^\circ$. For each k and $j = 1, \ldots, r_k$ we choose a point $z_{k,j} \in \partial F_{k,j}$. Set $F_k^* := \bigcup_{j=1}^{r_k} F_{k,j}$. Denote by $I_{k+1,j}$ the set of indexes $s = 1, \ldots, r_k$ such that $F_{k+1,s} \subset F_{k,j}$ and set $F_{k+1,j}^* := \bigcup_{s \in I_{k+1,j}} F_{k+1,s}$.

    In what follows by a tree we mean a connected polygonal line T such that $\mathbb{C} \setminus T$ is connected.

    Let us construct a sequence of trees $\{T_k\}_k$ with $T_{k-1} \subset T_k$ by induction. Take a point $z \notin X$ and choose a tree T_1 such that T_1 connects z with all points $z_{1,j}$, $j = 1, \ldots, r_1$ and such that the set $\mathbb{C} \setminus (F_1^* \cup T_1)$ is connected. Suppose now that the trees $T_1, \ldots, T_k$ are already constructed. Let us show how to construct the tree T_{k+1}. Since $F_{k,j}$ for $j = 1, \ldots, r_k$ contains a finite number of $\{F_{k+1,s}\}$ (where $s = 1, \ldots, r_{k+1}$), then we can choose a new tree $T_{k,k+1,j}$ that connects $z_{k,j}$ with all $z_{k+1,s}$ for $s \in I_{k+1,j}$ such that the domain $G_k := \mathbb{C} \setminus (T_k \cup Y_k)$, where $Y_k = \bigcup_{j=1}^{r_k} (F_{k+1,j}^* \cup T_{k,k+1,j})$ is simply connected. Now we put $T_{k+1} = T_k \cup \left(\bigcup_{j=1}^{r_k} T_{k,k+1,j}\right)$.

    Finally we take $T = \overline{\bigcup_{k=1}^{\infty} T_k}$ and let $Y = X \cup T$. Then Y is a compact set such that $X^\circ = Y^\circ$. Since all G_k are simply connected domains and $\overline{\mathbb{C}} \setminus Y = \bigcup_k G_k$, then Y is connected and finally, Y is a Carathéodory compact set because of $\partial Y = \partial X \cup T$. $\qquad\square$

Step 2. *If $\Omega \in \mathfrak{S}(X)$ and Φ is some conformal mapping $\mathbb{D}$ onto Ω, then $h_\Omega := (\widehat{\mu} \circ \Phi)\,\Phi' \in \mathrm{H}^1(\mathbb{D})$.*

Proof. By Step 1 we construct the Carathéodory connected compact set Y such that $X \subset Y$ and $X^\circ = Y^\circ$. Take a point $z_\Omega \in \Omega$ and consider the sequence $\{\Gamma_m\}_{m=1}^{\infty}$ of rectifiable contours such that $Y \subset D(\Gamma_m) \subset D(\Gamma_{m-1})$ and $\overline{D(\Gamma_m)}$ converges to Y as $m \to \infty$. Therefore, the kernel of the sequence $\{D(\Gamma_m)\}_m$ with respect to the point z_Ω is exactly Ω. From this moment the proof of the present step may be obtained by the word-to-word repetition of the proof of [CFP, Lemma 2.3] with the clear replacements in notations. $\qquad\square$

For each $j \in J$ we put $h_j := h_{\Omega_j}$ and define the measure μ_j by setting $\mu_j := \varphi_j(h_j \, d\xi) = (h_j \circ \psi_j)\, \omega_j$, where φ_j is some conformal mapping from $\mathbb{D}$ onto Ω_j, ψ_j is the inverse map for φ_j and $\omega_j = \varphi_j(d\xi)$ (see (3.2)).

Step 3.

 (i) $\widehat{\mu}_j(z) = \widehat{\mu}(z)$ *for all* $z \in \Omega_j$;

 (ii) $\widehat{\mu}_j(z) = 0$ *for all* $z \notin \overline{\Omega}_j$ *(which means that* $\mu_j \perp \mathcal{R}(\overline{\Omega}_j)$*).*

Proof. Take a point $z \notin \partial\Omega_j$. Then

$$\widehat{\mu}_j(z) = \frac{1}{2\pi i}\int_{\partial\Omega_j} \frac{h_j(\psi_j(\zeta))\, d\omega_j(\zeta)}{\zeta - z} = \frac{1}{2\pi i}\int_{\mathbb{T}} \frac{h_j(\xi)\, d\xi}{\varphi_j(\xi) - z}.$$

If $z \notin \Omega_j$, then the function $h_j(w)/(\varphi_j(w) - z)$ is a bounded analytic function in $\mathbb{D}$, so that

$$\int_{\mathbb{T}} \frac{h_j(\xi)\, d\xi}{\varphi_j(\xi) - z} = 0,$$

which gives (ii).

If $z \in \Omega_j$, then there exists $w_0 \in \mathbb{D}$ such that $z = \varphi_j(w_0)$ and the function

$$H(w, w_0) = \begin{cases} \dfrac{w - w_0}{\varphi_j(w) - \varphi_j(w_0)}, & w \neq w_0 \\[2ex] \dfrac{1}{\varphi_j'(w_0)}, & w = w_0 \end{cases}$$

is a bounded analytic function in $\mathbb{D}$. Therefore,

$$\frac{1}{2\pi i}\int_{\mathbb{T}} \frac{h_j(\xi)\, d\xi}{\varphi_j(\xi) - \varphi_j(w_0)} = \frac{1}{2\pi i}\int_{\mathbb{T}} \frac{h_j(\xi)\, H(\xi, w_0)\, d\xi}{\xi - w_0} = h_j(w_0)\, H(w_0, w_0) = \frac{h_j(w_0)}{\varphi_j'(w_0)}.$$

It gives, that for $z \in \Omega_j$ we have

$$\widehat{\mu}_j(z) = \frac{h_j(w_0)}{\varphi_j'(w_0)} = \frac{\widehat{\mu}(\varphi_j(w_0))\varphi_j'(w_0)}{\varphi_j'(w_0)} = \widehat{\mu}(\varphi_j(w_0)) = \widehat{\mu}(z). \qquad \square$$

The assertion (1) is the consequence of the following arguments. In such a case $X = \overline{\Omega}$ and the set of connected components of X° consists of only one domain $\Omega_1 := \Omega$ (and $J = \{1\}$). It follows from Step 3, that $\widehat{\mu}(z) = \widehat{\mu}_1(z)$ for all $z \notin \partial\Omega$, consequently $\mu - \mu_1 \perp \mathcal{R}(\partial\Omega)$. By Mergelyan's theorem (see [Me, Theorem 4.4]) one has $\mathrm{R}(\partial\Omega) = \mathrm{C}(\partial\Omega)$ if Ω is a Carathéodory domain, then $\mu = \mu_1$ which is exactly our assertion in (1).

Let us go to prove the assertion (2). Let $I \subset J$ be some finite set of indexes. Put

$$W_I := \bigcup_{j \in I} \Omega_j.$$

The following assertion is the consequence of [Bi2, Lemma 7] and Runge's theorem.

Step 4. *There exists a sequence of functions* $\{f_k\}_{k=1}^{\infty} \subset \mathcal{R}(X)$ *such that* $\|f_k\|_X \leq 1$, $f_k \rightrightarrows 1$ *on compact subsets of* W_I *and* $f_k \rightrightarrows 0$ *on compact subsets of* $X^\circ \setminus W_I$.

Therefore, for the sequence of measures $\{f_k\mu\}_{k=1}^{\infty}$ on ∂X we have $\|f_k\mu\| \leq \|\mu\|$. Hence, this sequence has a limit point μ_I in the weak-star topology (of the space of measures on ∂X). Since $\mu \perp \mathcal{R}(X)$ and $f_k \in \mathcal{R}(X)$, then $\mu_I \perp \mathcal{R}(X)$.

Step 5.

(i) $\widehat{\mu}_I(z) = \widehat{\mu}(z)$ *for all* $z \in W_I$;

(ii) $\widehat{\mu}_I(z) = 0$ *for all* $z \in X^\circ \setminus W_I$.

Proof. Considering if it is necessary some subsequence (for which we preserve the initial notation) of the sequence $\{f_k\mu\}_k$ we can say that the sequence $\{f_k\mu\}_k$ converges to μ_I in the weak-star topology of the space of measures on ∂X. Let $z \notin \partial X$. The desired assertion is the consequence of the following computation:

$$\widehat{\mu}_I(z) = \lim_{k\to\infty} \left(\frac{1}{2\pi i} \int \frac{(f_k(\zeta) - f_k(z))\, d\mu(\zeta)}{\zeta - z} + \frac{f_k(z)}{2\pi i} \int \frac{d\mu(\zeta)}{\zeta - z} \right) = \widehat{\mu}(z) \lim_{k\to\infty} f_k(z).$$

$\square$

It follows from Steps 3 and 5, that $\widehat{\mu}_I(z) = \sum_{j \in I} \widehat{\mu}_j(z)$ for all $z \notin \partial X$ and using the fact that $\mathrm{R}(\partial X) = \mathrm{C}(\partial X)$ (as in the proof of (1)) we conclude that

$$\mu_I = \sum_{j \in I} \mu_j. \tag{3.5}$$

Step 6 (see [Bi2, Lemma 10]). $\omega_j \perp \omega_k$ *for* $j \neq k$.

Taking into account (3.5), Step 6 and the fact that $\mu_j = (h_j \circ \psi_j)\omega_j \ll \omega_j$ we conclude, that $\mu_j \perp \mu_k$ for $j, k \in J$, $j \neq k$. Hence we have

$$\sum_{j \in I} \|\mu_j\| = \left\| \sum_{j \in I} \mu_j \right\| = \|\mu_I\| \leq \|\mu\|,$$

which means that $\sum_{j \in J} \|\mu_j\| < \infty$. Then, the series $\sum_{j \in J} \mu_j$ converges in the space of measures on ∂X to some measure σ. It is clear, that $\sigma \perp \mathcal{R}(X)$. For each $j \in J$ we have $\widehat{\sigma}(z) = \widehat{\mu}_j(z)$ for all $z \in \Omega_j$ and applying the result of Step 3 we conclude that $\widehat{\sigma}(z) = \widehat{\mu}(z)$ on X. Then, $\sigma = \mu$.

Take $k \in J$. Since $\mu_j \perp \mu_k$ for $j \in J \setminus \{k\}$, then for each Borel set $E \subset \partial X$ we have

$$\mu_{|\partial\Omega_k}(E) = \mu(E \cap \partial\Omega_k) = \sum_{j \in J} \mu_j(E \cap \partial\Omega_k) = \mu_k(E \cap \partial\Omega_k) = \mu_k(E). \qquad \square$$

Moreover, it is possible to find out in [Bi2] the following facts, concerning the objects that were introduced in the proof of Theorem 2. These facts do not needed for the proof of Theorem 2 but seems to be fairly interesting and important.

Remark 1. One has

$$\sum_{j \in J} \int_{\varphi_j(r\mathbb{T})} |\widehat{\mu}(\zeta)|\, d\zeta \leq C\|\mu\|,$$

for each $r \in (0,1)$ and

$$\sum_{j \in J} \|h_j\|_{\mathrm{H}^1(\mathbb{D})} \le C\|\mu\|.$$

where $C > 0$ is some absolute constant.

Next we obtain two propositions concerning the structure of all measures that are orthogonal to rational functions and which are supported on some special compacts sets.

In [CFP, Lemma 4.1] it was shown, that if Ω is a Jordan domain with rectifiable boundary and μ is a measure with $\mathrm{Supp}\,(\mu) \subset \Omega$, then the measure $\mu + \widehat{\mu}\,d\zeta_{|\partial\Omega}$ is orthogonal to $\mathcal{P}$. In view of the term $\widehat{\mu}\,d\zeta_{|\partial\Omega}$ it is not clear how to generalize this fact to domains with non rectifiable boundaries. The following proposition gives an appropriate generalization of [CFP, Lemma 4.1].

Proposition 3. *Let Ω be a Carathéodory domain in $\mathbb{C}$.*

(1) *Take a measure μ with $\mathrm{Supp}\,(\mu) \subset \Omega$ and put $\nu = \varphi^{-1}(\mu)$. Then the measure*

$$\mu^* := \mu + (\widehat{\nu} \circ \psi)\,\omega,$$

is orthogonal to $\mathrm{A}(\overline{\Omega})$.

(2) *Let $K \subset \Omega$ be some compact set and σ be a measure on $K \cup \partial\Omega$ such that $\sigma \perp \mathcal{R}(\overline{\Omega})$. Then there exists a function $h \in \mathrm{H}^1(\mathbb{D})$ such that*

$$\sigma = (\sigma_{|K})^* + (h \circ \psi)\,\omega.$$

Proof. (1) Put $M := \mathrm{Supp}\,(\nu)$. Since $\widehat{\nu}$ is analytic outside M, then $\widehat{\nu}\,d\xi$ is a measure on $\mathbb{T}$ and $\eta := \varphi(\widehat{\nu}\,d\xi)$ is a measure on $\partial\Omega$. Take $g \in \mathrm{A}(\overline{\Omega})$, so that $g \circ \varphi \in \mathrm{H}^\infty(\mathbb{D})$. Using Fubini and Cauchy theorems and the definition of $\widehat{\nu}$ we have:

$$\int g\,d\eta = \int_{\mathbb{T}} g(\varphi(\xi))\widehat{\nu}(\xi)\,d\xi = \int_M \left[\frac{1}{2\pi i} \int_{\mathbb{T}} \frac{g(\varphi(\xi))\,d\xi}{w - \xi}\right]d\nu(w) =$$

$$- \int_M g(\varphi(w))\,d\nu(w) = - \int g\,d\mu,$$

so that $\displaystyle\int g(z)\,d\mu^*(z) = \int g(z)\,d\mu(z) + \int g(\zeta)\,d\eta(\zeta) = 0.$

In order to prove (2) we need to observe that $\sigma - (\sigma_{|K})^*$ is a measure on $\partial\Omega$ orthogonal to $\mathcal{R}(\overline{\Omega})$. Then we can apply (3.3) which gives the desired result. $\square$

Corollary 3. *Let Y be a Carathéodory compact set, $Y^\circ \ne \varnothing$ and $K \subset Y^\circ$ be a compact set. Then for every measure μ on $K \cup \partial Y$ such that $\mu \perp \mathcal{R}(Y)$ one has:*

$$\mu = \sum_{\Omega \in \mathfrak{S}(Y)} \mu_\Omega,$$

where $\mu_\Omega = \mu_{|\overline{\Omega}} \perp \mathcal{R}(\overline{\Omega})$ and the series converges in norm in the space of measures on Y.

Proof. Since $K \subset Y^\circ$, then the set $\mathfrak{S}_K := \{\Omega \in \mathfrak{S}(Y) \text{ such that } K \cap \Omega \neq \varnothing\}$ is finite. Hence, applying Theorem 2(2) to the measure $\mu - \sum_{\Omega \in \mathfrak{S}_K} (\mu_{|K\cap\Omega})^*$ we obtain the desired result. $\qquad\qquad\square$

In order to formulate the next proposition we need the following topological definition:

Definition 1. Let U be a bounded open set in $\mathbb{C}$ and let $\gamma \subset \partial U$ be a closed Jordan arc. One says, that γ is a *gate* for U if there exists an open connected set V such that $V \setminus \gamma = V_1 \cup V_2$, where V_1 and V_2 are non empty open connected sets such that $V_1 \subset U$ and $V_2 \subset \mathbb{C} \setminus \overline{U}$.

Using the concept of a free arc (see [Po3]) we can say, that γ is a gate for a domain Ω if and only if γ is a free arc for Ω and for $\mathbb{C} \setminus \overline{\Omega}$. If Ω is a Jordan domain, then any closed subarc $\gamma \subset \partial\Omega$ is a gate for Ω.

Proposition 4. *Let U be a bounded not empty open set in $\mathbb{C}$ and μ be a measure, such that* $\mathrm{Supp}\,(\mu) \subset \overline{U}$ *and* $\mu \perp \mathcal{R}(\overline{U})$.

(1) *If there exists a gate γ for U, then $\mu_{|\gamma} \ll \omega(z, \cdot, U)_{|\gamma}$ for any point $z \in U$.*

(2) *If γ is a rectifiable gate for U, then $\mu_{|\gamma} \ll d\zeta_{|\gamma}$.*

Proof. Let $z \in U$ be an arbitrary point. Take a conformal mapping Φ from $\mathbb{D}$ onto $\overline{\mathbb{C}} \setminus \gamma$ with $\Phi(0) = \infty$. Since γ is rectifiable, then Φ has a continuous extension (denoted also by Φ) on $\overline{\mathbb{D}}$ and this extension can be chosen such that $\Phi(\lambda^\pm) = \gamma$, where $\lambda^\pm = \{\xi \in \mathbb{T} : \mathrm{Im}\,\xi \gtrless 0\}$. Let Ψ denotes the inverse mapping for Φ on $\overline{\mathbb{C}} \setminus \gamma$. Since γ is a gate for U, then the function $\Psi_{|U}$ has a continuous extension Ψ_0 on $U \cup \gamma$.

Take a Borel set $E \subset \gamma$ such that $\omega(z, E, U) = 0$ in the case (1), and $|d\zeta|(E) = 0$ in the case (2), and take a compact subset Y of E. We need to prove that $\mu(Y) = 0$.

One has that there exist two compact sets, says $Y^\pm \subset \mathbb{T}$, such that $Y^\pm \subset \lambda^\pm$ and $\Phi(Y^\pm) = Y$. Without loss of generality we suppose that $\Psi_0(Y) = Y^+$.

Let us show that $|d\xi|(Y^+) = 0$. In the case (2) γ is rectifiable and the desired property of Y^+ follows from [Po2, Theorem 10.11]. In the case (1) let us assume that $|d\xi|(Y^+) > 0$. Then there exists a compact set $Y_1 \subset Y^+$ such that $|d\xi|(Y_1) > 0$ and such that $Y_1 \subset \partial G$, where $G \subset \mathbb{D}$ is a domain whose boundary is the union of an arc of $\mathbb{T}$ and some circle orthogonal to this arc and such that $G \subset \Psi_0(U)$. Therefore,

$$0 = \omega(z, Y, U) \geq \omega(z, \Phi(Y_1), U) = \omega(\Psi_0(z), Y_1, \Psi_0(U)) \geq \omega(\Psi_0(z), Y_1, G) > 0.$$

This contradiction shows that $|d\xi|(Y^+) = 0$ in both cases under consideration.

Then, by [Ga, pag. 59], Y^+ is a peak set for the algebra $P(\overline{\mathbb{D}})$. Then there exists a function $f \in P(\overline{\mathbb{D}})$ such that

$$\begin{aligned}
f(w) &= 1, &&\text{as } w \in Y^+; \\
|f(w)| &< 1, &&\text{as } w \notin Y^+.
\end{aligned} \tag{3.6}$$

Denote by W the union of all bounded connected components of $\overline{\mathbb{C}} \setminus \overline{U}$ except the component V such that $\gamma \subset \partial V$ and consider the function

$$g = \begin{cases} f \circ \Psi_0 & \text{on } \overline{U}, \\ f \circ \Psi_{|W} & \text{on } W. \end{cases}$$

Since $\Psi_0 \in C(\overline{U})$ and $\Psi_{|W}$ is holomorphic in neighborhood of $\overline{W}$, then $g \in A(F)$, where $F = \overline{U} \cup W$. It follows from Mergelyan's theorem (see [Me, Theorem 4.4]) that $A(F) = R(F) \subset R(\overline{U})$. Then $g \in R(\overline{U})$. It follows from (3.6), that

$$g(z) = 1, \quad \text{as } z \in Y;$$
$$|g(z)| < 1, \quad \text{as } z \notin Y.$$

Applying the Lebesgue dominated convergence theorem to the sequence $\{g^m(\cdot)\}$ and taking into account $\mu \perp \mathcal{R}(\overline{U})$ we obtain:

$$0 = \lim_{m \to \infty} \int g^m(z) \, d\mu(z) = \int_Y d\mu(\zeta) = \mu(Y).$$

So, for all compact sets $Y \subset E$ one has $\mu(Y) = 0$. This means, that $|\mu|(E) = 0$. $\quad \square$

4. Approximation by polyanalytic functions

Let $X \subset \mathbb{C}$ be a compact set in $\mathbb{C}$ and $n \in \mathbb{N}$. Put $A_n(X) = C(X) \cap \operatorname{Hol}_n(X^\circ)$ and define the spaces $P_n(X)$ and $R_n(X, Y)$ (where $Y \supseteq X$ is some compact set in $\mathbb{C}$) as the closures in $C(X)$ of the subspaces $\{p_{|X} : p \in \mathcal{P}_n\}$ and $\{g_{|X} : g \in \mathcal{R}_n(Y)\}$ respectively. One has $P_n(X) \subset A_n(X)$.

We recall the concept of a Nevanlinna domain, that has been introduced in [Fe1, CFP], which is a useful tool for the study of uniform polyanalytic polynomial approximation.

Definition 2. Let Ω be a bounded simply connected domain in $\mathbb{C}$. One says that Ω is a *Nevanlinna domain* if there exist $u, v \in H^\infty(\Omega)$, $v \neq 0$ such that

$$\mathcal{F}(\zeta) := u(\zeta)/v(\zeta) = \overline{\zeta} \tag{4.1}$$

on $\partial \Omega$ almost everywhere in the sense of conformal mapping. This means that the following equality of *angular boundary values*

$$\overline{\varphi(\xi)} = (\mathcal{F} \circ \varphi)(\xi) = (u \circ \varphi)(\xi)/(v \circ \varphi)(\xi)$$

holds for almost all points $\xi \in \mathbb{T}$, where φ is some conformal mapping from $\mathbb{D}$ onto Ω.

Respectively, one says that Ω is a *locally Nevanlinna domain* if there exist a compact set $\Sigma \subset \Omega$ and $u, v \in H^\infty(\Omega \setminus \Sigma)$ such that the equality (4.1) holds on $\partial \Omega$ almost everywhere in the sense of conformal mapping.

We denote by $\mathcal{N}$ and $\mathcal{N}_{\mathrm{loc}}$ the sets of all Nevanlinna and locally Nevanlinna domains respectively. If Ω is a locally Nevanlinna domain, then there are many different possibilities to define the desired compact set Σ. We will write $\Omega(\mathcal{F}, \Sigma)$ in order to show which $\mathcal{F}$ and Σ we are deal with.

The following proposition is the direct consequence of the boundary uniqueness theorem of Luzin-Privalov [Pr, Chapter 4, Section 2.5]:

Proposition 5. *If* $\Omega \in \mathcal{N}_{\mathrm{loc}}$, $\Omega = \Omega(\mathcal{F}_1, \Sigma_1)$ *and* $\Omega = \Omega(\mathcal{F}_2, \Sigma_2)$, *then* $\mathcal{F}_1 = \mathcal{F}_2$ *in* $\Omega \setminus \widehat{\Sigma_1 \cup \Sigma_2}$.

It is always possible to assume, that Σ is minimal, which means that the function $\mathcal{F}$ cannot be meromorphically continued from $\Omega \setminus \Sigma$ to $\Omega \setminus \Sigma_1$ for any proper compact subset Σ_1 of Σ. The election of a such minimal Σ is also not unique, and we always need to find some appropriate Σ in order to apply the forthcoming results.

Remark 2. If Ω is Jordan domain with rectifiable boundary, then (4.1) holds $d\zeta$-almost everywhere on $\partial\Omega$ (see [Go, Chapter 10, Section 5]).

Moreover, if $\Omega \in \mathcal{N}_{\mathrm{loc}}$ and if there exists a rectifiable gate γ for Ω, then the functions u and v which are taken from Definition 2 have angular boundary values $u(\zeta)$ and $v(\zeta)$ for almost all $\zeta \in \gamma$ and (4.1) holds $d\zeta_{|\gamma}$-almost everywhere on γ.

Let Y be a Carathéodory compact set in $\mathbb{C}$ and $K \subseteq Y$ be a compact set. Put $X := K \cup \partial Y$ and take some integer $n \geq 2$. In this section we establish some conditions in order that $\mathrm{R}_n(X, Y) = \mathrm{A}_n(X)$ and $\mathrm{P}_n(X) = \mathrm{A}_n(X)$. We remaind, that in [CFP] the following situations have been considered:

(1) if Ω is a Carathéodory domain, $Y = \overline{\Omega}$ and $K = \varnothing$, then $\mathrm{R}_2(\partial\Omega, \overline{\Omega}) = \mathrm{C}(\partial\Omega)$ if and only if $\Omega \notin \mathcal{N}$ (see [CFP, Theorem 2.2(1)]);

(2) if $K = Y$, then $\mathrm{P}_n(Y) = \mathrm{A}_n(Y)$ if and only if each bounded connected component of $\mathbb{C} \setminus Y$ is not a Nevanlinna domain (see [CFP, Theorem 2.2(2)]).

The first main result in this section gives a useful generalization of [CFP, Theorem 4.3].

Theorem 3. *Let* Y *be a Carathéodory compact set in* $\mathbb{C}$, $Y^\circ \neq \varnothing$ *and* $K \subset Y^\circ$ *be a compact set. Put* $K_\Omega := K \cap \Omega$ *for each* $\Omega \in \mathfrak{S}(Y)$; $\mathfrak{S}_K := \{\Omega \in \mathfrak{S}(Y) : K_\Omega \neq \varnothing\}$.

If $\mathrm{P}_n(K) = \mathrm{A}_n(K)$ *for some integer* $n \geq 2$, $\Omega \notin \mathcal{N}$ *for each* $\Omega \in \mathfrak{S}(Y)$, *and for each* $\Omega \in \mathfrak{S}_K$ *one of the following assumptions:*

(1) $\Omega \notin \mathcal{N}_{\mathrm{loc}}$;

(2) $\Omega = \Omega(\mathcal{F}, \Sigma) \in \mathcal{N}_{\mathrm{loc}}$ *and* $\mathcal{F}$ *cannot be meromorphically continued from* $\Omega \setminus \widehat{\Sigma \cup K_\Omega}$ *to* $\Omega \setminus \widehat{K_\Omega}$;

holds, then for $X := K \cup \partial Y$ *one has* $\mathrm{A}_n(X) = \mathrm{R}_n(X, Y)$.

If, moreover, each bounded connected component of $\mathbb{C} \setminus Y$ *is not a Nevanlinna domain, then* $\mathrm{A}_n(X) = \mathrm{P}_n(X)$.

Define the function $\bar{\mathfrak{z}}$ as $\bar{\mathfrak{z}}(z) := \bar{z}$. The proof of Theorem 3 is essentially based on the following lemma.

Lemma 1. *Let Ω be a Carathéodory domain and $K \subset \Omega$ be a compact set. Assume that there exists a measure σ on $K \cup \partial\Omega$ such that $\sigma_{|\partial\Omega} \neq 0$ and $\sigma \perp \mathcal{R}_2(\overline{\Omega})$. Then there exist some compact set $\widetilde{\Sigma}$, $\widehat{K} \subset \widetilde{\Sigma} \subset \Omega$ and $\widetilde{u}, \widetilde{v} \in \mathrm{H}^\infty(\Omega \setminus \widetilde{\Sigma})$, $\widetilde{v} \neq 0$, such that $\Omega = \Omega(\widetilde{\mathcal{F}}, \widetilde{\Sigma}) \in \mathcal{N}_{\mathrm{loc}}$, where $\widetilde{\mathcal{F}} = \widetilde{u}/\widetilde{v}$ and such that $\widetilde{u}, \widetilde{v} \in \mathrm{Hol}(\Omega \setminus \widehat{K})$.*

Proof. For $s = 0, 1$ we define $\sigma_s := \overline{\mathfrak{z}}^s \sigma$, $\mu_s := \sigma_{s|K}$ and $\nu_s := \psi(\mu_s)$. Since $\sigma_s \perp \mathcal{R}(\overline{\Omega})$ for $s = 0, 1$, then, by Proposition 3, there exist two functions $h_0, h_1 \in \mathrm{H}^1(\mathbb{D})$ such that

$$\sigma_0 = \mu_0 + (\widehat{\nu}_0 \circ \psi + h_0 \circ \psi)\,\omega, \quad \sigma_1 = \mu_1 + (\widehat{\nu}_1 \circ \psi + h_1 \circ \psi)\,\omega$$

and in view of the relation $\sigma_1 = \overline{\mathfrak{z}}\sigma_0$ we have

$$\overline{\mathfrak{z}}(\widehat{\nu}_0 \circ \psi + h_0 \circ \psi)\,\omega = (\widehat{\nu}_1 \circ \psi + h_1 \circ \psi)\,\omega.$$

Since $\sigma_{0|\partial\Omega} \neq 0$, then $\widehat{\nu}_0 + h_0 \not\equiv 0$ in $\mathbb{D} \setminus \widehat{\psi(K)}$ and therefore

$$\overline{\varphi(\xi)} = \frac{\widehat{\nu}_1(\xi) + h_1(\xi)}{\widehat{\nu}_0(\xi) + h_0(\xi)}$$

for almost all $\xi \in \mathbb{T}$. Since $h_0, h_1 \in \mathrm{H}^1(\mathbb{D}) \subset \mathrm{N}(\mathbb{D})$ (the Nevanlinna class) and (according to [Pr, Chapter 2, Section 2.1]) each function in the Nevanlinna class is the ratio of two functions in $\mathrm{H}^\infty(\mathbb{D})$, then it is clear that there exist $u_1, v_1 \in \mathrm{Hol}(\mathbb{D} \setminus \widehat{\psi(K)})$, u_1, v_1 are bounded outside a neighborhood of $\widehat{\psi(K)}$ and such that

$$\frac{\widehat{\nu}_1(w) + h_1(w)}{\widehat{\nu}_0(w) + h_0(w)} = \frac{u_1(w)}{v_1(w)}$$

for all $w \in \mathbb{D} \setminus \widehat{\psi(K)}$. Observe, that $\varphi(\widehat{\psi(K)}) = \widehat{K}$. These facts allow us to define the functions $\widetilde{u}, \widetilde{v}$ and $\widetilde{\mathcal{F}}$ in $\Omega \setminus \widehat{K}$ by the formulas $\widetilde{u}(z) := u_1(\psi(z))$, $\widetilde{v}(z) := v_1(\psi(z))$ and $\widetilde{\mathcal{F}} := \widetilde{u}/\widetilde{v}$ respectively. $\qquad\square$

Proof of Theorem 3. At the beginning we prove the equality $\mathrm{A}_n(X) = \mathrm{R}_n(X, Y)$. Proceeding by contradiction, assume that $\mathrm{R}_n(X, Y) \neq \mathrm{A}_n(X)$. Then there exists a measure $\sigma \neq 0$ on X such that $\sigma \perp \mathrm{R}_n(X, Y)$ but $\sigma \not\perp \mathrm{A}_n(X)$.

Since $\sigma \perp \mathcal{R}_n(Y)$, then $\sigma, \overline{z}\sigma \perp \mathcal{R}(Y)$. By Corollary 3 one has

$$\sigma = \sum_{\Omega \in \mathfrak{S}(Y)} \sigma_\Omega \quad \text{and} \quad \overline{z}\sigma = \sum_{\Omega \in \mathfrak{S}(Y)} \sigma_{1,\Omega},$$

where $\sigma_\Omega = \sigma_{|\overline{\Omega}} \perp \mathcal{R}(\overline{\Omega})$, $\sigma_{1,\Omega} = (\overline{z}\sigma)_{|\overline{\Omega}} \perp \mathcal{R}(\overline{\Omega})$ for every $\Omega \in \mathfrak{S}(Y)$. Since σ_Ω is the restriction of σ to $\overline{\Omega}$ and $\sigma_{1,\overline{\Omega}}$ is the restriction of $\overline{z}\sigma$ to $\overline{\Omega}$, then $\sigma_{1,\Omega} = \overline{z}\sigma_\Omega$ and therefore, $\sigma_\Omega \perp \mathcal{R}_2(\overline{\Omega})$ for every $\Omega \in \mathfrak{S}(Y)$.

Since $K \subset Y^\circ$, then the set $\mathfrak{S}_K$ is finite. Take $\Omega \in \mathfrak{S}(Y) \setminus \mathfrak{S}_K$ so that $\Omega \notin \mathcal{N}$. We have σ_Ω is a measure on $\partial\Omega$ which gives (in view of [CFP, Theorem 2.2(1)] and the properties $\sigma_\Omega \perp \mathcal{R}_2(\overline{\Omega})$ and $\Omega \notin \mathcal{N}$) that $\sigma_\Omega \equiv 0$.

Therefore,

$$\sigma = \sum_{\Omega \in \mathfrak{S}_K} \sigma_\Omega = \sum_{\Omega \in \mathfrak{S}_K} \sigma_{|\overline{\Omega}}.$$

Put $\Gamma := \bigcup_{\Omega \in \mathfrak{S}_K} \partial\Omega$. If $\sigma_{|\Gamma} \equiv 0$, then σ is supported on K but it gives a contradiction with the assumption that $\mathrm{P}_n(K) = \mathrm{A}_n(K)$. Therefore, there exists at least one component $\Omega \in \mathfrak{S}_K$ such that $\sigma_{|\partial\Omega} \not\equiv 0$. Suppose that Ω is one of them. Applying Lemma 1 for Ω, K_Ω and σ_Ω we conclude, that $\Omega \in \mathcal{N}_{\mathrm{loc}}$. Since $\Omega \in \mathfrak{S}_K$, then Ω possesses the condition (1) or (2). If Ω possesses the condition (1), then $\Omega \notin \mathcal{N}_{\mathrm{loc}}$ and this contradiction implies, that $\sigma_\Omega \equiv 0$.

Let Ω possesses the condition (2) and let $\widetilde{\Sigma}$, $\widetilde{u}$ and $\widetilde{v}$ are taken from Lemma 1 for Ω, K_Ω and σ_Ω under consideration. By Proposition 5, $\mathcal{F} = \widetilde{\mathcal{F}}$ in $\widehat{\Omega \setminus \Sigma \cup \widetilde{\Sigma}}$ and since $\widetilde{u}, \widetilde{v} \in \mathrm{Hol}(\Omega \setminus \widehat{K_\Omega})$, then u/v can be continued meromorphically to $\Omega \setminus \widehat{K_\Omega}$. This fact contradicts to the condition (2) and therefore, $\sigma_\Omega \equiv 0$.

Hence, we have that $\sigma_\Omega \equiv 0$ for any $\Omega \in \mathfrak{S}_K$. Thus, $\sigma_\Omega \equiv 0$ for any $\Omega \in \mathfrak{S}(Y)$ which means that $\sigma \equiv 0$. But we have assumed that $\sigma \not\equiv 0$ and this contradiction ends the proof of the equality $\mathrm{R}_n(X, Y) = \mathrm{A}_n(X)$.

Suppose now that each bounded connected component of $\mathbb{C} \setminus Y$ is not a Nevanlinna domain. In order to prove that $\mathrm{P}_n(X) = \mathrm{A}_n(X)$ we need to show that $\mathrm{P}_n(X) = \mathrm{R}_n(X, Y)$. This fact may be verified repeating word-to-word the first part of the proof of [CFP, Theorem 2.2(2)] (from the beginning until Proposition 2.5). Therefore, the proof is completed. $\qquad\square$

Corollary 4. *Let Y be a Carathéodory compact set such that $Y^\circ \neq \varnothing$. Then $\mathrm{C}(\partial Y) = \mathrm{R}_2(\partial Y, Y)$ if and only if each connected component of Y° is not a Nevanlinna domain.*

Theorem 4. *Let Ω be a bounded simply connected domain in $\mathbb{C}$ and $\Gamma := \partial\Omega$.*

(1) *If $\Omega \in \mathcal{N}$, then $\mathrm{R}_n(\Gamma, \overline{\Omega}) \neq \mathrm{C}(\Gamma)$ for every integer $n \geq 1$.*

(2) *If $\Omega \in \mathcal{N}_{\mathrm{loc}} \setminus \mathcal{N}$ and there exists a rectifiable gate γ for Ω, then $\mathrm{R}_2(\Gamma, \overline{\Omega}) = \mathrm{C}(\Gamma)$.*

(3) *Let Ω and γ be as in (2). Let $X \subset \overline{\Omega}$ be a compact set such that $\Gamma \subseteq X$. Put $K := \overline{X \setminus \Gamma}$ and assume, that*

    (i) *$\Omega \setminus \widehat{K}$ is connected and γ is a gate for $\Omega \setminus \widehat{K}$.*

    (ii) *$\Omega = \Omega(\mathcal{F}, \Sigma)$ and $\mathcal{F}$ cannot be meromorphically continued from $\Omega \setminus \widehat{K \cup \Sigma}$ to $\Omega \setminus \widehat{K}$;*

    (iii) *$\mathrm{R}_n(K, \overline{\Omega}) = \mathrm{A}_n(K)$ for some integer $n \geq 2$.*

Then, $\mathrm{R}_n(X, \overline{\Omega}) = \mathrm{A}_n(X)$.

Proof. (1) Since $\Omega \in \mathcal{N}$, then there exist $u, v \in \mathrm{H}^\infty(\Omega)$, $v \neq 0$ such that (4.1) holds. Following the scheme using in the proof of the same implication in [CFP, Theorem 2.2(1)] it is possible to show that *for each rational function g having at least one pole in $\{z \in \Omega : v(z) \neq 0\}$ one has $g_{|\Gamma} \notin \mathrm{R}_n(\Gamma, \overline{\Omega})$ for every integer $n \geq 1$.*

(2) This case is included in (3) when $X = \Gamma$.

(3) Suppose that $R_n(X, \overline{\Omega}) \neq A_n(X)$. Then there exists a nonzero measure μ on X such that $\mu \perp R_n(X, \overline{\Omega})$ but $\mu \not\perp A_n(X)$. Consider the function

$$T_\mu(z) := \frac{1}{2\pi i} \int_X \frac{\overline{\zeta} - \overline{z}}{\zeta - z} d\mu(\zeta).$$

which has the following properties (see [TW]): T_μ is continuous except countable many points (that are the atoms of μ), $T_\mu(z) = 0$ for all $z \notin \overline{\Omega}$ and $\overline{\partial}^2 T_\mu = \frac{i}{2}\mu$ in the distributional sense.

By the continuity property of T_μ mentioned above $T_\mu(\zeta) = 0$ for all $\zeta \in \Gamma$ except the atoms of μ. Since γ is a rectifiable gate for Ω, then, by Proposition 4, $\mu_{|\gamma} \ll d\zeta_{|\gamma}$, so that μ has no atoms on γ and hence $T_\mu(\zeta) = 0$ for all $\zeta \in \gamma$. Put $\eta := \overline{3}\mu$ and observe that

$$T_\mu(z) = \widehat{\eta}(z) - \overline{z}\widehat{\mu}(z)$$

for all $z \in \mathbb{C} \setminus X$. So that, in order to obtain some further conclusions we need an additional information about behavior of holomorphic components $\widehat{\eta}$ and $\widehat{\mu}$ of the bianalytic function T_μ in $\mathbb{C} \setminus X$. It turns out, that $\widehat{\eta}$ and $\widehat{\mu}$ have for almost all $\zeta \in \gamma$ the finite angular limits $\widehat{\eta}(\zeta)$ and $\widehat{\mu}(\zeta)$ from Ω. This fact (it can be found in [Dn, Theorem 2.22]) is the result of a long-time difficult investigation of behavior of Cauchy transforms of measures (see [Dn, Ve]). Therefore, for almost all $\zeta \in \gamma$ we have

$$T_\mu(\zeta) = \widehat{\eta}(\zeta) - \overline{\zeta}\widehat{\mu}(\zeta) = \widehat{\eta}(\zeta) - \widetilde{\mathcal{F}}(\zeta)\widehat{\mu}(\zeta) = 0. \tag{4.2}$$

Let us show that $\widehat{\mu}(z) \not\equiv 0$ in $\Omega \setminus \widehat{K}$. Otherwise, $\widehat{\mu}(\zeta) = 0$ for almost all $\zeta \in \gamma$. By (4.2), $\widehat{\eta}(\zeta) = 0$ for almost all $\zeta \in \gamma$, so that $\widehat{\eta} = 0$ in $\Omega \setminus \widehat{K}$. This gives $T_\mu = 0$ in $\Omega \setminus \widehat{K}$. Since $\overline{\partial}^2 T_\mu = \frac{i}{2}\mu$ (in distributional sense), then μ is supported on K. Reminding the initial assumptions about μ we have $\mu \neq 0$, $\mu \perp R_n(K, \overline{\Omega})$ and $\mu \not\perp A_n(K)$ which is a contradiction with (iii).

Then $\widehat{\mu} \neq 0$ in $\Omega \setminus \widehat{K}$ and therefore $\widehat{\mu}(\zeta) \neq 0$ for almost all $\zeta \in \gamma$. This fact and (4.2) imply that for almost all $\zeta \in \gamma$ we have $\mathcal{F}(\zeta) = \widehat{\eta}(\zeta)/\widehat{\mu}(\zeta)$. Then, by Luzin-Privalov theorem, $\mathcal{F}$ must coincides with $\widehat{\eta}/\widehat{\mu}$ in $\Omega \setminus \widehat{K \cup \Sigma}$. Therefore, $\mathcal{F}$ can be meromorphically continued from $\Omega \setminus \widehat{K \cup \Sigma}$ to $\Omega \setminus \widehat{K}$ which gives the desired contradiction. $\qquad\square$

Let $\Gamma_{a,b}$ be the ellipse with semi-axes a and b focused at points ± 1. As it was shown in [CFP, § 3], $D(\Gamma_{a,b}) \in \mathcal{N}_{\mathrm{loc}} \setminus \mathcal{N}$ and the respective Σ may be chosen as $\Sigma = [-1, 1]$. By [CFP, Example 4.5] for $X_0 := \Gamma_{a,b} \cup [-1, 1]$ one has $P_2(X_0) \neq C(X_0) = P_3(X_0)$.

Example 3.

(1) Put $X_1 := \Gamma_{a,b} \cup \{z \in D(\Gamma_{a,b}) : |\mathrm{Im}\, z| \geq (b/2))\}$. Then, by Theorem 4(3), $P_n(X_1) = A_n(X_1)$ for every integer $n \geq 2$.

(2) Take $X_2 := \Gamma_{a,b} \cup [-a, a]$ and $X_3 := \Gamma_{a,b} \cup [-ib, ib]$. Since $X_0 \subset X_2$, then $P_2(X_2) \neq C(X_2)$, but, by Theorem 4(3), $P_2(X_3) = C(X_3)$.

Example 4. Paramonov have showed us the example of the domain $\Omega_1 = \varphi_1(\mathbb{D})$, where $\varphi_1(w) := (w + \sqrt{2})^4$. It follows from [CFP, Proposition 3.1], that $\Omega_1 \in \mathcal{N}$. Therefore, by Theorem 4(1) one has $R_n(\partial\Omega_1, \overline{\Omega}_1) \neq C(\partial\Omega_1)$ for every integer $n \geq 1$. Following the same idea we consider the domain $\Omega_2 = \varphi_2(D(\Gamma_{\sqrt{2},1}))$, where $\varphi_2(w) = (w + \sqrt{3})^4$, which is in $\mathcal{N}_{\mathrm{loc}} \setminus \mathcal{N}$ and hence, by Theorem 4(2), $R_2(\partial\Omega_2, \overline{\Omega}_2) = C(\partial\Omega_2)$. Observe, that Ω_1 and Ω_2 are not a Carathéodory domains so that we need to refer Theorem 4 in order to conclude the desired approximability properties.

We end this section establishing one rigidity property of Nevanlinna domains. Usually, an algebraic curve is the set in $\mathbb{C}^2$ of the form $\{(w_1, w_2) \in \mathbb{C}^2 : Q(w_1, w_2) = 0\}$, where $Q \in \mathbb{C}[w_1, w_2]$, but here we mean by an algebraic planar curve the set $\{z \in \mathbb{C} : (z, \overline{z}) \in L\}$, where L is an algebraic curve in the usual sense.

Recall that an arc γ is *analytic* if it is the image of $[0, 1]$ under a map which is conformal in some neighborhood of $[0, 1]$. In such a case there exist a neighborhood U of γ and a function $S \in \mathrm{Hol}(U)$ such that $\overline{\zeta} = S(\zeta)$ on γ. This function S is called the *Schwarz function* of γ (see [Da]).

Theorem 5. *Let Ω be a bounded simply connected domain and L be an algebraic planar curve. If*

(i) $\Omega \in \mathcal{N}$ *and*

(ii) *there exists an arc $\lambda \subset L \cap \partial\Omega$ such that $\omega(z, \lambda, \Omega) > 0$ (for some $z \in \Omega$),*

then $\partial\Omega$ is analytic and $\partial\Omega \subset L$.

Proof. Since L is an algebraic planar curve, then there exists $Q \in \mathbb{C}[w_1, w_2]$ such that $L = \{z \in \mathbb{C} : Q(z, \overline{z}) = 0\}$. By [Se, Corollary 6.4], L has at most a finite number of singular points. Then it is possible to find a subarc λ_1 of λ such that $\omega(z, \lambda_1, \Omega) > 0$ and all points of λ_1 are regular for L. Take a point $z_0 \in \lambda_1$. Without loss of generality we assume that $\partial Q/\partial w_2(z_0, \overline{z_0}) \neq 0$. By the implicit function theorem there exists an open set $U \ni z_0$ and a function $S \in \mathrm{Hol}(U)$ such that

$$U \cap \lambda_1 = \{z \in U : \overline{z} = S(z), \text{ and } S'(z) \neq 0\}$$

which means that there exists an arc $\lambda_2 \subset U \cap \lambda_1$ such that λ_2 is an analytic arc having the Schwarz function S and $\omega(z, \lambda_2, \Omega) > 0$. In particular, L is analytic except at most finite number of points.

If $\mathcal{F}$ is taken for Ω from Definition 2, then for almost all $\zeta \in \lambda_2$ we have $\mathcal{F}(\zeta) = \overline{\zeta} = S(\zeta)$ and, consequently,

$$Q(\zeta, \mathcal{F}(\zeta)) = Q(\zeta, \overline{\zeta}) = Q(\zeta, S(\zeta)) = 0.$$

Since $\omega(z, \lambda_2, \Omega) > 0$, then, by Luzin-Privalov theorem, the meromorphic function $Q(z, \mathcal{F}(z)) = 0$ for all $z \in \Omega$. Hence, for all $\zeta \in \varphi(E)$, where $E = F(\varphi) \cap F(\varphi \circ u) \cap F(\varphi \circ v)$ and u and v are taken for Ω from Definition 2, one has $Q(\zeta, \overline{\zeta}) = Q(\zeta, \mathcal{F}(\zeta)) = 0$. Since $|d\xi|(E) = 2\pi$, then $\overline{\varphi(E)} = \partial\Omega$ and then $Q(\zeta, \overline{\zeta}) = 0$ for all

$\zeta \in \partial\Omega$. This means that $\partial\Omega \subset L$, so that $\partial\Omega$ is analytic except at most finite number of singular points.

For each such point a, by [Ba, Corollary 2.1], there exists an open set $U \ni a$ such that $L \cap U$ is a finite union of at least two analytic arcs passing through a. Take two such arcs having the Schwarz functions S_1 and S_2. Since $\Omega \in \mathcal{N}$, then S_1 and S_2 are analytically continuations of each other which is impossible because the respective arcs are intersected. Then, $\partial\Omega$ is locally analytic and $\Omega \in \mathcal{N}$ implies that $\partial\Omega$ is analytic. $\qquad\qquad\Box$

If in Theorem 5 we additionally assume, that L is a contour in $\mathbb{C}$, then we have $\Omega = D(L)$.

References

[Ba] M.B. Balk. *Polyanalytic Functions*, Mathematical Research, **63**, Akademie Verlag, Berlin, 1991.

[Bi1] E. Bishop. *The structure of certain measures*, Duke Math. J., 1958, **25** (2), 283–289.

[Bi2] E. Bishop. *Boundary measures of analytic differentials*, Duke Math. J., 1960, **27** (3), 331–340.

[BGP1] A. Boivin, P.M. Gauthier and P.V. Paramonov. *Approximation on closed sets by analytic or meromorphic solutions of elliptic equations and applications*, Canadian J. of Math., 2002, **54** (5), 945–969.

[BGP2] A. Boivin, P.M. Gauthier and P.V. Paramonov. *Uniform approximation on closed subsets of $\mathbb{C}$ by polyanalytic functions*, Izvestia RAN, 2004, **68** (3), 15–28 (in Russian).

[Ca] J.J. Carmona. *Mergelyan approximation theorem for rational modules*, J. Approx. Theory., 1985, **44**, 113–126.

[CFP] J.J. Carmona, K.Yu. Fedorovskiy and P.V. Paramonov. *On uniform approximation by polyanalytic polynomials and Dirichlet problem for bianalytic functions*, Sb. Math., 2002, **193** (10), 1469–1492.

[Da] P. Davis. *The Schwarz functions and its applications*, Carus Mathematical Monographs, **17**, 1974.

[Dn] E.M. Dyn'kin. *Methods of the theory of singular integrals: Hilbert transform and Calderon-Zygmund theory*, Commutative harmonic analysis, I, 167–259, Encyclopaedia Math. Sci., **15**, Springer, Berlin, 1991.

[Ga] T.W. Gamelin *Uniform algebras*, Prentice-Hall, Inc., Englewood Cliffs, N. J., 1969.

[Go] G.M. Goluzin. *Geometric theory of functions of a complex variable. 2-nd edition*, "Nauka", Moscow, 1966; English translation: Amer. Math. Soc., Providence, R.I., 1969.

[Fe1] K.Yu. Fedorovski. *Uniform n-analytic polynomial approximations of functions on rectifiable contours in $\mathbb{C}$*, Math. Notes., 1996, **59** (4), 435–439.

[Fe2] K.Yu. Fedorovski. *Approximation and boundary properties of polyanalytic functions*, Proc. Steklov Inst. Math., 2001, **235**, 251–260.

[Ma] S. Mazurkiewicz. *Über erreichbare Punkte*, Fund. Math., 1936, **26**, 150–155.

[Me] S.N. Mergelyan. *Uniform approximation to functions of a complex variables*, Uspekhi Mat. Nauk (Russia Math. Surveys), 1952, **7** (2), 31–122.

[Po1] Ch. Pommerenke. *Boundary behavior of the conformal maps*, Springer-Verlag, Berlin, 1992.

[Po2] Ch. Pommerenke. *Univalent functions*, Vandenhoek & Ruprecht in Göttingen, 1973.

[Po3] Ch. Pommerenke. *Conformal maps at the boundary*, Handbook of Complex analysis: Geometric Function Theory. Volume 1, 2002.

[Pr] I.I. Privalov. *Boundary properties of analytic functions*, M.-L., Gostehizdat, 1950. (Russian).

[Se] A. Seidenberg. *Elements of the theory of algebraic curves*, Addison Wesley, 1968.

[TW] T. Trent and J.L-M. Wang. *Uniform approximation by rational modules on nowhere dense sets*, Proc. Amer. Math. Soc., 1981, **81** (1), 62–64.

[Ve] J. Verdera. L^2 *boundedness of the Cauchy integral and Menger curvature*, Harmonic analysis and boundary value problems, 139–158, Contemp. Math., **277**, Amer. Math. Soc., Providence, RI, 2001.

[Wa] J.L. Wang. *A localization operator for rational modules*, Rocky Mountain J. of Math., 1989, **19** (4), 999–1002.

[Za] A.B. Zajtsev. *On uniform approximation of functions by polynomial solutions of second-order elliptic equations on plane compact sets*, Izvestia RAN, 2004, **68** (6), 85–98 (in Russian).

Joan Josep Carmona
Departament de Matemàtiques
Universitat Autònoma de Barcelona
e-mail: jcar@mat.uab.es

Konstantin Yu. Fedorovskiy
Department of the Computer Science
Russian State University of Management
e-mail: const@imis.ru, const@fedorovski.mccme.ru

Operator Theory:
Advances and Applications, Vol. 158, 131–139
© 2005 Birkhäuser Verlag Basel/Switzerland

Capacities of Generalized Cantor Sets

V.Ya. Eiderman

In memoriam of my teacher Semyon Yakovlevich Khavinson

1. Introduction

Our main goal is to obtain sharp estimates of capacities of Cantor sets in $\mathbb{R}^m$, $m \geq 1$, that are obtained as the Cartesian product of one-dimensional (in general, different) Cantor sets. As an application we deduce the criterion for vanishing of the capacity for such sets.

Let $K(t)$, $t \in (0, \infty)$, be a nonnegative, non-increasing and continuous function. The capacity $C_K(E)$ of a bounded Borel set $E \subset \mathbb{R}^m$ is defined by

$$C_K(E) = \sup \mu(E),$$

where the supremum is taken over all nonnegative measures μ supported in E such that $U^\mu(x) := \int_{\mathbb{R}^m} K(|x - y|)\, d\mu(y) \leq 1$, $x \in E$. This definition is meaningful if $K(t) \to +\infty$ as $t \to 0+$ and

$$\int_0 K(t) t^{m-1} dt < \infty. \tag{1.1}$$

Now we define Cantor sets. Suppose that sequences $\{k_j^{(s)}\}_{j=0}^\infty$, $s = 1, \ldots, m$, of positive integers and sequences $\{l_j^{(s)}\}_{j=0}^\infty$, $s = 1, \ldots, m$, of positive numbers are given with $k_0^{(s)} = 1$, $s = 1, \ldots, m$, and

$$k_j^{(s)} \geq 2, \quad k_j^{(s)} l_j^{(s)} \leq l_{j-1}^{(s)}, \quad j \geq 1, \quad s = 1, \ldots, m.$$

Fix s and let $E_0^{(s)} = [0, l_0^{(s)}]$. Suppose that the set $E_{n-1}^{(s)}$, $n \geq 1$, has already been constructed and this set consists of $k_0^{(s)} \cdots k_{n-1}^{(s)}$ closed intervals of length $l_{n-1}^{(s)}$.

The author was partially supported by the Russian Foundation of Basic Research (Grant no 01-01-00608).

For $n \geq 1$, the set $E_n^{(s)}$ is obtained from $E_{n-1}^{(s)}$ by replacement of each of these intervals (say, $[a_t, b_t]$) by the union of $k_n^{(s)}$ equidistant intervals of length $l_j^{(s)}$:

$$[a_t, a_t + l_j^{(s)}], \ldots, [b_t - l_j^{(s)}, b_t].$$

It is possible that $E_n^{(s)} = E_{n-1}^{(s)}$, but the intersection of the intervals of $E_n^{(s)}$ can only contain some of the endpoints of these intervals. We set

$$E_n = E_n(\{k_j^{(s)}\}, \{l_j^{(s)}\}) = E_n^{(1)} \times E_n^{(2)} \times \cdots \times E_n^{(m)},$$

and

$$E = \bigcap_{n=0}^{\infty} E_n. \tag{1.2}$$

If $E_n^{(1)} = \cdots = E_n^{(m)}$ for all n the set E is called a generalized symmetric Cantor set. In this case we omit the upper index (s) in $k_j^{(s)}, l_j^{(s)}$.

Estimation of the capacity $C_K(E)$ of Cantor sets is an important ingredient of investigation in various problems in analysis (see for example [2], [4], [6], [7]). Ohtsuka [8] obtained the criterion for vanishing of $C_K(E)$ which is equivalent to the following theorem:

Theorem A. [8] *Let E be a generalized symmetric Cantor set and let either $K(t) = t^{-\alpha}$, $0 < \alpha < m$, or $K(t) = \log(1/t)$. Then,*

$$C_K(E) = 0 \iff \sum_{j=0}^{\infty} (k_0 \cdot \cdots \cdot k_j)^{-m} K(l_j) = \infty. \tag{1.3}$$

In [3, Lemma 3.3] it was shown that the criterion (1.3) does not take place in general (even under condition (1.1)). On the other hand, if $k_j \equiv 2$ then (1.3) holds without any additional conditions on K [2]. For $m = 1$ and $k_j \equiv$ const the same criterion was established in [1]. In [3, Corollary 3.1] we proved that (1.3) holds for all K and for bounded sequences $\{k_j\}$. A refinement of these results was given in [5]. In this paper we obtain the estimates and the criterion for vanishing of the capacity $C_K(E)$ which is valid without any additional assumptions on a kernel K and sequences $\{k_j^{(s)}\}$. As a corollary we deduce the results mentioned above.

Remark in passing that Monterie [7] considered a more general class of Cantor sets than the Cartesian product of one-dimensional Cantor sets, and obtained certain necessary as well as sufficient conditions of positivity of the capacity in the case $K(t) = \log \frac{1}{t}$. In the special cases there have also been obtained some necessary and sufficient conditions. These results and our theorems do not imply each other.

2. Main results

For each $s = 1, 2, \ldots, m$ we introduce the function

$$
\varphi_{s,n}(t) = \begin{cases}
(k_0^{(s)} \cdots \cdot k_n^{(s)} l_n^{(s)})^{-1} t, & 0 \le t < l_n^{(s)}, \\
(k_0^{(s)} \cdots \cdot k_{j+1}^{(s)})^{-1}, & l_{j+1}^{(s)} \le t \le l_j^{(s)}/k_{j+1}^{(s)}, \\
(k_0^{(s)} \cdots \cdot k_j^{(s)} l_j^{(s)})^{-1} t, & l_j^{(s)}/k_{j+1}^{(s)} \le t \le l_j^{(s)}, \\
& j = n-1, n-2, \ldots, 0, \\
1, & l_0^{(s)} \le t < \infty.
\end{cases}
$$

Clearly, $\varphi_{s,n}(t)$ is a continuous non-decreasing function for $0 \le t < \infty$. We set

$$
\Phi_n(t) = \varphi_{1,n}(t) \cdots \cdot \varphi_{m,n}(t).
$$

Theorem 2.1. *Suppose that $K(t)$, $t > 0$, is a nonnegative, non-increasing and continuous function satisfying (1.1) and such that $K(t) \to +\infty$ as $t \to 0+$. There exists a constant $A > 1$ depending only on the dimension m, for which*

$$
A^{-1} \left[\int_0^\infty K(t)\, d\Phi_n(t) \right]^{-1} \le C_K(E_n) \le A \left[\int_0^\infty K(t)\, d\Phi_n(t) \right]^{-1}. \tag{2.1}
$$

In other words,

$$
C_K(E_n) \approx \left[\int_0^\infty K(t)\, d\Phi_n(t) \right]^{-1}.
$$

Proof. Let μ_n be the probability measure uniformly distributed on E_n. Fix $x = (x^{(1)}, \ldots, x^{(m)}) \in E_n$ and $t > 0$ and show that

$$
\mu_n(B(x,t)) \approx \Phi_n(t), \quad 0 < t < \infty, \tag{2.2}
$$

where $B(x,t) = \{\zeta \in \mathbb{R}^m : |\zeta - x| < t\}$. Fix $s \in [1,m]$ and suppose that $l_{j+1}^{(s)} \le t < l_j^{(s)}$ with some $j \in [0, n-1]$. Let $M_{j+1}^{(s)}$, $N_{j+1}^{(s)}$ be the numbers of intervals of length $l_{j+1}^{(s)}$ forming $E_{j+1}^{(s)}$ and intersecting or contained in the interval $[x^{(s)} - t, x^{(s)} + t]$, respectively. Then

$$
M_{j+1}^{(s)} \approx N_{j+1}^{(s)} \approx \begin{cases}
1, & l_{j+1}^{(s)} \le t \le l_j^{(s)}/k_{j+1}^{(s)}, \\
t k_{j+1}^{(s)}/l_j^{(s)}, & l_j^{(s)}/k_{j+1}^{(s)} \le t \le l_j^{(s)},
\end{cases}
$$

with absolute constants of comparison. Every interval from $E_{j+1}^{(s)}$, $j+1 = 0, 1, \ldots,$ $n-1$, contains $k_{j+2}^{(s)} \cdots \cdot k_n^{(s)}$ intervals of length $l_n^{(s)}$. Denoting by $\mathcal{H}^p(G)$ the p-dimensional Hausdorff measure of a set G, we have

$$
H^{(s)}(t) := \mathcal{H}^1(E_n^{(s)} \cap [x^{(s)} - t, x^{(s)} + t]) \approx N_{j+1}^{(s)} k_{j+2}^{(s)} \cdots \cdot k_n^{(s)} l_n^{(s)}
$$

$$
= k_0^{(s)} \cdots \cdot k_n^{(s)} l_n^{(s)} \varphi_{s,n}(t) = \mathcal{H}^1(E_n^{(s)}) \varphi_{s,n}(t), \quad l_{j+1}^{(s)} \le t < l_j^{(s)}.
$$

Obviously,

$$
H^{(s)}(t) \approx \begin{cases}
t = \mathcal{H}^1(E_n^{(s)}) \varphi_{s,n}(t), & 0 \le t < l_n^{(s)}, \\
k_0^{(s)} \cdots \cdot k_n^{(s)} l_n^{(s)} = \mathcal{H}^1(E_n^{(s)}) \varphi_{s,n}(t), & l_0^{(s)} \le t < \infty.
\end{cases}
$$

134
V.Ya. Eiderman

Thus,

$$H^{(s)}(t) \approx \mathcal{H}^1(E_n^{(s)})\varphi_{s,n}(t), \quad 0 \le t < \infty.$$

Clearly,

$$\mathcal{H}^m(E_n \cap B(x,t)) \approx H^{(1)}(t) \cdots \cdots H^{(m)}(t),$$

where constants of comparison depend only on m. The density of the measure μ_n is equal to $1/\mathcal{H}^m(E_n) = [\mathcal{H}^1(E_n^{(1)}) \cdots \cdots \mathcal{H}^1(E_n^{(m)})]^{-1}$. Therefore,

$$\mu_n(B(x,t)) = \mathcal{H}^m(E_n \cap B(x,t))/\mathcal{H}^m(E_n) \approx \Phi_n(t),$$

and (2.2) is proved.

Without loss of generality we may assume that $K(t) \to 0$ as $t \to +\infty$. The condition (1.1) implies that $K(t)t^m \to 0$ as $t \to 0+$. Hence,

$$\lim_{t\to 0+} K(t)\mu_n(B(x,t)) = \lim_{t\to 0+} K(t)\Phi_n(t) = 0,$$

$$\lim_{t\to\infty} K(t)\mu_n(B(x,t)) = \lim_{t\to\infty} K(t)\Phi_n(t) = 0.$$

Integrating by parts we obtain

$$U^{\mu_n}(x) := \int_{E_n} K(|x-y|)\,d\mu_n(y) = \int_0^\infty K(t)\,d\mu_n(B(x,t))$$

$$= -\int_0^\infty \mu_n(B(x,t))\,dK(t) \approx -\int_0^\infty \Phi_n(t)\,dK(t) = \int_0^\infty K(t)\,d\Phi_n(t).$$

$$(2.3)$$

This relation directly implies the first inequality in (2.1).

In order to obtain the second inequality in (2.1) we consider a measure ν_n concentrated on E_n and such that $U^{\nu_n}(x) \le 1$ on E_n. By (2.3),

$$1 < cP_n U^{\mu_n}(x) \quad \forall x \in E_n,$$

where $P_n = [\int_0^\infty K(t)\,d\Phi_n(t)]^{-1}$ and the constant c depends only on m. Hence,

$$\nu_n(E_n) < cP_n \int_{E_n} U^{\mu_n}\,d\nu_n = cP_n \int_{E_n} U^{\nu_n}\,d\mu_n \le cP_n.$$

It follows from this estimate that $C_K(E_n) \le cP_n$, and the proof of Theorem 2.1 is complete. $\square$

Corollary 2.2. *Suppose that $K(t)$ satisfies the conditions of Theorem 2.1 and that E is a Cantor set defined by (1.2). Then*

$$C_K(E) \approx \left[\int_0^\infty K(t)\,d\Phi(t)\right]^{-1},$$

$$(2.4)$$

where $\Phi(t) = \varphi_1(t) \cdots \cdots \varphi_m(t)$ and

$$\varphi_s(t) = \begin{cases} (k_0^{(s)} \cdots \cdots k_{j+1}^{(s)})^{-1}, & l_{j+1}^{(s)} \le t \le l_j^{(s)}/k_{j+1}^{(s)}, \\ (k_0^{(s)} \cdots \cdots k_j^{(s)} l_j^{(s)})^{-1}t, & l_j^{(s)}/k_{j+1}^{(s)} \le t \le l_j^{(s)}, \\ & j = 0,1,\dots, \\ 1, & l_0^{(s)} \le t < \infty. \end{cases}$$

Proof. By definitions of functions Φ_n and Φ,

$$\Phi_n(t) \le \Phi(t) \quad \text{for } t > 0 \text{ and } n = 1, 2, \ldots$$

Hence,

$$P_n := \left[\int_0^\infty K(t)\, d\Phi_n(t) \right]^{-1} \ge \left[\int_0^\infty K(t)\, d\Phi(t) \right]^{-1} =: P.$$

This inequality and the first inequality in (2.1) yield

$$A^{-1} P \le C_K(E).$$

Indeed, it is enough to take the equilibrium measures ν_n^* of compact sets E_n and to extract a weakly convergent subsequence.

Let ν be a measure supported in E such that $U^\nu(x) \le 1$ for all $x \in E$. By the generalized maximum principle [9] $U^\nu(x) \le d_m \ \forall x \in \mathbb{R}^m$, where d_m depends only on m. Hence,

$$C_K(E) \le d_m C_K(E_n), \quad n = 0, 1, \ldots$$

Moreover, $\lim_{n \to \infty} P_n = P$. These relations and the second inequality in (2.1) imply

$$C_K(E) \le A d_m P,$$

as required. $\qquad\qquad\qquad\qquad\qquad\qquad\qquad\qquad\qquad\qquad\qquad\qquad\square$

Corollary 2.3. *Under assumptions of Corollary 2.2*

$$C_K(E) = 0 \iff \int_0^\infty K(t)\, d\Phi(t) = \infty. \tag{2.5}$$

3. The case of generalized symmetric Cantor sets

In this section we assume that

$$k_j^{(1)} = \cdots = k_j^{(m)} =: k_j, \quad l_j^{(1)} = \cdots = l_j^{(m)} =: l_j, \quad j = 0, 1, \ldots \tag{3.1}$$

Theorem 2.1 and Corollary 2.2 immediately give the following

Theorem 3.1. *Suppose that $K(t)$ satisfies the conditions of Theorem 2.1 and (3.1) holds. Then*

$$C_K(E_n) \approx \left[(k_0 \cdots\cdots k_n l_n)^{-m} \int_0^{l_n} K(t)\, dt^m \right.$$

$$\left. + \sum_{j=0}^{n-1} (k_0 \cdots\cdots k_j l_j)^{-m} \int_{l_j/k_{j+1}}^{l_j} K(t)\, dt^m \right]^{-1}, \tag{3.2}$$

$$C_K(E) \approx \left[\sum_{j=0}^{\infty} (k_0 \cdots\cdots k_j l_j)^{-m} \int_{l_j/k_{j+1}}^{l_j} K(t)\, dt^m \right]^{-1}. \tag{3.3}$$

Since $l_{j+1} \le l_j/k_{j+1}$, trivial estimates of integrals in (3.2), (3.3) imply the following inequalities.

136   V.Ya. Eiderman

Corollary 3.2. *Under assumptions of Theorem* 3.1

$$c^{-1}\left[(k_0\cdots\cdot k_n l_n)^{-m}\int_0^{l_n} K(t)\,dt^m + \sum_{j=0}^{n-1}(k_0\cdots\cdot k_j)^{-m}K(l_{j+1})\right]^{-1}$$

$$\le C_K(E_n)\le c\left[\sum_{j=0}^{n}(k_0\cdots\cdot k_j)^{-m}K(l_j)\right]^{-1}, \tag{3.4}$$

$$c^{-1}\left[\sum_{j=0}^{\infty}(k_0\cdots\cdot k_j)^{-m}K(l_{j+1})\right]^{-1}\le C_K(E)\le c\left[\sum_{j=0}^{\infty}(k_0\cdots\cdot k_j)^{-m}K(l_j)\right]^{-1}, \tag{3.5}$$

where c depends only on m.

Corollary 3.3. *Suppose that*

$$\int_0^r K(t)\,dt^m \le \delta K(r)r^m, \quad 0 < r \le r_0 \tag{3.6}$$

for some $\delta > 0$ and $r_0 > 0$. If $l_0 \le r_0$ then

$$\delta^{-1}c^{-1}L_n \le C_K(E_n)\le cL_n, \quad where \quad L_n = \left[\sum_{j=0}^{n}(k_0\cdots\cdot k_j)^{-m}K(l_j)\right]^{-1}, \tag{3.7}$$

and c depends only on m.

Proof. Inequalities (3.7) follow directly from (3.2) and (3.6). $\square$

Corollary 3.4. *Suppose that at least one of the following conditions holds:*
1. *a kernel $K(t)$ satisfies (3.6);*
2. *a sequence $\{k_j\}$ is bounded: $k_j \le M$, $j = 1, 2, \ldots$*
   Then

$$QL \le C_K(E)\le cL, \quad where \quad L = \left[\sum_{j=0}^{\infty}(k_0\cdots\cdot k_j)^{-m}K(l_j)\right]^{-1}, \tag{3.8}$$

$Q = \delta^{-1}c^{-1}$ *in Case 1,* $Q = M^{-m}c^{-1}$ *in Case 2, and c depends only on m.*

Proof. In Case 1 inequalities (3.8) follow from (3.3). In Case 2 we have

$$\sum_{j=0}^{\infty}(k_0\cdots\cdot k_j)^{-m}K(l_{j+1})\le M^m\sum_{j=0}^{\infty}(k_0\cdots\cdot k_{j+1})^{-m}K(l_{j+1})$$

$$< M^m\sum_{j=0}^{\infty}(k_0\cdots\cdot k_j)^{-m}K(l_j).$$

This estimate and (3.5) imply (3.8) with $Q = M^{-m}c^{-1}$. $\square$

From Corollary 3.4 we immediately deduce the criterion for vanishing of the capacity for generalized symmetric Cantor sets.

Corollary 3.5. *Suppose that at least one of the conditions 1, 2 in Corollary 3.4 holds. Then*

$$C_K(E) = 0 \iff \sum_{j=0}^{\infty}(k_0 \cdots \cdots k_j)^{-m}K(l_j) = \infty. \tag{3.9}$$

Since the functions $K(t) = t^{-\alpha}$, $0 < \alpha < m$, and $K(t) = \log(1/t)$ satisfy (3.6), Corollary 3.5 generalizes Theorem A and other related results mentioned in the introduction. As we noted above, the criterion (3.9) is not correct without additional conditions on $K(t)$ or $\{k_j\}$. Hence the same remark concerns Corollary 3.4 as well. In the general situation we have the weaker estimates (3.5).

4. Another application of Theorem 2.1

An important ingredient of the proof of the main result in [6] is the following lemma which is of independent interest.

Lemma B. [6, Lemma 2.2] *Let $E^{(1)}$ be a linear Cantor set with $k_j^{(1)} = 2$, $j = 1, 2, \ldots$ Then there exists a linear Cantor set $E^{(2)}$ such that $C_{\log(1/t)}(E^{(2)}) = 0$ and $C_{\log(1/t)}(E^{(1)} \times E^{(2)}) > 0$.*

As an application of our results in Section 2 we deduce a generalization of this statement. Our proof will be simpler than the original proof in [6].

Lemma 4.1. *For every m-dimensional Cantor set E, $m \geq 1$, defined by (1.2) and for every kernel $K(t)$ satisfying conditions of Theorem 2.1 with $m = 1$ there exists a linear Cantor set G with $k_j = 2$, $j = 1, 2, \ldots$, such that $C_K(G) = 0$ and $C_K(E \times G) > 0$.*

Proof. Let $\Phi(t)$ and $\varphi(t)$ be the functions associated with E and G respectively and defined in Corollary 2.2. Clearly, $\Phi(t) \downarrow 0$ as $t \to 0+$. It is enough to construct a sequence $\{l_j\}$ such that $0 < l_{j+1} \leq 2l_j$,

$$\sum_{j=0}^{\infty} K(l_j)2^{-j} = \infty, \tag{4.1}$$

$$\sum_{j=0}^{\infty} K(l_{j+1})\Phi(l_j)2^{-j} < \infty. \tag{4.2}$$

Indeed, (4.1) implies that $C_K(G) = 0$ (see (3.9) with $m = 1$ and $k_j = 2$, $j = 1, 2, \ldots$). Moreover,

$$\int_0^{\infty} K(t)\,d[\Phi(t)\varphi(t)] = \sum_{j=0}^{\infty}\int_{l_{j+1}}^{l_j} K(t)\,d[\Phi(t)\varphi(t)]$$

$$< \sum_{j=0}^{\infty} K(l_{j+1})\Phi(l_j)\varphi(l_j) < \infty,$$

since $\varphi(l_j) = 2^{-j}$. The criterion (2.5) yields $C_K(E \times G) > 0$.

In view of the condition $\int_0 K(t)\,dt < \infty$ (i.e., (1.1) with $m = 1$),

$$\sum_{j=0}^{\infty} K(2^{-j})2^{-j} < \infty \quad \text{and} \quad tK(t) \to 0 \text{ as } t \to 0+ \, .$$

Fix $x \in (0,1]$. Let $k \geq 0$ be such that $2^{-k-1} < x \leq 2^{-k}$. Since $K(t)$ is a non-increasing function, for each $m \in \mathbb{N}$ we have

$$\sum_{j=m}^{\infty} K(x2^{-j})x2^{-j} \leq \sum_{j=m}^{\infty} K(2^{-k-1}2^{-j})2^{-k}2^{-j} = 2 \sum_{p=k+m+1}^{\infty} K(2^{-p})2^{-p}$$

$$\leq 2 \sum_{p=m}^{\infty} K(2^{-p})2^{-p} := \varepsilon_m \to 0 \quad \text{as} \quad m \to \infty.$$

Hence, the series $\sum_{j=0}^{\infty} K(x2^{-j})x2^{-j}$ converges uniformly for $x \in [0,1]$. We see that

$$S(x) = \sum_{j=0}^{\infty} K(x2^{-j})2^{-j}$$

is a continuous function for $x \in (0,1]$. Obviously, $S(x) \uparrow \infty$ as $x \to 0+$. We choose j_1 such that

$$\sum_{j=0}^{j_1} K(2^{-j})2^{-j} \geq \frac{1}{2}S(1), \quad \Phi(2^{-j_1}) < \frac{1}{2}.$$

Set $j_0 = -1$ and $l_j = 2^{-j}$, $j_0 + 1 \leq j \leq j_1$. By continuity of $S(x)$ there exists a positive number l_{j_1+1} such that

$$2^{-j_1-1}S(l_{j_1+1}) = \sum_{j=j_1+1}^{\infty} K(l_{j_1+1}2^{j_1+1-j})2^{-j} = S(1).$$

Since

$$2^{-j_1-1}S(2^{-j_1-1}) = \sum_{j=j_1+1}^{\infty} K(2^{-j_1-1}2^{j_1+1-j})2^{-j} \leq S(1),$$

we have $l_{j_1+1} \leq 2^{-j_1-1} = l_{j_1}/2$. Continuing in this way, we take j_2 for which

$$\sum_{j=j_1+1}^{j_2} K(l_{j_1+1}2^{j_1+1-j})2^{-j} \geq \frac{1}{2}S(1), \quad \Phi(2^{-j_2}) < \frac{1}{2^2},$$

set $l_j = l_{j_1+1}2^{j_1+1-j}$, $j_1 + 1 \leq j \leq j_2$, etc. We have

$$\sum_{j=0}^{\infty} K(l_j)2^{-j} = \sum_{k=0}^{\infty} \sum_{j=j_k+1}^{j_{k+1}} K(l_j)2^{-j} \geq \sum_{k=0}^{\infty} \frac{1}{2}S(1) = \infty,$$

$$\sum_{j=0}^{\infty} K(l_{j+1})\Phi(l_j)2^{-j} = 2\sum_{j=1}^{\infty} K(l_j)\Phi(l_{j-1})2^{-j} \approx \sum_{k=0}^{\infty} \frac{1}{2^k} < \infty.$$

Thus, (4.1), (4.2) are satisfied and Lemma 4.1 is proved. $\square$

References

[1] A.F. Beardon, *The generalized capacity of Cantor sets*, Quart. J. Math. Oxford (2) **19** (1968), 301–304.

[2] L. Carleson, *Selected problems on exceptional sets*, Van Nostrand, Princeton, N. J., 1967.

[3] V.Ya. Eiderman, *On the comparison of Hausdorff measure and capacity*, Algebra i Analiz **3** (1991), no. 6, 173–188. (Russian); Transl. in St. Petersburg Math. J. **3** (1992), no. 6, 1367–1381.

[4] V.Ya. Eiderman, *Estimates for potentials and δ-subharmonic functions outside exceptional sets*, Izvestiya RAN: Ser. Mat. **61** (1997), no. 6, 181–218. (Russian); Transl. in Izvestiya: Mathematics **61** (1997), no. 6, 1293–1329.

[5] V.Ya. Eiderman, *Metric properties of exceptional sets*, Complex Analysis and Differential Equations. Proceedings of the Marcus Wallenberg Symposium in Honor of Matts Essén, Held in Uppsala, Sweden, June 15–18, 1997. Uppsala Univ., 1999, 131–145.

[6] N. Levenberg, T.J. Ransford, J. Rostand, Z. Słodkowski *Countability via capacity*, Math. Z. **242** (2002), 399–406.

[7] M.A. Monterie, *Capacities of certain Cantor sets*, Indag. Math. (N.S.), **8** (1997), 247–266.

[8] M. Ohtsuka, *Capacité d'ensembles de Cantor généralisés*, Nagoya Math. J. **11** (1957), 151–160.

[9] T. Ugaheri, *On the general potential and capacity*, Jap. J. Math. **20** (1950), 37–43.

V.Ya. Eiderman
Moscow State Civil Engineering University
Yaroslavskoe Shosse 26
Moscow 129337, Russia
e-mail: `eiderman@orc.ru`

Operator Theory:
Advances and Applications, Vol. 158, 141–158

On $A^p_{\omega,\gamma}$ Spaces in the Half-Plane

A.M. Jerbashian

To Semyon Yakovlevich Khavinson, a great man and mathematician

Abstract. This paper is devoted to investigation of some Banach spaces $A^p_{\omega,\gamma}$ of functions holomorphic in the upper half-plane. The functions from the considered spaces can have arbitrary growth near the finite points of the real axis. The canonical representations of functions from these classes are established by an approach based on the use of Fourier–Laplace transform apparatus.

Mathematics Subject Classification (2000). Primary 30H05; Secondary 47B38.

Keywords. Banach spaces, holomorphic functions, representations.

It is well known that rotation-invariant growth conditions and the Fourier–Taylor expansions apparatus are the most natural ones in problems related to classes and spaces of functions regular in the unit disc. The geometric feature of the half-plane is the existence of the non-finite boundary point ∞, which makes similar problems in the half-plane essentially different. Particularly, parallel shift-invariant growth conditions and the Fourier–Laplace transform apparatus turn out to be the most natural ones in the half-plane. One has to note that the known shift-invariant growth conditions mainly consider the finite boundary points equivalently and mean a different requirement at ∞. This is evident in view of the theory of Hardy spaces in the half-plane, Nevanlinna's factorization and uniqueness theorem and Phragmén–Lindelöf principle. On the other hand, the growth description of the whole set of subharmonic functions possessing nonnegative harmonic majorants in the half-plane [4] and some other results show that some times it is natural to complement a shift-invariant growth condition by a local condition in the neighborhood of ∞. This makes natural the below definition of the general spaces $A^p_{\omega,\gamma}$ in the half-plane.

1. The spaces $A^p_{\omega,\gamma}$

1.1. We define $A^p_{\omega,\gamma}$ $(0 < p < +\infty, \ -\infty < \gamma \leq 2)$ as the set of those functions $f(z)$ holomorphic in the upper half-plane $G^+ = \{z : \operatorname{Im} z > 0\}$, which for sufficiently

142   A.M. Jerbashian

small $\rho > 0$ satisfy the Nevanlinna condition

$$\liminf_{R \to +\infty} \frac{1}{R} \int_{\beta}^{\pi-\beta} \log^+ |f(Re^{i\vartheta})| \left(\sin \frac{\pi(\vartheta - \beta)}{\pi - 2\beta} \right)^{1-\pi/\kappa} d\vartheta = 0 \qquad (1.1)$$

where $\beta = \arcsin \frac{\rho}{R} = \frac{\pi}{2} - \kappa$ and, simultaneously,

$$\|f\|_{p,\omega,\gamma}^p \equiv \iint_{G^+} |f(z)|^p \frac{d\mu_\omega(z)}{(1+|z|)^\gamma} < +\infty, \qquad (1.2)$$

where $d\mu_\omega(x+iy) = dx d\omega(2y)$ and it is supposed that $\omega(t) \in \Omega_\alpha$ $(-1 \le \alpha < +\infty)$, i.e., $\omega(t)$ is given in $[0, +\infty)$ and such that

(i) $\omega(t)$ $\nearrow$ (is non-decreasing) in $(0, +\infty)$, $\omega(0) = \omega(+0)$ and there exists a sequence $\delta_k \downarrow 0$ such that $\omega(\delta_k) \downarrow$ (is strictly decreasing);

(ii) $\omega(t) \asymp t^{1+\alpha}$ for $\Delta_0 \le x < +\infty$ and some $\Delta_0 \ge 0$

($f(t) \asymp g(t)$ means that $m_1 f(t) \le g(t) \le m_2 f(t)$ for some constants $m_{1,2} > 0$). One can see that if $\omega(t) \in \Omega_\alpha$ $(\alpha \ge -1)$ then (ii) is true for any $\Delta \in (0, \Delta_0]$.

We shall assume that $L_{\omega,\gamma}^p$ is the Lebesgue space defined solely by (1.2).

Remark 1.1. It is obvious that

$$A_{\omega,\gamma}^p = (i + z)^{\gamma/p} A_{\omega,0}^p.$$

Remark 1.2. For $\omega(t) = t^{1+\alpha}$ $(\alpha > -1)$, $\gamma = 0$ and $p \ge 1$ the spaces $A_{\omega,\gamma}^p$ coincide with the well-known A_α^p in the half-plane (see [1], [2], [3]). In this case (1.2) implies (1.1), and this implication is true even in the somehow more general case when $\omega(t)$ is continuously differentiable in $(0, +\infty)$ and such that $\omega'(t) \ge Mt^\alpha$ $(\alpha > -1)$ for almost all $t > 0$, where $M > 0$ is a constant.

Indeed, if $f(z) \in A_{\omega,0}^p$, where $p \ge 1$, $\omega(t)$ is continuously differentiable in $(0, +\infty)$ and $\omega'(t) \ge Mt^\alpha$ $(\alpha > -1)$ for almost all $t > 0$, then by (1.2)

$$\iint_{G^+} |f(\zeta)|^p (\operatorname{Im} \zeta)^\alpha d\sigma(\zeta) < +\infty,$$

where $\sigma(z)$ is the Lebesgue surface measure, i.e., $f(z) \in A_\alpha^p$. Thus, the well-known representation is true:

$$f(z) = \frac{1}{2\pi} \iint_{G^+} \frac{f(\zeta)(\operatorname{Im} \zeta)^\alpha}{[i(\overline{\zeta} - z)]^{2+\alpha}} d\sigma(\zeta), \quad z \in G^+,$$

where $\sigma(\zeta)$ is Lebesgue's area measure. Hence (1.1) holds since for $\operatorname{Im} z \ge \rho > 0$ and $p > 1$

$$|f(z)|^p \le C_1 \|f\|_{A_\alpha^p}^p \left\{ \iint_{G^+} \frac{(\operatorname{Im} \zeta)^\alpha d\sigma(\zeta)}{(|\operatorname{Re} \zeta - x| + \operatorname{Im} \zeta + \rho)^{(2+\alpha)q}} \right\}^{1/q}$$

$$\le C_2 \|f\|_{A_\alpha^p}^p \left\{ \int_0^{+\infty} \frac{\eta^\alpha d\eta}{(\eta + \rho)^{(2+\alpha)q-1}} \right\}^{1/q} = C_3 \|f\|_{A_\alpha^p}^p \rho^{-(2+\alpha)/p},$$

where $C_{1,2,3}$ are constants and $1/p + 1/q = 1$. For $p = 1$ the proof is more simple.

Note that in the general case the condition (1.1) can not be derived from (1.2). We shall return to the necessity of (1.2) in the last section of the paper.

1.2. Before analyzing the spaces $A^p_{\omega,\gamma}$, we recall some properties [4] of the holomorphic Hardy spaces

$$H^p_\gamma \equiv H^p\left(\frac{dx}{(1+|x|)^\gamma}\right) = (z+i)^{\gamma/p} H^p_0 \qquad (1.3)$$

$(0 < p < +\infty, \ -\infty < \gamma \leq 2)$, where H^p_0 is the Hille-Tamarkin's Hardy space in the upper half-plane $G^+ = \{z : \operatorname{Im} z > 0\}$, defined as the set of those $f(z)$ for which

$$\sup_{y>0} \int_{-\infty}^{+\infty} |f(x+iy)|^p dx < +\infty.$$

H^p_γ coincides with the set of those functions $f(z)$ holomorphic in G^+, for which $|f(z)|^p$ have harmonic majorants in G^+ (i.e., $f(z)$ is from the conformal image of Hardy's H^p in $|z| < 1$) and

$$f(x) \in L^p\left(\frac{dx}{(1+|x|)^\gamma}\right) \equiv L^p_\gamma$$

on the real axis. H^p_γ $(1 \leq p < +\infty, -\infty < \gamma \leq 2)$ is a Banach space with the norm $\|f(z)\|_{H^p_\gamma} = \|f(x)\|_{L^p_\gamma}$. For $\gamma = 2$ the space H^p_γ coincides with the conformal image of Hardy's H^p and for $\gamma = 0$ with H^p_0. Besides, it follows from the results of [4] that H^p_γ $(0 < p < +\infty, -\infty < \gamma \leq 2)$ coincides with the set of those functions holomorphic in G^+, which satisfy (1.1) for any $\rho > 0$ and

$$\liminf_{y\to+0} \int_{-\infty}^{+\infty} |f(x+iy)|^p \frac{dx}{(1+|x|)^\gamma} < +\infty. \qquad (1.4)$$

One can observe that in [4] the half-plane G^+ could be exhausted by disc segments, and hence under (1.4) the condition (1.1) is equivalent to

$$\liminf_{R\to+\infty} \frac{1}{R} \int_\beta^{\pi-\beta} |f(Re^{i\vartheta})|^p \left(\sin\frac{\pi(\vartheta-\beta)}{\pi-2\beta}\right)^{1-\pi/\kappa} d\vartheta < +\infty \qquad (1.1')$$

for any $\rho > 0$. Besides, if $f(z) \in H^p_\gamma$ $(-\infty < \gamma \leq 2, 0 < p < +\infty)$, then for any $0 < M < +\infty$

$$\sup_{0<y<M} \int_{-\infty}^{+\infty} |f(x+iy)|^p \frac{dx}{(1+|x|)^\gamma} < +\infty. \qquad (1.5)$$

Checking (1.4), one can easily show that for any $p > 1$ and $\gamma < 1$

$$H^p_\gamma \subset H^1_{\gamma'}, \quad 1 - \frac{1-\gamma}{p} < \gamma' < 1. \qquad (1.6)$$

1.3. Note that if $f(z) \in A^p_{\omega,\gamma}$ $(0 < p < +\infty, \omega \in \Omega_\alpha, -1 \leq \alpha < +\infty, -\infty < \gamma < 1)$, then $f(z+i\rho)$ belongs to H^p_γ for any $\rho > 0$. Moreover, the following assertion is true.

144   A.M. Jerbashian

Proposition 1.1. *For any $p > 0$ and $\gamma \in (-\infty, 2]$ the sum $\bigcup_{\omega \in \Omega_\alpha} A^p_{\omega, \gamma}$ coincides with the set of all those functions which belong to H^p_γ in any half-plane $G^+_\rho = \{z : \text{Im } z > \rho\}$ $(\rho > 0)$.*

Proof. Assuming that $f(z) \in H^p_\gamma$ in any half-plane G^+_ρ $(\rho > 0)$, one can put

$$\omega_f(t) = \int_0^{t/2} \frac{d\rho}{1 + M_p(\rho)}, \quad M_p(\rho) = \int_{-\infty}^{+\infty} |f(x + i\rho)|^p \frac{dx}{(1 + |x|)^\gamma},$$

for $0 < t \leq 1$ and define $\omega_f(t) \equiv \omega_f(1)$ for $1 \leq t < +\infty$, then (1.2) obviously becomes true. Thus, $f(z) \in A^p_{\omega, \gamma}$. $\qquad\square$

Proposition 1.2. *$A^p_{\omega, \gamma}$ $(1 \leq p < +\infty, -\infty < \gamma < 1, \omega \in \Omega_\alpha, \alpha \geq -1)$ is a Banach space with the norm (1.2).*

Proof. As $L^p_{\omega, \gamma}$ (which we have defined solely by (1.2)) is a Banach space, it suffices to show that $A^p_{\omega, \gamma}$ is a closed subspace of $L^p_{\omega, \gamma}$. Thus, assuming that $\{f_n\}^\infty_1 \subset A^p_{\omega, \gamma}$ is a sequence convergent in the norm of $L^p_{\omega, \gamma}$ $(f_n \to f \in L^p_{\omega, \gamma})$ we shall prove that $f \in A^p_{\omega, \gamma}$. Obviously

$$\int_0^{1/2} d\omega(2y) \int_{-\infty}^{+\infty} |f_n(x + iy) - f(x + iy)|^p \frac{dx}{(1 + |x|)^\gamma} \to 0$$

as $n \to \infty$. Hence, by Fatou's lemma $\int_0^1 g(t) d\omega(t) = 0$ for

$$g(2y) \equiv \liminf_{n \to \infty} \int_{-\infty}^{+\infty} |f_n(x + iy) - f(x + iy)|^p \frac{dx}{(1 + |x|)^\gamma}. \qquad (1.7)$$

As $\omega(t) \in \Omega_\alpha$, there exists a sequence $\eta_k \downarrow 0$ such that $\omega(\eta_{k+1}) < \omega(\eta_k)$. Introducing the measure $\nu(E) = \bigvee_E \omega$ we conclude that $\nu([\eta_{k+1}, \eta_k]) > 0$ for any $k \geq 1$ and obviously $g(t) = 0$ in $[\eta_{k+1}, \eta_k]$ almost everywhere with respect to the measure ν. On the other hand, $f(x + it) \in L^p((1 + |x|)^{-\gamma} dx)$ for almost every $t > 0$ in respect to the measure ν. Thus, there exists at least a sequence $y_k \downarrow 0$ such that simultaneously $g(2y_k) = 0$ and $f(x + iy_k) \in L^p((1 + |x|)^{-\gamma} dx)$.

Now choose a subsequence of $\{f_n\}$, for which the limit (1.7) is attained for $y = y_1$. From this subsequence choose another one, for which (1.7) is attained for $y = y_2$, etc. Then, by diagonal operation, choose a subsequence (for which we keep the same notation $\{f_n\}$) over which

$$g(2y_k) \equiv \lim_{n \to \infty} \int_{-\infty}^{+\infty} |f_n(x + iy_k) - f(x + iy_k)|^p \frac{dx}{(1 + |x|)^\gamma} = 0 \qquad (1.8)$$

for all $k \geq 1$. Then observe that $f_n(z + i\rho) \in H^p_\gamma$ $(n \geq 1)$ for any $\rho > 0$ and particularly for $\rho = y_k$ $(k = 1, 2, \ldots)$. By (1.8), for any fixed $k \geq 1$ the sequence $\{f_n(z + iy_k)\}^\infty_{n=1}$ is fundamental in H^p_γ and consequently $f_n(z + iy_k)$ (as $n \to \infty$) tends to some F form H^p_γ taken over $G^+_{y_k}$, and hence $f_n(z)$ tends to $F(z)$ uniformly inside G^+ and $F \in H^p_\gamma$ in any half-plane G^+_ρ. Thus, we conclude that for $F(z)$ (1.1) is true and, in addition, for any number $A > 0$

$$\iint_{|x|<A,\ \frac{1}{A}<y<A} |F(z) - f(z)|^p \frac{d\mu_\omega(z)}{(1+|z|)^\gamma}$$

$$\leq 2^{p-1} \left\{ \iint_{|x|<A,\ \frac{1}{A}<y<A} |F(z) - f_n(z)|^p \frac{d\mu_\omega(z)}{(1+|\mathrm{Re}z|)^\gamma} \right.$$

$$\left. + \iint_{|x|<A,\ \frac{1}{A}<y<A} |f(z) - f_n(z)|^p \frac{d\mu_\omega(z)}{(1+|\mathrm{Re}z|)^\gamma} \right\} \to 0$$

as $n \to \infty$. Letting $A \to +\infty$ we conclude that $\|F - f\|_{L^p_{\omega,\gamma}} = 0$. $\qquad\square$

2. Representation over a strip

2.1. Assuming that $\omega(t) \in \Omega_\alpha$ ($\alpha \geq -1$), we shall deal with the following continuous analog of M.M. Djrbashian's Cauchy-type kernel:

$$C_\omega(z) = \int_0^{+\infty} e^{itz} \frac{dt}{I_\omega(t)}, \quad I_\omega(t) = \int_0^{+\infty} e^{-tx} d\omega(x). \tag{2.1}$$

The function $C_\omega(z)$ is holomorphic in G^+ since for any $k \geq 1$ the integral (2.1) is uniformly convergent in $G^+_{\delta_k}$ in virtue of the obvious estimate

$$I_\omega(t) \geq t \int_{\delta_k}^{+\infty} e^{-tx} [\omega(x) - \omega(0)]dx \geq e^{-\delta_k t} [\omega(\delta_k) - \omega(0)], \quad k \geq 1.$$

Note that for the first time the kernel (2.1) has been used in [5] (see also [6]), where it was constructed in the multidimensional case of tube domains. Besides, one can see that for the simple scale $\omega(t) = t^{1+\alpha}$ ($-2 < \alpha < +\infty$)

$$I_\omega(t) = \Gamma(2+\alpha)t^{-(1+\alpha)} \quad \text{and} \quad C_\omega(z) = (-iz)^{-(2+\alpha)}.$$

2.2. For proving the canonical representations of $A^p_{\omega,\gamma}$ a new approach in using Fourier-Laplace transforms and differing from S.M.Gindikin's [9] one is revealed. We shall need an auxiliary statement on representation by Cauchy integral in H^p_γ and a somehow simple estimate for $C_\omega(z)$.

Lemma 2.1. *If $f(z) \in H^p_\gamma$ ($p \geq 1$, $\gamma < 1$), then for any $z \in G^+$*

$$f(z) = \frac{1}{2\pi i} \int_0^{2\pi} \frac{f(t)}{t-z} dt \quad \text{and} \quad \frac{1}{2\pi i} \int_0^{2\pi} \frac{f(t)}{t-\bar{z}} dt = 0. \tag{2.2}$$

Proof. By (1.6), it suffices to consider only the case $p = 1$. Fixing any $z \in G^+$ and any $\rho > 0$, for arbitrary $R > |z|$ one can write a representation over the boundary of a shifted semidisc:

$$f(z + i\rho) = \frac{1}{2\pi i} \int_{\partial[G^+(R)+i\rho]} \frac{f(\zeta)d\zeta}{\zeta - (z + i\rho)} \tag{2.3}$$

$$= \frac{1}{2\pi i} \int_{-R}^{R} \frac{f(t + i\rho)}{t - z} dt + \frac{1}{2\pi i} \int_0^{\pi} \frac{f(Re^{i\vartheta} + i\rho)}{Re^{i\vartheta} - z} dRe^{i\vartheta}$$

$$\equiv I_1(R) + I_2(R),$$

where $G^+(R) = \{z \in G^+ : |z| < R\}$. It is obvious that

$$\lim_{R \to +\infty} I_1(R) = \frac{1}{2\pi i} \int_{-\infty}^{+\infty} \frac{f(t + i\rho)}{t - z} dt. \tag{2.4}$$

For proving that $I_2(R) \to 0$ as $R \to +\infty$, observe that $f(z) = (z + i)^\gamma f_1(z)$, and both formulas of (2.2) are true for the function $f_1(z) \in H^1(dx)$. Hence

$$I_2(R) = \frac{1}{2\pi i} \int_{\vartheta=0}^{\pi} \frac{(Re^{i\vartheta} + i(\rho + 1))^\gamma}{Re^{i\vartheta} - z} \left(\frac{1}{2\pi i} \int_{-\infty}^{+\infty} \frac{f_1(t)dt}{t - (Re^{i\vartheta} + i\rho)} \right) dRe^{i\vartheta}$$

$$= \frac{1}{4\pi^2} \int_{-\infty}^{+\infty} f_1(t)dt \int_{\vartheta=0}^{\pi} \frac{(Re^{i\vartheta} + i(\rho + 1))^\gamma dRe^{i\vartheta}}{(Re^{i\vartheta} - z)(Re^{i\vartheta} + i\rho - t)} \tag{2.5}$$

$$\equiv \frac{1}{4\pi^2} \int_{-\infty}^{+\infty} f_1(t) J(R, t) dt,$$

where for large enough R

$$|J(R,t)| \le CR^{\gamma-1} \int_0^{\pi} \frac{d\vartheta}{\left| e^{i\vartheta} - \frac{t-i\rho}{R} \right|} \equiv A(R,t),$$

and $C > 0$ is a constant independent of t and R. For evaluating of the latter integral, one can use the inequalities $\frac{2}{\pi}x \le \sin x$ $(0 < x < \frac{\pi}{2})$ and $a^2 + b^2 \ge \frac{1}{2}(a + b)^2$ and derive $\left| e^{i\vartheta} - \frac{t-i\rho}{R} \right| > \frac{1}{2} \left(\left|1 - \frac{t}{R}\right| + \frac{2}{\pi}\sqrt{\frac{t}{R}}\vartheta + \sqrt{2}\frac{\rho}{R} \right)$ for $t > 0$ and any $\vartheta \in (0, \pi)$. Hence

$$|J(R,t)| \le \pi CR^{\gamma-1}\sqrt{\frac{R}{t}} \log \frac{\left|1 - \frac{t}{R}\right| + \sqrt{2}\frac{\rho}{R} + 2\sqrt{\frac{t}{R}}}{\left|1 - \frac{t}{R}\right| + \sqrt{2}\frac{\rho}{R}} \quad (t > 0). \tag{2.6}$$

If $t > R$, then

$$|J(R,t)| \le \pi CR^{\gamma-1} \log \left(\max_{x \ge 1} g(x) \right), \quad g(x) = \frac{x - 1 + 2\sqrt{x} + \sqrt{2}\frac{\rho}{R}}{x - 1 + \sqrt{2}\frac{\rho}{R}},$$

where $g(1) = 1 + \sqrt{2}\frac{R}{\rho}$, $g(+\infty) = 1$ and $g'(x) < 0$ for large enough R. Hence, for any $\varepsilon > 0$

$$|J(R,t)| < \pi CR^{\gamma-1} \log \left(1 + \sqrt{2}\frac{R}{\rho} \right) < 2\pi CR^{\gamma-1+\varepsilon} \quad (t > R).$$

If $0 < t < R$, then using (2.6) for $x = t/R \in (1/2, 1)$ we find that

$$|J(R,t)| < \sqrt{2}\pi C R^{\gamma-1} \log \left(1 + \frac{2\sqrt{x}}{1 - x + \sqrt{2}\frac{\rho}{R}} \right)$$

$$< \sqrt{2}\pi C R^{\gamma-1} \log \left(1 + \sqrt{2}\frac{R}{\rho} \right) < 4\pi R^{\gamma-1+\varepsilon}$$

for any $\varepsilon > 0$ and large enough R. The same estimate is true also for $x = t/R \in (0, 1/2)$. Observing that $A(R,t) = A(R, |t|)$, for $t < 0$ we conclude that $|J(R,t)| < 4\pi C R^{\gamma-1+\varepsilon}$ for any $\varepsilon \in (0, 1 - \gamma)$ provided R is large enough. Hence $I_2(R) \to 0$ as $R \to +\infty$ since $f_1(t) \in L^1(dx)$ in (2.5), and by (2.3) and (2.4)

$$f(z + i\rho) = \frac{1}{2\pi i} \int_{-\infty}^{+\infty} \frac{f(t + i\rho)}{t - z} dt, \quad z \in G^+, \; \rho > 0.$$

Letting $\rho \to 0$ we arrive at the first formula of (2.2). Indeed, for any fixed $z \in G^+$

$$\left| \frac{1}{2\pi i} \int_{-\infty}^{+\infty} \frac{f(t + i\rho)}{t - z} dt - \frac{1}{2\pi i} \int_{-\infty}^{+\infty} \frac{f(t)}{t - z} dt \right|$$

$$\leq C' \frac{1}{2\pi i} \int_{-\infty}^{+\infty} \frac{|f(t + i\rho) - f(t)|}{1 + |t|} dt \leq \frac{C'}{2\pi} \|f(z + i\rho) - f(z)\|_{H^1_\gamma} = o(1).$$

The second formula of (2.2) is proved in the same way, starting by (2.3) with $\bar{z}$ instead of z and zero at the left-hand side. $\qquad\square$

Lemma 2.2. *Let $\omega(t)$ satisfy the condition (i) (Subsection 1.1) and let $\omega(t) = \omega(\Delta)$ for some $\Delta > 0$ and all $\Delta < t < +\infty$. Then for large enough k there exists some constant $M \equiv M_k$ depending solely on k, such that for any $\delta \in [0, \delta_{k+1}]$*

$$|C_{\omega_\delta}(z)| \leq M|z|^{-1}, \quad z \in G^+_{3\delta_k}, \tag{2.7}$$

where $\omega_\delta(t) = \omega(t + \delta)$ $(0 < t < +\infty)$.

Proof. If k is large enough to provide $2\delta_k < \Delta$, then for any $\delta \in [0, \delta_{k+1}]$

$$I_{\omega_\delta}(t) \equiv I_\delta(t) = \int_0^{\Delta-\delta} e^{-tx} d[\omega(x + \delta) - \omega(\delta)]$$

$$\geq e^{-t(\Delta-\delta)}[\omega(\Delta) - \omega(\delta)] + t \int_{\delta_k}^{\Delta-\delta} e^{-tx}[\omega(x + \delta) - \omega(\delta)]dx$$

$$\geq e^{-t(\Delta-\delta)}[\omega(\Delta) - \omega(\delta)] + [e^{-t\delta_k} - e^{-t(\Delta-\delta)}][\omega(\delta_k + \delta) - \omega(\delta)]$$

$$\geq e^{-t\delta_k}[\omega(\delta_k + \delta) - \omega(\delta)] \geq e^{-t\delta_k}[\omega(\delta_k) - \omega(\delta_{k+1})] > 0.$$

Besides,

$$|I'_\delta(t)| = \int_0^{\Delta-\delta} e^{-tx} x d\omega(x + \delta) \leq \int_0^{\Delta-\delta} x d\omega(x + \delta) \leq \Delta\omega(\Delta).$$

Consequently, for $z = x + iy \in G^+_{3\delta_k}$ $(y > 3\delta_k)$

$$C_{\omega\delta}(z) = \frac{1}{iz}\int_0^{+\infty}\frac{de^{itz}}{I_\delta(t)} = \frac{1}{iz}\frac{e^{itz}}{I_\delta(t)}\bigg|_{t=0}^{+\infty} + \frac{1}{iz}\int_0^{+\infty}e^{itz}\frac{I'_\delta(t)}{[I_\delta(t)]^2}dt \equiv A + B,$$

where

$$|A| \leq \frac{1}{|z|[\omega(\delta_k) - \omega(\delta_{k+1})]} \quad \text{and} \quad |B| \leq \frac{\Delta\omega(\Delta)}{|z|\delta_k[\omega(\delta_k) - \omega(\delta_{k+1})]^2}.$$

$\square$

2.3. Theorem 2.1. *Let $f(z) \in A^p_{\omega,\gamma}(G^+)$ for some $1 \leq p < +\infty$, $-\infty < \gamma < 1$ and $\omega(x)$ satisfying the condition (i) and such that $\omega(t) = \omega(\Delta) < +\infty$ $(\Delta < t < +\infty)$ for some $\Delta > 0$. Then*

$$f(z) = \frac{1}{2\pi}\iint_{G^+} f(w)C_\omega(z - \overline{w})d\mu_\omega(w), \quad z \in G^+, \tag{2.8}$$

$$f(z) = \frac{1}{\pi}\iint_{G^+} \{\text{Re } f(w)\}C_\omega(z - \overline{w})d\mu_\omega(w), \quad z \in G^+, \tag{2.9}$$

where both integrals are absolutely and uniformly convergent inside G^+.

Proof. First, note that in our assumption $d\omega(2y) = 0$, $\Delta/2 < y < +\infty$, similar to (1.6)

$$A^p_{\omega,\gamma} \subset A^1_{\omega,\gamma'}, \quad 1 - \frac{1-\gamma}{p} < \gamma' < 1,$$

for any $p > 1$ and $\gamma < 1$. Thus, the uniform convergence of the integrals (2.8) and (2.9) in any compact lying inside G^+ is obvious by (2.7). Besides, by the above inclusion it suffices to prove the representations (2.8) and (2.9) only for $p = 1$. So, let $f(z) \in A^1_{\omega,\gamma}(G^+)$ where $\omega(t)$ and γ are as required. Then $f(z + i\rho) \in H^1_\gamma$ for any $\rho > 0$. Hence, by Lemma 2.1

$$f(z + i\rho) = \frac{1}{2\pi i}\int_{-\infty}^{+\infty}\frac{f(\xi + i\rho)}{\xi - z}d\xi$$

for any fixed point $z = x + iy \in G^+$. Consequently, for any $\delta \in (0, y_0)$

$$f(z + i\rho) = \lim_{\substack{a\to-\infty\\b\to+\infty}}\frac{1}{2\pi}\int_0^{+\infty}e^{i\tau z}d\tau\int_a^b e^{-i\tau\xi}f(\xi + i\rho)d\xi$$

$$= \lim_{\substack{a\to-\infty\\b\to+\infty}}\frac{1}{2\pi}\int_0^{+\infty}e^{i\tau z}\left(\int_0^{\Delta-\delta}e^{-\tau x}d\omega(x + \delta)\right)\left(\int_a^b e^{-i\tau\xi}f(\xi + i\rho)d\xi\right)\frac{d\tau}{I_\delta(\tau)}$$

$$= \lim_{\substack{a\to-\infty\\b\to+\infty}}\frac{1}{2\pi}\int_{\delta/2}^{\Delta/2}d\omega(2v)\int_0^{+\infty}e^{i\tau(z+2iv-i\delta)}\frac{d\tau}{I_\delta(\tau)}\int_a^b e^{-i\tau\xi}f(\xi + i\rho)d\xi$$

$$\equiv \lim_{\substack{a\to-\infty\\b\to+\infty}}\int_{\delta/2}^{\Delta/2}J_{a,b}(v)d\omega(2v),$$

where

$$J_{a,b}(v) = \frac{1}{2\pi} \int_0^{+\infty} \left(e^{-\tau v} \int_a^b e^{-i\tau\xi} f(\xi + i\rho)d\xi \right) \overline{\left(\frac{e^{-i\tau(\bar{z}-v+i\delta)}}{I_\delta(\tau)} \right)} d\tau$$

$$\equiv \frac{1}{2\pi} \int_{-\infty}^{+\infty} A_{a,b}(\tau)\overline{B(\tau)}d\tau,$$

and it is assumed that $A_{a,b}(\tau) \equiv B(\tau) \equiv 0$ for $\tau \leq 0$. One can see that $A_{a,b}(\tau)$ and $B(\tau)$ are bounded functions of $L_1(-\infty, +\infty)$, which are continuous in $(-\infty, +\infty) \setminus \{0\}$. Therefore, the following well-known formula on Fourier transforms of such functions is valid (see, for instance, [6], Ch. I, §3, p. 39, Theorem 1.12(2°))

$$J_{a,b}(v) = \frac{1}{2\pi} \int_{-\infty}^{+\infty} \mathcal{F}[A_{a,b}](u)\,\overline{\mathcal{F}[B](u)}\,du,$$

where $\mathcal{F}[\varphi]$ stands for the Fourier transform of $\varphi \in L_1(-\infty, +\infty)$. Using this formula we conclude that

$$J_{a,b}(v) = \frac{1}{2\pi} \int_{-\infty}^{+\infty} \left[\frac{1}{2\pi} \int_a^b f(\xi + i\rho)d\xi \int_0^{+\infty} e^{i\tau(-u+iv-\xi)}d\tau \right]$$

$$\times \left[\int_0^{+\infty} e^{it(u+z+iv-i\delta)} \frac{dt}{I_\delta(t)} \right] du$$

$$= \frac{1}{2\pi} \int_{-\infty}^{+\infty} \left(\frac{1}{2\pi i} \int_a^b \frac{f(\xi + i\rho)}{\xi - (u+iv)} d\xi \right) C_{\omega\delta}(z - (u - iv) - i\delta)du.$$

The latter integral is absolutely convergent even for $a = -\infty$ and $b = +\infty$ provided δ is small enough. Indeed, choosing k great enough so that $y > 4\delta_k$ and assuming that $\delta \in (0, \delta_{k+1})$, for $v \in (\delta/2, \Delta/2)$ we shall have $y + v - \delta \geq y - \delta_k/2 > 3\delta_k$ and consequently by (2.7)

$$I_1 \equiv \int_{-\infty}^{+\infty} \left(\int_{-\infty}^{+\infty} \frac{|f(\xi + i\rho)|}{|\xi - (u+iv)|}d\xi \right) |C_{\omega\delta}(z - (u - iv) - i\delta)|du$$

$$\leq 2M \int_{-\infty}^{+\infty} |f(\xi + i\rho)|\Phi(\xi)d\xi,$$

where

$$\Phi(\xi) = \int_{-\infty}^{+\infty} \frac{d\sigma}{(|\sigma| + 3\delta_k)(|\xi - x - \sigma| + \delta/2)}$$

is a continuous function in $(-\infty, +\infty)$. Besides, one can show that

$$\Phi(\xi) \leq \frac{C}{1 + |\xi|} \log(1 + |\xi|) < \frac{C'}{(1 + |\xi|)^\gamma}, \qquad -\infty < \xi < +\infty,$$

150   A.M. Jerbashian

where C and C' are some constants depending on x, δ and γ. Hence $I_1 < +\infty$, and consequently

$$\lim_{\substack{a\to-\infty\\b\to+\infty}} J_{a,b}(v) = \frac{1}{2\pi}\int_{-\infty}^{+\infty}\left(\frac{1}{2\pi i}\int_{-\infty}^{+\infty}\frac{f(\xi+i\rho)d\xi}{\xi-(u+iv)}\right)C_{\omega_\delta}(z-(u-iv)-i\delta)du$$

$$= \frac{1}{2\pi}\int_{-\infty}^{+\infty}f(u+iv+i\rho)C_{\omega_\delta}(z-(u-iv)-i\delta)du$$

$$= \frac{1}{2\pi}\int_{-\infty}^{+\infty}\left(\frac{1}{2\pi i}\int_{-\infty}^{+\infty}\frac{f(t+iv)dt}{t-(u+i\rho)}\right)C_{\omega_\delta}(z-(u-iv)-i\delta)du$$

$$= \frac{1}{2\pi}\int_{-\infty}^{+\infty}f(u+iv)\left(\frac{1}{2\pi i}\int_{-\infty}^{+\infty}\frac{C_{\omega_\delta}(t+z+iv-i\delta)}{t-(-u+i\rho)}dt\right)du$$

$$= \frac{1}{2\pi}\int_{-\infty}^{+\infty}f(u+iv)C_{\omega_\delta}(z-(u-iv)+i\rho-i\delta)du$$

since

$$I_2 \equiv \int_{-\infty}^{+\infty}|f(u+iv)|du\int_{-\infty}^{+\infty}\frac{|C_{\omega_\delta}(z-(t-iv)-i\delta)|}{|u-t-i\rho|}dt < +\infty$$

and $C_{\omega_\delta}(z+i(y+v-\delta)) \in H^1_\gamma$ in G^+. The latter inclusion simply follows from (2.7), and $I_2 < +\infty$ holds in the same way as $I_1 < +\infty$ above. Thus, for any fixed $z \in G^+$ and small enough $\delta > 0$

$$f(z) = \frac{1}{2\pi}\iint_{\delta/2<v<\Delta/2} f(w)C_{\omega_\delta}(z-\overline{w}-i\delta)d\mu_\omega(w).$$

For letting $\delta \to +0$, we fix any $\varepsilon > 0$ and use the following two estimates

$$I_1 \equiv \frac{1}{2\pi}\iint_{\substack{\delta/2<v<\Delta/2\\|u|>A}}|f(w)||C_{\omega_\delta}(z-\overline{w}-i\delta)|d\mu_\omega(w) < \frac{\varepsilon}{4},$$

$$I_2 \equiv \frac{1}{2\pi}\iint_{\substack{\delta/2<v<\Delta/2\\|u|>A}}|f(w)||C_{\omega_\delta}(z-\overline{w})|d\mu_\omega(w) < \frac{\varepsilon}{4}$$

which are true for large enough $A > 0$. Besides, we use the following two estimates which are true for fixed A and small enough $\delta > 0$:

$$I_3 \equiv \frac{1}{2\pi}\iint_{0<v<\delta/2}|f(w)||C_{\omega_\delta}(z-\overline{w})|d\mu_\omega(w) < \frac{\varepsilon}{4},$$

$$I_4 \equiv \frac{1}{2\pi}\iint_{\substack{\delta/2<v<\Delta/2\\|u|<A}}|f(w)||C_{\omega_\delta}(z-\overline{w}-i\delta)-C_{\omega_\delta}(z-\overline{w})|d\mu_\omega(w) < \frac{\varepsilon}{4}.$$

Hence we conclude that for small enough $\delta > 0$

$$\left|f(z) - \frac{1}{2\pi}\iint_{0<v<\Delta/2} f(w)C_\omega(z-\overline{w})d\mu_\omega(w)\right| < \varepsilon,$$

and the representation (2.8) holds. Note that the above inequalities for $I_{1,2,3}$ obviously follow from (2.7) (where M does not depend on δ). For proving the remaining estimate for I_4, one can use the estimate of I_δ given in the Proof of Lemma 2.2 and obtain that

$$I_4 \leq \frac{\omega(\delta) - \omega(0)}{2\pi} \iint_{\substack{\delta/2 < v < \Delta/2 \\ |u| < A}} |f(w)| \left(\int_0^{+\infty} e^{-t(y+v)} \frac{dt}{[I_\delta(t)]^2} \right) d\mu_\omega(w)$$

$$\leq \frac{\omega(\delta) - \omega(0)}{[\omega(\delta_k) - \omega(\delta_{k+1})]^2} \frac{(1+A)^\gamma}{\pi y} \|f\|_{1,\omega,\gamma} \to 0 \quad \text{as} \quad \delta \to +0.$$

For proving (2.9), it suffices to repeat the above proof starting by

$$\frac{1}{2\pi i} \int_{-\infty}^{+\infty} \frac{f(\xi + i\rho)}{\xi - \overline{z}} d\xi = \frac{1}{2\pi i} \int_{-\infty}^{+\infty} \frac{\overline{f(\xi + i\rho)}}{\xi - z} d\xi \equiv 0, \quad z \in G^+.$$

This will lead to the identity

$$0 \equiv \frac{1}{2\pi} \iint_{G^+} \overline{f(w)} C_\omega(z - \overline{w}) d\mu(w), \quad z \in G^+.$$

Adding this identity to (2.8) we come to (2.9). $\qquad\qquad\qquad\qquad\qquad\square$

3. General representations

3.1. The following two lemmas will be used for letting $\Delta \to +\infty$ in (2.8) and (2.9) and obtaining a representation of $A^p_{\omega,\gamma}$ by some integrals which can be taken over the whole half-plane. In the below Lemma 3.1 D^{-a} will stand for the Riemann–Liouville's integro-differentiation:

$$D^{-a}u(x) \equiv \frac{1}{\Gamma(a)} \int_0^x (x - t)^{a-1} u(t) dt, \quad a > 0,$$

$$D^0 u(x) \equiv u(x),$$

$$D^a u(x) = \frac{d^p}{dx^p} D^{-(p-a)} u(x), \quad a > 0, \quad p - 1 < a \leq p.$$

Lemma 3.1. *Let $\omega(x) \in \Omega_\alpha$ for some $\alpha \geq -1$. Then for any non-integer $\beta \in ([\alpha] - 1, \alpha)$*

$$C_\omega(z) = \frac{1}{(-iz)^{2+\beta}} \int_0^{+\infty} e^{itz} \left\{ D^{2+\beta} [I_\omega(t)]^{-1} \right\} dt, \quad z \in G^+, \qquad (3.1)$$

where the right-hand side integral is absolutely convergent. Besides, for any $\rho > 0$ there exists a constant $M \equiv M_{\rho,\beta} > 0$ such that

$$|C_\omega(z)| \leq \frac{M}{|z|^{2+\beta}}, \quad z \in G^+_\rho. \qquad (3.2)$$

Proof. Let $m \geq 1$ be the integer deduced from $m-1 < 2+\alpha \leq m$ $(m-1 < \beta < m)$. Then the function $D^{-(m-2+\beta)}[I_\omega(t)]^{-1}$ is infinitely differentiable in $(0, +\infty)$ (i.e., is of $C^\infty(0, +\infty)$). Indeed, the function $I_\omega(t)$ $(I_\omega(t) > 0,\ I_\omega(t) \searrow,\ I_\omega(+\infty) = 0)$ has holomorphic continuation in $\mathrm{Re}\, z > 0$. Hence $[I_\omega(z)]^{-1}$ is holomorphic in a neighborhood of $(0, +\infty)$ and the function

$$D^{-(m-2-\beta)}[I_\omega(t)]^{-1} = \frac{t^{m-2-\beta}}{\Gamma(m-2-\beta)} \int_0^1 (1-x)^{m-3-\beta}[I_\omega(tx)]^{-1}dx$$

is of $C^\infty(0, +\infty)$. Consequently, integration by parts leads to (3.1):

$$\int_0^{+\infty} e^{itz}\left\{ D^{2+\beta}[I_\omega(t)]^{-1} \right\}dt = \int_0^{+\infty} e^{itz}d\left\{ D^{m-1}D^{-(m-\beta-2)}[I_\omega(t)]^{-1} \right\} \quad (3.3)$$

$$= e^{itz}\sum_{n=0}^{m-1}(-iz)^n D^{m-1-n}D^{-(m-\beta-2)}[I_\omega(t)]^{-1}\Big|_{t=0}^{+\infty}$$

$$+ (-iz)^m \int_0^{+\infty} e^{itz}\left\{ D^{-(m-\beta-2)}[I_\omega(t)]^{-1} \right\}dt$$

$$= \frac{(-iz)^m}{\Gamma(m-\beta-2)} \int_0^{+\infty} \frac{d\lambda}{I_\omega(\lambda)} \int_\lambda^{+\infty} e^{itz}(t-\lambda)^{m-\beta-3}dt = (-iz)^{\beta+2}C_\omega(z)$$

provided all integrals which arise in the above operations are convergent and

$$e^{-ty}D^n D^{-(m-\beta-2)}[I_\omega(t)]^{-1}\Big|_{t=0,\,+\infty} = 0 \quad (3.4)$$

for any $y > 0$ and $0 \leq n \leq m-1$. For proving (3.4), first we shall verify that our condition $\omega(x) \asymp x^{1+\alpha}$ $(0 \leq \Delta_0 < x < +\infty)$ implies

$$\left| \frac{d^k}{dt^k}[I_\omega(t)]^{-1} \right| \leq M_1 e^{1+\alpha-k}, \quad 0 < t < 1,\ 0 \leq k \leq m, \quad (3.5)$$

where M_1 is a constant. Indeed, for $k = 0$ we obviously have

$$I_\omega(t) \asymp t^{-(1+\alpha)} \int_{\Delta_0 t}^{+\infty} e^{-x}x^{1+\alpha}dx, \quad 0 < t < 1,$$

and it suffices to see that successively differentiating any summand in the expression of $\left(d^\lambda / dt^\lambda\right)[I_\omega(t)]^{-1}$ $(\lambda \geq 0)$ we either differentiate the nominator (and this adds the multiplier t^{-1} to the estimate of the summand) or we multiply this summand by $(\lambda+1)I_\omega'(t)/I_\omega(t) = O(t^{-1})$ $(t \to +0)$. By (3.5), for $0 < t < +\infty$

$$D^n D^{-(m-\beta-2)}[I_\omega(t)]^{-1} = \frac{d^n}{dt^n}\frac{t^{m-\beta-2}}{\Gamma(m-\beta-2)} \int_0^1 (1-\sigma)^{m-\beta-3}[I_\omega(t\sigma)]^{-1}d\sigma \quad (3.6)$$

$$= \sum_{k=0}^{n} C_n^k \frac{m-\beta-2}{\Gamma(m-\beta-1-n+k)} t^{m-\beta-2-n+k}$$

$$\times \int_0^1 (1-\sigma)^{m-\beta-3}\frac{\partial^k}{\partial t^k}[I_\omega(t\sigma)]^{-1}d\sigma, \quad 0 \leq n \leq m,$$

where C^k_n are the binomial coefficients and all integrals are absolutely convergent. Hence the case $t \to +0$ of (3.4) follows. Further, it is obvious that for any $k \geq 0$ the derivative $\left|\left(d^k / dt^k\right) I_\omega(t)\right|$ is bounded as $1 \leq t < +\infty$ and for any $y > 0$ and any $k \geq 0$

$$I_\omega(t) \geq t \int_{\frac{y}{2(k+1)}}^{1+\frac{y}{2(k+1)}} e^{-tx}[\omega(x) - \omega(0)]dx$$

$$\geq \left(1 - \frac{1}{e}\right)\left[\omega\left(\frac{y}{2(k+1)}\right) - \omega(0)\right]^{-\frac{ty}{2(k+1)}}, \qquad 1 \leq t < +\infty.$$

Consequently, for any $y > 0$, $k \geq 0$ and $1 \leq t < +\infty$

$$\left|\frac{d^k}{dt^k}[I_\omega(t)]^{-1}\right| \leq M_2 e^{ty/2}\left[\omega\left(\frac{y}{2(k+1)}\right) - \omega(0)\right]^{-(k+1)}, \tag{3.7}$$

where M_2 is a constant independent of t and y. The case $t \to +\infty$ of (3.4) follows from (3.6) and (3.7).

It remains to see that in virtue of (3.6) and (3.5), (3.7) for any $y > 0$ and any $n(0 \leq n \leq m)$ there exists a constant M_3 such that

$$\left|D^n D^{m-\beta-2}[I_\omega(t)]^{-1}\right| \leq M_3 e^{ty/2}t^{\alpha-\beta-1}, \qquad 0 < t < +\infty. \tag{3.8}$$

Hence the absolute convergence of all integrals appearing in the operations of (3.3) holds. The estimate (3.2) follows from (3.1) and (3.8). $\qquad\square$

Lemma 3.2. *If $\omega \in \Omega_\alpha$ for some $\alpha \geq -1$,*

$$\omega_\Delta(t) = \begin{cases} \omega(t) & for \quad 0 \leq t \leq \Delta \\ \omega(\Delta) & for \quad \Delta < t < +\infty \end{cases} \qquad and \quad C_\Delta(z) \equiv C_{\omega_\Delta}(z), \tag{3.9}$$

then there exists a constant $M \equiv M_\omega$ depending solely on ω and such that

$$|C_\omega(z) - C_\Delta(z)| \leq \frac{M}{\Delta^{2+\alpha}}, \qquad z \in G^+, \quad 1 < \Delta < +\infty. \tag{3.10}$$

Proof. It is obvious that for $z = x + iy$ $(y > 0)$

$$|C_\omega(z) - C_\Delta(z)| \leq \int_0^{+\infty} e^{-yt} \frac{I_1(t)}{I_2(t)[I_1(t) + I_2(t)]}dt, \tag{3.11}$$

where

$$I_1(t) = \int_\Delta^{+\infty} e^{-tx}d\omega(x) \quad and \quad I_2(t) = \int_0^\Delta e^{-tx}d\omega(x).$$

Besides, one can see that

$$I_1(t) = e^{-t\Delta}\int_0^{+\infty} e^{-tx}[\omega(x + \Delta) - \omega(\Delta)]dt < M'_\omega e^{-t\Delta}t \int_0^{+\infty} e^{-tx}(x + \Delta)^{1+\alpha}dx$$

$$= M'_\omega e^{-t\Delta}t\Delta^{2+\alpha}\int_0^{+\infty} e^{-tx\Delta}(1 + x)^{1+\alpha}dx,$$

where the constant M'_ω depends only on ω. Hence, there exists another constant M''_ω such that

$$I_1(t) < M''_\omega e^{-t\Delta}\Delta^{1+\alpha}, \quad t\Delta > 1.$$

Further,

$$I_2(t) = e^{-t\Delta}[\omega(\Delta) - \omega(0)] + t\Delta \int_0^\Delta e^{-tx\Delta}[\omega(x\Delta) - \omega(0)]dx.$$

Hence, for $t\Delta > 1$

$$I_2(t) > t\Delta \int_{1/4}^1 e^{-tx\Delta}[\omega(x\Delta) - \omega(0)]dx \geq \left[\omega\left(\frac{\Delta}{4}\right) - \omega(0)\right] e^{-\frac{t\Delta}{4}}\left(1 - e^{-\frac{3t\Delta}{4}}\right)$$

$$> M'''_\omega e^{-\frac{t\Delta}{4}}\Delta^{1+\alpha}\frac{3t\Delta}{4 + 3t\Delta} > \frac{3}{7}M'''_\omega e^{-\frac{t\Delta}{4}}\Delta^{1+\alpha}$$

and $I_2(t) > e^{-t\Delta}[\omega(\Delta) - \omega(0)] > M^{IV}_\omega \Delta^{1+\alpha}$ for $0 < t\Delta < 1$. By these inequalities,

$$\frac{I_1(t)}{I_2(t)[I_1(t) + I_2(t)]} < \begin{cases} \dfrac{I_1(t)}{[I_2(t)]^2} < M^V_\omega e^{-\frac{t\Delta}{2}}\Delta^{-(1+\alpha)}, & t\Delta > 1 \\[2mm] \dfrac{1}{I_2(t)} < \dfrac{\Delta^{-(1+\alpha)}}{M^{IV}_\omega} < M^{VI}_\omega e^{-\frac{t\Delta}{2}}\Delta^{-(1+\alpha)}, & 0 < t\Delta < 1. \end{cases}$$

Hence by (3.11) we come to (3.10). $\qquad\qquad\qquad\qquad\qquad\qquad\qquad\qquad\square$

3.2. Theorem 3.1. *Let* $f(z) \in A^p_{\omega,\gamma}$ *for some* $1 \leq p < +\infty$ *and* $-\infty < \gamma < 1 - (1+\alpha)(p-1)$. *Then*

$$f(z) = \frac{1}{2\pi} \iint_{G^+} f(w)C_\omega(z - \overline{w})d\mu_\omega(w), \quad z \in G^+, \tag{3.12}$$

$$f(z) = \frac{1}{\pi} \iint_{G^+} \{\mathrm{Re}\, f(w)\}C_\omega(z - \overline{w})d\mu_\omega(w), \quad z \in G^+, \tag{3.13}$$

where both integrals are absolutely and uniformly convergent inside G^+.

Proof. The absolute and uniform convergence of the integrals in (3.12) and (3.13) inside G^+ is obvious by (3.2). Further, one can verify that under our assumptions

$$A^p_{\omega,\gamma} \subset A^1_{\omega,\gamma'}, \quad \max\left\{1 - \frac{1-\gamma}{p}, \frac{\gamma + (2+\alpha)(p-1)}{p}\right\} < \gamma' < 1. \tag{3.14}$$

Thus, it suffices to prove the representation (3.12) only for $p = 1$. We shall omit the proof of (3.13) since it holds similarly, by the same passage $\Delta \to +\infty$ in (2.9).

We start by proving that for any $\Delta > \Delta_0$, $\rho > 0$ and any $\delta \in (0, 1]$

$$|C_\Delta(z)| \leq \frac{M_\rho}{|z|^{1-\delta}}, \quad z \in G^+_\rho, \tag{3.15}$$

where the constant M_ρ depends only on ρ and δ, and C_Δ is that of (3.9). Indeed, $\omega_\Delta(t) \in \Omega_{-1}$. Hence, if $0 < \delta < 1$, then by (3.1)

$$C_\Delta(z) = \frac{1}{(-iz)^{1-\delta}} \int_0^{+\infty} e^{itz} \varphi_{\Delta,\delta}(t) dt, \quad z \in G^+, \tag{3.16}$$

where $\varphi_{\Delta,\delta}(t) = \frac{d}{dt} D^{-\delta}[I_\Delta(t)]^{-1} \geq 0$ and the right-hand side integral of (3.16) is convergent. On the other hand, for any $y > 0$

$$C_\Delta(iy) = \frac{1}{y^{1-\delta}} \int_0^{+\infty} e^{-ty} \varphi_{\Delta,\delta}(t) dt = \int_0^{+\infty} e^{-ty} \frac{dt}{I_\Delta(t)},$$

where the right-hand side integral decreases by Δ (for any fixed y and δ). Hence

$$\int_0^{+\infty} e^{-ty} \varphi_{\Delta,\delta}(t) dt$$

has the same property. Consequently, for $y > \rho$ and $\Delta > \Delta_0$

$$\int_0^{+\infty} e^{-ty} \varphi_{\Delta,\delta}(t) dt \leq \int_0^{+\infty} e^{-t\rho} \varphi_{\Delta_0,\delta}(t) dt \equiv M_\rho < +\infty,$$

and (3.15) holds by (3.16). For $\delta = 1$ (3.15) is obvious.

Now fix any $z \in G^+$ and choose a number A_1 enough large to provide

$$I_1 \equiv \frac{1}{2\pi} \iint_{\substack{0<v<\Delta/2 \\ |u|>A_1}} |f(w)||C_\Delta(z-\overline{w})| d\mu_\omega(w) < \frac{\varepsilon}{4}, \quad \Delta > \Delta_0. \tag{3.17}$$

This choice of A_1 is possible by (3.15). Indeed, using (3.15) with $1 - \delta = \gamma^+$ we come to the following inequalities which imply (3.17):

$$I_1 \leq \frac{M_y}{2\pi} \iint_{\substack{0<v<\Delta/2 \\ |u|>A_1}} |f(w)| \frac{d\mu_\omega(w)}{(1+|w|)^{\gamma^+}} \leq \frac{M_y}{2\pi} \iint_{\substack{0<v<+\infty \\ |u|>A_1}} |f(w)| \frac{d\mu_\omega(w)}{(1+|w|)^{\gamma}}.$$

Using (3.2), choose $A_2 > A_1$ enough large to provide that

$$I_2 \equiv \frac{1}{2\pi} \iint_{\substack{0<v<+\infty \\ |u|>A_2}} |f(w)||C_\omega(z-\overline{w})| d\mu_\omega(w) < \frac{\varepsilon}{4}, \quad \Delta > \Delta_0. \tag{3.18}$$

Now, using (3.10) choose $\Delta_1 > \Delta_0$ enough large to provide that

$$I_3 \equiv \frac{1}{2\pi} \iint_{\substack{0<v<\Delta/2 \\ |u|<A_2}} |f(w)||C_\omega(z-\overline{w}) - C_{\Delta_1}(z-\overline{w})| d\mu_\omega(w) < \frac{\varepsilon}{4}. \tag{3.19}$$

This choice of Δ_1 is possible since by (3.10)

$$I_3 \leq \frac{1}{2\pi} \frac{M}{\Delta_1^{2+\alpha}} \iint_{\substack{0<v<\Delta_1/2 \\ |u|<A_2}} |f(w)| d\mu_\omega(w) \leq \frac{M(A_2+\Delta_1)^{\gamma^+}}{\Delta_1^{2+\alpha}} \|f\|_{1,p,\gamma}.$$

At last, using (3.2) choose $\Delta > \Delta_1$ enough large to provide that

$$I_4 = \frac{1}{2\pi} \iint_{\substack{\Delta/2<v<+\infty \\ |u|<A_2}} |f(w)||C_\omega(z-\overline{w})| d\mu_\omega(w) < \frac{\varepsilon}{4}. \tag{3.20}$$

156
A.M. Jerbashian

Taking Δ enough large in the representation (2.8) of $f(z)$, by (3.17)-(3.20) we get

$$\left| f(z) - \frac{1}{2\pi} \iint_{G^+} f(w) C_\omega(z - \overline{w}) d\mu_\omega(w) \right| < \varepsilon,$$

and (3.12) holds by the arbitrariness of ε. $\qquad\square$

4. Final remarks

Theorem 4.1. *Let $f(z) \in L^p_{\omega,\gamma}$ ($1 \le p < +\infty$, $-\infty < \gamma < 1$), and let $\omega(x)$ be as required in Theorem 2.1 or Theorem 3.1. Then the right-hand side integrals in formulas (2.8), (2.9) or (3.12), (3.13) represent functions which are holomorphic in G^+ and satisfy (1.1).*

Proof. Under the mentioned requirements, the right-hand side integrals in (2.8), (2.9) and (3.12), (3.13) obviously are absolutely and uniformly convergent inside G^+ and hence, these integrals represent holomorphic functions in G^+.

Let $\omega(x)$ be as required in Theorem 3.1, and let

$$F(z) = \frac{1}{2\pi} \iint_{G^+} f(w) C_\omega(z - \overline{w}) d\mu_\omega(w), \quad z \in G^+.$$

If $z = Re^{i\vartheta}$ and $\operatorname{Im} z = R \sin \vartheta \ge \rho$, where $\rho > 0$ is a fixed number, then by (3.2)

$$|F(Re^{i\vartheta})| \le M \iint_{G^+} |f(w)| \frac{d\mu_\omega(w)}{|Re^{i\vartheta} - \overline{w}|^{2+\beta}},$$

where $\beta \in ([\alpha]-1, \alpha)$ is any fixed non-integral number and $M \equiv M_{\rho,\beta}$ is a constant depending on ρ and β. One can verify that in the above inequality

$$|Re^{i\vartheta} - \overline{w}| > \frac{1}{\sqrt{2}} \left(|R - |w|| + \frac{\rho}{\sqrt{R}} \sqrt{|w|} \right).$$

Hence

$$|F(Re^{i\vartheta})| \le M' \iint_{G^+} \frac{|f(w)| d\mu_\omega(w)}{\left(|R - |w|| + \frac{\rho}{\sqrt{R}} \sqrt{|w|} \right)^{2+\beta}}, \tag{4.1}$$

where $f(w) \in L^p_{\omega,\gamma} \subset L^1_{\omega,\gamma'}$ with some $\gamma' \in (0,1)$ (see (3.14)).

If $|w| \le R/2$, then $1 \le (1 + |w|)^{-(1+\beta/2)} R^{(\gamma')^+}$, and

$$\iint_{\substack{w \in G^+ \\ |w| \le R/2}} \frac{|f(w)| d\mu_\omega(w)}{\left(|R - |w|| + \frac{\rho}{\sqrt{R}} \sqrt{|w|} \right)^{2+\beta}} \tag{4.2}$$

$$\le \iint_{\substack{w \in G^+ \\ |w| \le R/2}} |f(w)| \frac{d\mu_\omega(w)}{(|R - |w||)^{2+\beta}}$$

$$\le \left(\frac{2}{R} \right)^{2+\beta} \iint_{\substack{w \in G^+ \\ |w| \le R/2}} |f(w)| d\mu_\omega(w) \le \frac{2^{2+\beta}}{R^{2+\beta-(\gamma')^+}} \|f\|_{L^1_{\omega,\gamma'}}.$$

If $|w| > R/2$ and $R > 2$, then

$$\frac{1}{|w|^{1+\beta/2}} < \frac{2^{1+\beta/2}}{(1+|w|)^{1+\beta/2}} \quad \text{and} \quad \frac{1}{(1+|w|)^{1+\beta/2-\gamma'}} < \frac{R^{(1+\beta/2-\gamma')^-}}{2^{(1+\beta/2-\gamma')^+}},$$

and

$$\iint_{\substack{w \in G^+ \\ |w| > R/2}} \frac{|f(w)| d\mu_\omega(w)}{\left(|R-|w|| + \frac{\rho}{\sqrt{R}}\sqrt{|w|}\right)^{2+\beta}} \tag{4.3}$$

$$\leq \frac{R^{1+\beta/2}}{\rho^{2+\beta}} \iint_{\substack{w \in G^+ \\ |w| > R/2}} \frac{|f(w)|}{|w|^{1+\beta/2}} d\mu_\omega(w)$$

$$\leq 2^{1+\beta/2-(1+\beta/2-\gamma')^+} R^{1+\beta/2+(1+\beta/2-\gamma')^-} \rho^{-(2+\beta)} \|f\|_{L^1_{\omega,\gamma'}}.$$

The validity of (1.1) for the left-hand side integral of (3.12) follows from (4.1), (4.2) and (4.3). This proof makes obvious the validity of (1.1) for the left-hand side integral of (3.13).

If $\omega(x)$ is as required in Theorem 2.1, then the above argument is remains valid for any $\beta \in (-2, -1)$, and again we come to (1.1) for the right-hand side integrals in (2.8) and (2.9). $\qquad\square$

Remark 4.1. In view of the above Theorem 4.1, the representations (2.8), (2.9) and (3.12), (3.13) along with the kernel estimates of [7] can be used in proving projection theorems from $L^p_{\omega,\gamma}$ to $A^p_{\omega,\gamma}$ and revealing the duals of $A^p_{\omega,\gamma}$ spaces, in the same manner as it is done in [8] for the case of the unit disc.

References

[1] R.R. Coifman, R. Rochberg, *Representation theorems for holomorphic and harmonic functions in L^p*, Astérisque, **77** (1980), 12–66.

[2] F. Ricci, M. Taibleson, *Boundary values of harmonic functions in mixed norm spaces and their atomic structure*, Annali Scuola Normale Superiore – Pisa, Classe di Scienze, Ser. IV, **X** (1983), no. 1, 1–54.

[3] M.M. Djrbashian, A.E. Djrbashian, *Integral representation for some classes of analytic functions in a half-plane*, Dokl. Akad. Nauk USSR, **285** (1985), no. 3, 547–550.

[4] A.M. Jerbashian, *On some classes of subharmonic functions which have nonnegative harmonic majorants in a half-plane*, J. Contemp. Math. Anal., **28** (1993), no. 4, 42–61.

[5] A.H. Karapetyan, *Some questions of integral representations in multidimensional complex analysis*, Candidate Thesis, Yerevan, 1988. (Russian)

[6] M.M. Djrbashian, *Integral transforms and representations of functions in the complex domain*, Nauka, Moscow, 1966. (Russian)

[7] A.M. Jerbashian, *Evaluation of M.M.Djrbashian ω-kernels*, Archives of Inequalities and Applications, **1** (2003), 399–412.

[8] A.M. Jerbashian, *Weighted classes of regular functions area integrable over the unit disc*, Preprint 2002-01, Institute of Mathematics, National Acad. Sci. of Armenia, Yerevan, 2002.

[9] S.M. Gindikin, *Analysis in homogeneous domains*, Uspehi Mat. Nauk, **19**, no.4 (1964), 3–92; English transl. in: Russian Math. Surveys, **19**, no.4 (1964).

A.M. Jerbashian
24-b Marshal Baghramian Avenue
Institute of Mathematics
National Academy of Sciences of Armenia
375019 Yerevan
Armenia
e-mail: `armen_jerbashian@yahoo.com`

Operator Theory:
Advances and Applications, Vol. 158, 159–176

Estimate of the Cauchy Integral over Ahlfors Regular Curves

Mark Melnikov and Xavier Tolsa

Abstract. We obtain the complete characterization of those domains $G \subset \mathbb{C}$ which admit the so-called estimate of the Cauchy integral, that is to say, $\left| \int_{\partial G} f(z) \, dz \right| \leq C(G) \, \|f\|_\infty \, \gamma(E)$ for all $E \subset G$ and $f \in H^\infty(G \setminus E)$, where $\gamma(E)$ is the analytic capacity of E. The corresponding result for continuous functions f and the continuous analytic capacity $\alpha(E)$ is also proved.

1. Introduction

The problem of estimating the Cauchy integral over the boundary of a domain was posed by Vitushkin in connection with the theory of uniform rational approximation on compact subsets of the complex plane. The problem consists of characterizing those bounded domains $G \subset \mathbb{C}$ with rectifiable boundary for which there exists a constant $C_1(G)$ such that for any compact set $E \subset G$ and any function f bounded and holomorphic on $G \setminus E$ the following estimate holds:

$$\left| \int_{\partial G} f(z) \, dz \right| \leq C_1(G) \, \|f\|_\infty \gamma(E), \tag{1}$$

where $\gamma(E)$ is the analytic capacity of E (see next section for the precise meaning of the notions appearing in this section). Vitushkin also raised the analogous question for functions f continuous on $\bar{G}$ and holomorphic on $G \setminus E$, changing $\gamma(E)$ by the continuous analytic capacity $\alpha(E)$.

The estimate (1) was proved in 1966 in [Me1] for G being a disk, and more generally, for G with analytic boundary. Later on, Vitushkin [Vi1] proved the estimate (1) for domains G bounded by piecewise Lyapunov curves (i.e., piecewise $C^{1+\varepsilon}$ curves). In [De], Davie generalized the result to the case of hypolyapunov curves (i.e., curves satisfying a Dini type condition).

In [Vi2, Section III.1], Vitushkin showed an example of a domain G with rectifiable boundary such that the estimate of the integral (1) does not hold (that

The authors were partially supported by grants MTM2004-00519, HPRN-2000-0116, and 2001-SGR-00431.

is, rectifiability of the boundary alone does not imply (1)), and he conjectured that (1) holds if G is a Jordan domain such that

$$\gamma(F) \geq C_2(G)\,\mathcal{H}^1(F) \quad \text{for all closed subsets } F \subset \partial G, \tag{2}$$

for some constant $C_2(G) > 0$ ($\mathcal{H}^1$ denotes the one-dimensional Hausdorff measure, or arc length). In the present paper we prove that this conjecture is true.

Let us remark that when ∂G is a curve, condition (2) is equivalent to the fact ∂G is Ahlfors regular, that is to say,

$$\mathcal{H}^1(B(x,r) \cap \partial G) \leq Cr \quad \text{for all } x \in \mathbb{C},\, r > 0.$$

This follows from a theorem of David [Dd] which asserts that the Cauchy integral operator is bounded in L^2 on Ahlfors regular curves.

Now we state our result in detail.

Theorem 1. *Let G be a bounded open set in $\mathbb{C}$ whose boundary ∂G is a finite disjoint union of Jordan rectifiable (closed) curves. Then, the following conditions are equivalent:*

(a) *There exists some constant $C_2(G) > 0$ such that for any closed set $F \subset \partial G$,*

$$\gamma(F) \geq C_2(G)\,\mathcal{H}^1(F).$$

(b) *The Cauchy integral operator $\mathcal{C}_{\mathcal{H}^1|\partial G}$ is bounded on $L^2(\mathcal{H}^1_{|\partial G})$.*

(c) *There exists some constant $C_1(G)$ such that for any compact set $E \subset G$ and any function $f \in H^\infty(G \setminus E)$, we have*

$$\left| \int_{\partial G} f(z)\,dz \right| \leq C_1(G)\|f\|_\infty \gamma(E). \tag{3}$$

(d) *There exists some constant $C_3(G)$ such that for any compact set $E \subset G$ and any function $f \in H^\infty(G \setminus E) \cap C(\bar{G})$, we have*

$$\left| \int_{\partial G} f(z)\,dz \right| \leq C_3(G)\|f\|_\infty \alpha(E). \tag{4}$$

The constants $C_1(G)$ and $C_3(G)$ depend only on $C_2(G)$, or equivalently, on the $L^2(\mathcal{H}^1_{|\partial G})$ norm of $\mathcal{C}_{\mathcal{H}^1|\partial G}$, and conversely.

Notice that in Theorem 1 we consider a more general setting than G being a Jordan domain. Moreover, we show that Vitushkin's condition (2) is not only sufficient, but also necessary for the estimate of the Cauchy integral.

Vitushkin's original motivation for studying estimates like (1) was to obtain necessary and sufficient conditions for uniform rational approximation. Let $X \subset \mathbb{C}$ be compact and, as usual, let $R(X)$ be the algebra of uniform on X limits of rational functions with poles out of X, and $A(X)$ the algebra of functions continuous on X and holomorphic on $\overset{\circ}{X}$ (this is the interior of X). Given $f \in A(X)$ one asks when f also belongs to $R(X)$. A direct consequence of Theorem 1 is that such function f has to satisfy the following estimate, for any Jordan domain G with Ahlfors regular boundary:

$$\left| \int_{\partial G} f(z)\, dz \right| \le C(G)\, \omega_f\big(\mathrm{diam}(G)\big)\, \gamma(G \setminus X), \tag{5}$$

where $C(G)$ depends only on the Ahlfors regularity constant of ∂G and $\omega_f(\cdot)$ is the modulus of continuity of f. Conversely, condition (5) is also sufficient for f to be in $R(X)$. Indeed, it is enough that (5) holds either for all the *squares* $G \subset \mathbb{C}$ [Vi2] or for all the *disks* $G \subset \mathbb{C}$ [Pa].

In the proof of Theorem 1 we use the quite recent results and ideas involved in the proof of the semiadditivity of γ [To4] and α [To5].

The plan of the paper is the following. In Section 2 we recall some preliminary definitions and results. In Section 3 we prove the equivalence (a) $\Leftrightarrow$ (b) of Theorem 1. The main implications are (b) $\Rightarrow$ (c) and (b) $\Rightarrow$ (d), and they are proved in Sections 4 and 5 respectively. In Section 6 we show that (c) $\Rightarrow$ (a) and, finally, in Section 7 that (d) $\Rightarrow$ (a).

2. Preliminaries

Throughout all the paper, the letter C will stand for an absolute constant that may change at different occurrences. Constants with subscripts, such as C_1, will retain its value, in general.

The notation $A \approx B$ means that there exists an absolute constant $C > 0$ such that $C^{-1}A \le B \le CA$.

2.1. Analytic capacity, the Cauchy transform, and curvature of measures

The *analytic capacity* of a compact set $E \subset \mathbb{C}$ is

$$\gamma(E) = \sup |f'(\infty)|, \tag{6}$$

where the supremum is taken over all holomorphic functions $f : \mathbb{C} \setminus E \longrightarrow \mathbb{C}$ with $|f| \le 1$ on $\mathbb{C} \setminus E$, and $f'(\infty) = \lim_{z \to \infty} z(f(z) - f(\infty))$.

The continuous analytic capacity of E is

$$\alpha(E) = \sup |f'(\infty)|,$$

where the supremum is taken over all complex functions which are continuous in $\mathbb{C}$, holomorphic on $\mathbb{C} \setminus E$, and satisfy $|f(z)| \le 1$ for all $z \in \mathbb{C}$.

A positive Radon measure μ is said to have linear growth if there exists some constant C such that $\mu(B(x,r)) \le Cr$ for all $x \in \mathbb{C}$, $r > 0$. The linear density of μ at $x \in \mathbb{C}$ is (if it exists)

$$\Theta_\mu(x) = \lim_{r \to 0} \frac{\mu(B(x,r))}{r}.$$

Given a complex Radon measure ν on $\mathbb{C}$, the *Cauchy transform* of ν is

$$\mathcal{C}\nu(z) = \int \frac{1}{\xi - z}\, d\nu(\xi).$$

This definition does not make sense, in general, for $z \in \text{supp}(\nu)$, although one can easily see that the integral above is convergent at a.e. $z \in \mathbb{C}$ (with respect to Lebesgue measure). This is the reason why one considers the *truncated Cauchy transform* of ν, which is defined as

$$\mathcal{C}_\varepsilon \nu(z) = \int_{|\xi - z| > \varepsilon} \frac{1}{\xi - z} \, d\nu(\xi),$$

for any $\varepsilon > 0$ and $z \in \mathbb{C}$. Given a μ-measurable function f on $\mathbb{C}$ (where μ is some fixed positive Radon measure on $\mathbb{C}$), the Cauchy integral operator $\mathcal{C}_\mu$ is defined by

$$\mathcal{C}_\mu f := \mathcal{C}(f \, d\mu).$$

The ε-truncated version of $\mathcal{C}_\mu$ is $\mathcal{C}_{\mu,\varepsilon} f := \mathcal{C}_\varepsilon(f \, d\mu)$. We say the Cauchy integral operator is bounded on $L^2(\mu)$ if the operators $\mathcal{C}_{\mu,\varepsilon}$ are bounded on $L^2(\mu)$ uniformly on $\varepsilon > 0$.

The maximal Cauchy transform of a complex measure ν is

$$\mathcal{C}_* \nu(x) = \sup_{\varepsilon > 0} |\mathcal{C}_\varepsilon \nu(x)|.$$

We also set $\mathcal{C}_{\mu,*}(f) = \mathcal{C}_*(f \, d\mu)$.

If in the supremum (6) which defines $\gamma(E)$, additionally we ask the functions f to be Cauchy transforms of *positive* measures supported in E, we get the capacity γ_+ of E. The definition of α_+ is analogous.

Given three pairwise different points $x, y, z \in \mathbb{C}$, their *Menger curvature* is

$$c(x, y, z) = \frac{1}{R(x, y, z)},$$

where $R(x, y, z)$ is the radius of the circumference passing through x, y, z (with $R(x, y, z) = \infty$, $c(x, y, z) = 0$ if x, y, z lie on the same line). If two among these points coincide, we let $c(x, y, z) = 0$. For a positive Radon measure μ, we set

$$c_\mu^2(x) = \iint c(x, y, z)^2 \, d\mu(y) d\mu(z),$$

and we define the *curvature of μ* as

$$c^2(\mu) = \int c_\mu^2(x) \, d\mu(x) = \iiint c(x, y, z)^2 \, d\mu(x) d\mu(y) d\mu(z). \tag{7}$$

The notion of curvature of measures was introduced in [Me2], where some estimates of analytic capacity in terms of curvature were obtained.

Curvature of measures is connected to the Cauchy transform too. Indeed, Melnikov and Verdera [MV] proved that if μ has linear growth, then

$$\|\mathcal{C}_\varepsilon \mu\|_{L^2(\mu)}^2 = \frac{1}{6} c_\varepsilon^2(\mu) + O(\mu(\mathbb{C})), \tag{8}$$

where $c_\varepsilon^2(\mu)$ is an ε-truncated version of $c^2(\mu)$ (defined as in the right-hand side of (7), but with the triple integral over $\{x, y, z \in \mathbb{C} : |x - y|, |y - z|, |x - z| > \varepsilon\}$).

In [To4], the following result has been proved:

Theorem A. *For any compact set E, we have*

$$\gamma(E) \approx \gamma_+(E)$$

$$\approx \sup\left\{\mu(E) : \operatorname{supp}(\mu) \subset E,\ \mu(B(x,r)) \leq r\,\forall x \in E,\ r > 0\ \text{and}\ c^2(\mu) \leq \mu(E)\right\}$$

$$\approx \sup\left\{\mu(E) : \operatorname{supp}(\mu) \subset E,\ \mu(B(x,r)) \leq r\,\forall x \in E,\ r > 0\ \text{and}\right.$$

$$\left.\|\mathcal{C}_\mu\|_{L^2(\mu),L^2(\mu)} \leq 1\right\}, \tag{9}$$

with absolute constants.

The corresponding result for α was obtained in [To5].

Theorem B. *For any compact set E, we have*

$$\alpha(E) \approx \alpha_+(E)$$

$$\approx \sup\left\{\mu(E) : \operatorname{supp}(\mu) \subset E,\ \Theta_\mu(x) = 0\ \forall x \in E,\right.$$

$$\left.\mu(B(x,r)) \leq r\,\forall x \in E, r > 0\ \text{and}\ c^2(\mu) \leq \mu(E)\right\} \tag{10}$$

$$\approx \sup\left\{\mu(E) : \operatorname{supp}(\mu) \subset E,\ \Theta_\mu(x) = 0\ \forall x \in E,\right.$$

$$\left.\mu(B(x,r)) \leq r\,\forall x \in E, r > 0\ \text{and}\ \|\mathcal{C}_\mu\|_{L^2(\mu),L^2(\mu)} \leq 1\right\},$$

with absolute constants.

A direct consequence of the Theorems A and B is that γ and α are semiadditive. That is, for all compact sets $E, F \subset \mathbb{C}$,

$$\gamma(E \cup F) \leq C\big(\gamma(E) + \gamma(F)\big),$$

and

$$\alpha(E \cup F) \leq C\big(\alpha(E) + \alpha(F)\big).$$

The following potential was introduced by Verdera in [Ve]:

$$U_\mu(x) := M\mu(x) + c_\mu(x), \tag{11}$$

where M is the maximal radial Hardy-Littlewood operator:

$$M\mu(x) = \sup_{r>0} \frac{\mu(B(x,r))}{r},$$

and $c_\mu(x) = \big(c_\mu^2(x)\big)^{1/2}$. We remark that γ can be characterized in terms of this potential:

$$\gamma(E) \approx \sup\{\mu(E) : \operatorname{supp}(\mu) \subset E,\ U_\mu(x) \leq 1\,\forall x \in \mathbb{C}\}.$$

An analogous characterization for α and α_+ exists. See [To5] for the details.

2.2. Vitushkin's localization operator V_φ

Given $f \in L^1_{loc}(\mathbb{C})$ and $\varphi \in C^\infty$ compactly supported, we set

$$V_\varphi f := \varphi f - \frac{1}{\pi} \mathcal{C}(f\,\bar{\partial}\varphi) = \frac{1}{\pi} \mathcal{C}(\varphi\,\bar{\partial}f).$$

Here $\bar{\partial}f$ should be understood in the sense of distributions. Recall that if the support φ is contained in a ball of radius r, $\|\varphi\|_\infty \leq C_4$, and $\|\nabla\varphi\|_\infty \leq C_4/r$, then $V_\varphi f$ has the following properties (see [Ga, Lemma VIII-7.1], for example):

- $\|V_\varphi f\|_\infty \leq C_5\|f\|_\infty$,
- $\|V_\varphi f\|_\infty \leq C_6\,\omega_f(r)$, where ω_f stands for the modulus of continuity of f,
- $V_\varphi f$ is holomorphic outside $\operatorname{supp}(\bar{\partial}f) \cap \operatorname{supp}(\varphi)$,
- if f is continuous on $\mathbb{C}$, then $V_\varphi f$ is also continuous in $\mathbb{C}$.

The constants C_5 and C_6 depend only on C_4.

3. Proof of the equivalence (a) $\Leftrightarrow$ (b)

The implication (b) $\Rightarrow$ (a) is well known. It follows from the fact that the L^2 boundedness of the Cauchy integral operator implies that it is of weak type $(1,1)$ (see [NTV1] or [To2], for example), and from a dualization of the weak $(1,1)$ inequality. It can also be proved using the estimate of analytic capacity in terms of curvature obtained in [Me2].

We consider now the other implication. So we assume that

$$\gamma(F) \geq C^{-1}\mathcal{H}^1(F) \quad \text{for all closed sets } F \subset \partial G. \tag{12}$$

This condition implies that $\mathcal{H}^1_{|\partial G}$ has linear growth, because for any closed ball $\bar{B}(x,r)$ we have

$$\mathcal{H}^1(\bar{B}(x,r) \cap \partial G) \leq C\gamma(\bar{B}(x,r) \cap \partial G) \leq Cr.$$

In [To3] it has been shown that there is an absolute constant C_7 such that for any complex Radon measure ν and any $\lambda > 0$ the following holds:

$$\gamma_+\left\{x \in \mathbb{C} : \mathcal{C}_*\nu(x) > \lambda\right\} \leq C_7\frac{\|\nu\|}{\lambda}. \tag{13}$$

Then, by (12) and the comparability between γ and γ_+, we get

$$\mathcal{H}^1\left\{x \in \partial G : \mathcal{C}_*\nu(x) > \lambda\right\} \leq C\gamma\left\{x \in \partial G : \mathcal{C}_*\nu(x) > \lambda\right\} \leq C\frac{\|\nu\|}{\lambda}.$$

Thus the Cauchy integral operator is bounded from the space of complex Radon measures $M(\mathbb{C})$ into $L^{1,\infty}(\mathcal{H}^1_{|\partial G})$. In particular, it is of weak type $(1,1)$ (with respect to arc length measure on ∂G). This is equivalent to the L^2 boundedness on ∂G (see [To1] or [NTV1], for example).

Let us remark that one can prove the implication (a) $\Rightarrow$ (b) by different arguments. For example, instead of using the inequality (13), one can use the local $T(b)$ theorem of Nazarov, Treil and Volberg [NTV2].

4. Proof of the implication (b) $\Rightarrow$ (c)

4.1. A preliminary lemma

The following result will play a key role in the proof of (b) $\Rightarrow$ (c).

Lemma 2. *Let $E \subset \mathbb{C}$ be compact. There exists an open set Ω containing E with a Whitney decomposition $\Omega = \bigcup_{i \in I} Q_i$, where $\{Q_i\}_{i \in I}$ are Whitney squares such that $\gamma_+(\Omega) \approx \gamma_+(E)$ and*

$$\sum_{i \in I} \gamma_+(E \cap 2Q_i) \le C\gamma_+(E).$$

Let us remark that, unless stated otherwise, we assume that all the squares are closed and have sides parallel to the axes.

Proof. In (a) and (b) in Lemma 5.1 of [To4] the same result has been proved assuming that E is a finite union of segments. To prove it for this type of sets, it has been shown in [To4, Lemma 4.1] that there exists a measure σ supported on E, with linear growth, such that $\sigma(E) \approx \gamma_+(E)$ and $U_\sigma(x) \ge 1$ for all $x \in E$. If E is an arbitrary compact set, then we have

$$\gamma_+(E) \approx \inf\{\mu(\mathbb{C}) : \mu \in M_+(\mathbb{C}),\ U_\mu(x) \ge 1 \,\forall x \in E\},$$

where $M_+(\mathbb{C})$ stands for the set of all positive Radon measures, by [To3, Theorem 3.3]. As a consequence, in this case there also exists some measure σ (which in general will not be supported on E and will not have linear growth) such that $\sigma(E) \approx \gamma_+(E)$ and $U_\sigma(x) \ge 1$ for all $x \in E$.

Now the same arguments used to prove (a) and (b) of [To4, Lemma 5.1] work since it can be checked that the assumptions concerning the support and linear growth of σ are not necessary. $\qquad\square$

4.2. Proof of the implication (b) $\Rightarrow$ (c)

Suppose that the Cauchy integral operator $\mathcal{C}_{\mathcal{H}^1_{|\partial G}}$ is bounded on $L^2(\mathcal{H}^1_{|\partial G})$. We want to show that (3) holds. We assume that f vanishes on $\mathbb{C} \setminus \bar{G}$ and also, by homogeneity, that $\|f\|_\infty \le 1$.

Let Ω be the open set containing E in Lemma 2, and $\Omega = \bigcup_{i \in I} Q_i$ a Whitney decomposition into squares satisfying the conditions mentioned in the same lemma. Now we consider a partition of unity: let $\{\varphi_i\}_{i \in I}$ be a family of C^∞ functions such that $0 \le \varphi_i \le 1$, $\|\nabla \varphi_i\|_\infty \le C/\ell(Q_i)$ and $\operatorname{supp}(\varphi_i) \subset \frac{3}{2}Q_i$ for each $i \in I$, so that $\sum_{i \in I} \varphi_i = 1$ on Ω.

Consider the (finite) subfamily of squares $\{Q_j\}_{j \in J}$, $J \subset I$, such that $2Q_j \cap E \ne \varnothing$. Notice that $\psi := \sum_{j \in J} \varphi_j = 1$ on a neighborhood of E. Moreover, ψ is a compactly supported C^∞ function because J is finite. Then we have

$$f = \sum_{j \in J} V_{\varphi_j} f + V_{1-\psi} f.$$

The function $V_{1-\psi}f$ is bounded on G, although its L^∞ norm may depend on $\#J$. Also, it is holomorphic in G because

$$\bar{\partial}(V_{1-\psi}f) = (1-\psi)\,\bar{\partial}f = 0 \quad \text{in } G,$$

since $1-\psi$ vanishes in a neighborhood of E and f is holomorphic in $G\setminus E$. Thus,

$$\int_{\partial G} f(z)\,dz = \sum_{j\in J}\int_{\partial G} V_{\varphi_j}f(z)\,dz. \tag{14}$$

Let $J_1 \subset J$ be the set of indices such that $2Q_j \cap \partial G = \varnothing$ if $j \in J_1$, and $J_2 = J\setminus J_1$. By the definition of γ, for $j\in J_1$ we have

$$\left|\int_{\partial G} V_{\varphi_j}f(z)\,dz\right| \le C\gamma(E\cap 2Q_j),$$

because $E\cap 2Q_i \subset G$ and $\|V_{\varphi_j}f\|_\infty \le C$. Therefore, using Lemma 2 and the fact that $\gamma\approx\gamma_+$, we obtain

$$\left|\sum_{j\in J_1}\int_{\partial G} V_{\varphi_j}f(z)\,dz\right| \le C\sum_{j\in J_1}\gamma(E\cap 2Q_j) \le C\gamma(E). \tag{15}$$

For $j\in J_2$ we will show below that

$$\left|\int_{\partial G} V_{\varphi_j}f(z)\,dz\right| \le C\big(\gamma(E\cap 2Q_j) + \mathcal{H}^1(\partial G\cap 3Q_j)\big). \tag{16}$$

Before proving this estimate, let us see that (3) follows from (14), (15), and (16). Indeed, using Lemma 2 again, by the finite overlap of the Whitney squares Q_j, we get

$$\left|\sum_{j\in J_2}\int_{\partial G} V_{\varphi_j}f(z)\,dz\right| \le C\Big(\sum_{j\in J}\gamma(E\cap 2Q_j) + \sum_{j\in J}\mathcal{H}^1(\partial G\cap 3Q_j)\Big)$$
$$\le C\big(\gamma(E) + \mathcal{H}^1(\partial G\cap\Omega)\big).$$

Because of the $L^2(\mathcal{H}^1_{|\partial G})$ boundedness of the Cauchy integral operator, we have

$$\mathcal{H}^1(\partial G\cap\Omega) \le C\gamma_+(\partial G\cap\Omega) \le C\gamma_+(\Omega) \le C\gamma_+(E).$$

Thus,

$$\left|\sum_{j\in J_2}\int_{\partial G} V_{\varphi_j}f(z)\,dz\right| \le C\gamma(E),$$

which, jointly with (15), yields (3).

It only remains to prove (16) for $j\in J_2$. Let D_j be a disk with radius $\ell(Q_j)/4$ whose center z_j coincides with the center of Q_j. Consider the following measure:

$$\nu_j := \frac{-1}{\pi\mathcal{H}^2(D_j)}\left(\int f\bar{\partial}\varphi_j\,dm\right)\mathcal{H}^2|D_j.$$

We want to compare $V_{\varphi_j}f$ with the function $g_j := \mathcal{C}\nu_j$. Notice that

$$\frac{1}{\pi}\int f\bar{\partial}\varphi_j\,dm = (V_{\varphi_j}f)'(\infty) = g_j'(\infty).$$

Then, using the definition of γ, the fact that $\|V_{\varphi_j} f\|_\infty \leq C$, and the semiadditivity of γ, we get

$$\left| \frac{1}{\pi} \int f \bar{\partial} \varphi_j \, dm \right| \leq C\gamma\big((E \cup \partial G) \cap 2Q_j\big) \leq C\big(\gamma(E \cap 2Q_j) + \gamma(\partial G \cap 2Q_j)\big)$$

$$\leq C\big(\gamma(E \cap 2Q_j) + \mathcal{H}^1(\partial G \cap 2Q_j)\big). \tag{17}$$

As a consequence,

$$\|g_j\|_\infty \leq \left| \frac{C}{\ell(Q_j)} \int f \bar{\partial} \varphi_j \, dm \right| \leq C \frac{\gamma(E \cap 2Q_j) + \mathcal{H}^1(\partial G \cap 2Q_j)}{\ell(Q_j)} \leq C. \tag{18}$$

In the last inequality we used the fact that the arc length measure on ∂G has linear growth, because the Cauchy integral operator is bounded on $L^2(\mathcal{H}^1_{|\partial G})$. Since g_j and $V_{\varphi_j} f$ are bounded by C and $(V_{\varphi_j} f)'(\infty) = g'_j(\infty)$, the following estimate holds for $z \notin 3Q_j$:

$$|V_{\varphi_j} f(z) - g_j(z)| \leq \frac{C\ell(2Q_j)\gamma\big((E \cup \partial G) \cap 2Q_j\big)}{\operatorname{dist}(z, 2Q_j)^2}$$

$$\leq \frac{C\ell(Q_j)\big[\gamma(E \cap 2Q_j) + \mathcal{H}^1(\partial G \cap 2Q_j)\big]}{|z - z_j|^2}. \tag{19}$$

Let us estimate the integral

$$\int_{\partial G} |V_{\varphi_j} f(z) - g_j(z)| \, d\mathcal{H}^1(z) = \int_{\partial G \cap 3Q_j} + \int_{\partial G \setminus 3Q_j} =: I_1 + I_2.$$

From the uniform boundedness of g_j and $V_{\varphi_j} f$, it follows immediately that $I_1 \leq C\mathcal{H}^1(\partial G \cap 3Q_j)$. To deal with I_2 we use (19) and the fact that the arc length measure on ∂G has linear growth. Then we get

$$I_2 \leq \int_{\partial G \setminus 3Q_j} \frac{C\ell(Q_j)\big[\gamma(E \cap 2Q_j) + \mathcal{H}^1(\partial G \cap 2Q_j)\big]}{|z - z_j|^2} \, d\mathcal{H}^1(z)$$

$$\leq C\big(\gamma(E \cap 2Q_j) + \mathcal{H}^1(\partial G \cap 2Q_j)\big).$$

As a consequence,

$$\left| \int_{\partial G} V_{\varphi_j} f(z) \, dz \right| \leq C\big(\gamma(E \cap 2Q_j) + \mathcal{H}^1(\partial G \cap 3Q_j)\big) + \left| \int_{\partial G} g_j(z) \, dz \right|.$$

By Fubini and (17) we obtain

$$\left| \int_{\partial G} g_j(z) \, dz \right| = \left| \int_{\partial G} C\nu_j(z) \, dz \right| = C|\nu_j(G)| \leq C\big(\gamma(E \cap 2Q_j) + \mathcal{H}^1(\partial G \cap 2Q_j)\big).$$

So (16) holds, and the theorem follows.

5. Proof of the implication (b) $\Rightarrow$ (d)

5.1. The capacities γ^h

The capacity γ^h of a compact set $E \subset \mathbb{C}$, introduced in [To5], is defined as follows. We consider a continuous function $h : (0, +\infty) \to (0, +\infty)$ such that $h(r)/r$ is non-decreasing in r,

$$h(r) \leq r \quad \text{and} \quad h(2r) \leq 4h(r) \quad \text{for all } r > 0, \tag{20}$$

and moreover

$$\lim_{r \to 0+} \frac{h(r)}{r} = 0. \tag{21}$$

We set

$$\gamma^h(E) = \sup |f'(\infty)|,$$

where the supremum is taken over all functions $f \in L^\infty(\mathbb{C})$ which are holomorphic in $\mathbb{C} \setminus E$, with $f(\infty) = 0$, $\|f\|_{L^\infty(\mathbb{C})} \leq 1$, such that

$$\left| \int f \, \bar{\partial} \varphi \, d\mathcal{L}^2 \right| \leq h(r) r \|\nabla \varphi\|_\infty \tag{22}$$

for any real function $\varphi \in \mathcal{C}_c^\infty$ supported on some ball of radius r. If f satisfies all these properties we say that f is admissible for γ^h and E, and we write $f \in A^h(E)$.

The capacity $\gamma_+^h(E)$ is defined in an analogous way, but we ask an additional condition on the functions in the supremum above. Namely, f should be the Cauchy transform of some positive Radon measure supported on E.

Let us remark that the doubling property $h(2r) \leq 4h(r)$, for $r > 0$, implies that

$$h(\lambda t) \leq 4\lambda^2 h(t) \qquad \text{for all } \lambda > 1 \text{ and } t > 0. \tag{23}$$

In [To5], the following result has been proved:

Theorem C. *For any compact set E, we have*

$$\alpha(E) \approx \sup_h \gamma^h(E),$$

with the supremum over all continuous functions $h : (0, +\infty) \longrightarrow (0, +\infty)$ satisfying (20) and (21), with $h(r)/r$ non-decreasing. Moreover, for any fixed h fulfilling these properties,

$$\gamma^h(E) \approx \gamma_+^h(E)$$

$$\approx \sup\left\{\mu(E) : \text{supp}(\mu) \subset E, \ \mu(B(x,r)) \leq h(r) \ \forall x \in E, \ r > 0 \text{ and } c^2(\mu) \leq \mu(E)\right\}$$

$$\approx \sup\left\{\mu(E) : \text{supp}(\mu) \subset E, \ \mu(B(x,r)) \leq h(r) \ \forall x \in E, \ r > 0 \text{ and }\right.$$

$$\left. \|\mathcal{C}_\mu\|_{L^2(\mu), L^2(\mu)} \leq 1\right\}, \tag{24}$$

with absolute constants (independent of E and h, in particular). As a consequence, γ^h and α are countably semiadditive.

The next lemma asserts that the localization operator V_φ behaves well with respect to the capacity γ^h:

Lemma 3 ([To5]). *Let $E \subset \mathbb{C}$ be compact and $f \in A^h(E)$. Let φ be a C^∞ function supported on $\bar{B}(x_0, r)$, such that $\|\varphi\|_\infty \leq C_8$ and $\|\nabla\varphi\|_\infty \leq C_8 r^{-1}$. Then there exists some constant C depending on C_8 such that $C^{-1}V_\varphi f \in A^h(E \cap \bar{B}(x_0, r))$.*

5.2. Estimate of the Cauchy integral in terms of γ^h

We will prove the following result, which may have some independent interest.

Theorem 4. *Let G be a bounded open set in $\mathbb{C}$ such that its boundary ∂G is a finite disjoint union of Jordan rectifiable curves. Suppose that the Cauchy integral operator $C_{\mathcal{H}^1|\partial G}$ is bounded on $L^2(\mathcal{H}^1_{|\partial G})$. Then there exists a constant $C_9(G)$, which only depends on the $L^2(\mathcal{H}^1_{|\partial G})$ norm of $C_{\mathcal{H}^1|\partial G}$, such that for any compact set $E \subset \mathbb{C}$ and any function $f \in C(\bar{\mathbb{C}}) \cap H^\infty(G \setminus E)$, we have*

$$\left| \int_{\partial G} f(z)\, dz \right| \leq C_9(G)\gamma^h(E), \tag{25}$$

where $h(r) = r\,\omega_f(r)$.

The implication (b) $\Rightarrow$ (d) in Theorem 1 follows easily from the preceding theorem. Indeed, given $f \in C(\bar{G}) \cap H^\infty(G \setminus E)$, we may extend it continuously to the whole complex plane without increasing $\|f\|_\infty$. Then we apply Theorem 4 to $f/\|f\|_\infty$, and we get

$$\left| \int_{\partial G} f(z)\, dz \right| \leq C_9(G)\|f\|_\infty\gamma^h(E) \leq C\|f\|_\infty\alpha(E)$$

(observe that $C_9(G)$ does not depend on h).

Let us remark that the condition $f \in C(\bar{\mathbb{C}}) \cap H^\infty(G \setminus E)$, implies that $C^{-1}f \in A^h(E)$ for some absolute constant C and $h(r) = r\,\omega_f(r)$ (see [To5, Lemma 4.1]. The converse implication seems to be false in general (as far as we know).

5.3. Preliminary lemmas for the proof of Theorem 4

The following lemma will play an essential role in the proof of Theorem 4.

Lemma 5. *Let $E \subset \mathbb{C}$ be compact. There exists an open set Ω containing E with a Whitney decomposition $\Omega = \bigcup_{i \in I} Q_i$, where $\{Q_i\}_{i \in I}$ are Whitney squares such that $\gamma^h_+(\Omega) \approx \gamma^h_+(E)$ and*

$$\sum_{i \in I} \gamma^h_+(E \cap 2Q_i) \leq C\gamma^h_+(E).$$

The proof of this result is similar to the one of Lemma 2. See Lemma 7.2 of [To5] for the details.

We will also need next lemma.

Lemma 6. *Consider $f \in C(\bar{\mathbb{C}})$ and set $h(r) = r\omega_f(r)$. Let φ be a C^∞ function supported on a square Q such that $\|\varphi\|_\infty \leq 1$ and $\|\nabla\varphi\|_\infty \leq C_{10}/\ell(Q)$. Let Γ be an Ahlfors regular closed Jordan curve which intersects Q. We have*

$$\left| \int_\Gamma V_\varphi f(z)\,dz \right| \leq C_{11} h(\mathrm{diam}(\Gamma \cap 2Q)),$$

where C_{11} only depends on C_{10} and the Ahlfors regularity constant of Γ.

Proof. We consider first the case $\Gamma \subset 2Q$. We take $z_0 \in \Gamma$. Notice that $V_\varphi f = V_\varphi(f - f(z_0))$, because $V_\varphi 1 = 0$. So if we set $\widetilde{f}(z) := f(z) - f(z_0)$, we have

$$\int_\Gamma V_\varphi f(z)\,dz = \int_\Gamma V_\varphi \widetilde{f}(z)\,dz = \int_\Gamma \varphi(z)\widetilde{f}(z)\,dz - \frac{1}{\pi} \int_\Gamma \mathcal{C}(\widetilde{f}\bar\partial\varphi)(z)\,dz =: I_1 + I_2.$$

First we estimate I_1:

$$\begin{aligned} |I_1| &\leq \int_\Gamma |\varphi(z)\widetilde{f}(z)|\,d\mathcal{H}^1(z) \leq \|\widetilde{f}\|_{\infty,\Gamma}\, \mathcal{H}^1(\Gamma) \\ &\leq C\omega_f(\mathrm{diam}(\Gamma))\mathcal{H}^1(\Gamma) \leq Ch(\mathrm{diam}(\Gamma)). \end{aligned}$$

Now we turn our attention to I_2. Observe that by Cauchy's formula we have

$$I_2 = C \int_{\mathrm{Int}(\Gamma)} \widetilde{f}(z)\bar\partial\varphi(z)\,d\mathcal{H}^2(z),$$

where $\mathrm{Int}(\Gamma)$ stands for the bounded component of $\mathbb{C} \setminus \Gamma$. Then we get

$$|I_2| \leq \frac{C}{\ell(Q)}\|\widetilde{f}\|_{\infty,\mathrm{Int}(\Gamma)}\,\mathcal{H}^2(\mathrm{Int}(\Gamma))^2 \leq \frac{C\,\omega_f(\mathrm{diam}(\Gamma))\,\mathrm{diam}(\Gamma)^2}{\ell(Q)} \leq Ch(\mathrm{diam}(\Gamma)).$$

Thus the lemma holds in this case.

Suppose now that $\Gamma \not\subset 2Q$. From the fact that $\Gamma \cap Q \neq \varnothing$, we deduce

$$\mathrm{diam}(\Gamma \cap 2Q) \geq \frac{\ell(Q)}{2}. \tag{26}$$

Let z_Q be the center of Q. Notice that the ball $B(z_Q, 2\ell(Q))$ contains $\mathrm{supp}(\varphi)$. We consider the curve $\widetilde{\Gamma} := \partial\big(B(z_Q, 2\ell(Q)) \cap \mathrm{Int}(\Gamma)\big)$. Since $V_\varphi f$ is holomorphic outside $B(z_Q, 2\ell(Q))$, by Cauchy's theorem we have

$$\int_\Gamma V_\varphi f(z)\,dz = \int_{\widetilde{\Gamma}} V_\varphi f(z)\,dz. \quad \text{Therefore,} \quad \left| \int_\Gamma V_\varphi f(z)\,dz \right| \leq C\|V_\varphi f\|_\infty \mathcal{H}^1(\widetilde{\Gamma}).$$

Recall that from the identity

$$V_\varphi f(z) = \frac{1}{\pi} \int \frac{f(z) - f(\xi)}{z - \xi}\,\bar\partial\varphi(\xi)\,d\mathcal{H}^2(\xi),$$

one easily infers that $\|V_\varphi f\|_\infty \leq C\omega_f(\ell(Q)) = Ch(\ell(Q))/\ell(Q)$. On the other hand, from the Ahlfors regularity of Γ, it easily follows that $\mathcal{H}^1(\widetilde{\Gamma}) \leq C\ell(Q)$. Thus, $\left|\int_\Gamma V_\varphi f(z)\,dz\right| \leq Ch(\ell(Q))$ and so the lemma also holds in this situation, by (26). $\qquad\square$

5.4. Proof of Theorem 4

Let Ω be the open set containing E described in Lemma 5, and $\Omega = \bigcup_{i \in I} Q_i$ the corresponding Whitney decomposition into squares Q_i. We consider the same family of C^∞ functions φ_i, $i \in I$, used in the proof of the implication (b) $\Rightarrow$ (c) of Theorem 1, with $\mathrm{supp}(\varphi_i) \subset \frac{3}{2}Q_i$, and also the same finite subfamily of squares Q_j, $j \in J \subset I$ (recall that these are the squares such that $2Q_j \cap E \neq \varnothing$). Arguing as in the proof of (b) $\Rightarrow$ (c), we have again

$$\int_{\partial G} f(z)\,dz = \sum_{j \in J} \int_{\partial G} V_{\varphi_j} f(z)\,dz. \tag{27}$$

Let $J_1 \subset J$ be the set of indices such that $2Q_j \cap \partial G = \varnothing$ if $j \in J_1$, and $J_2 = J \setminus J_1$. By Lemma 3 and the definition of γ^h, for $j \in J_1$ we have

$$\left| \int_{\partial G} V_{\varphi_j} f(z)\,dz \right| \leq C\gamma^h(E \cap 2Q_j),$$

since $E \cap 2Q_i \subset G$ and $C^{-1}V_{\varphi_j} f \in A^h(E \cap 2Q_j)$, for some constant C. Therefore, by Lemma 5 and the fact that $\gamma^h \approx \gamma^h_+$,

$$\left| \sum_{j \in J_1} \int_{\partial G} V_{\varphi_j} f(z)\,dz \right| \leq C \sum_{j \in J_1} \gamma^h(E \cap 2Q_j) \leq C\gamma^h(E). \tag{28}$$

Now we have to deal with the squares Q_j such that $2Q_j \cap \partial G = \varnothing$. Let Γ_k, $k \in K$, denote the family of disjoint closed Jordan curves such that $\partial G = \bigcup_{k \in K} \Gamma_k$. Using Lemma 6 and the definition of γ^h, for $j \in J_2$, we have

$$\left| \int_{\partial G} V_{\varphi_j} f(z)\,dz \right| \leq \sum_{k:\Gamma_k \cap 2Q_j \neq \varnothing} \left| \int_{\Gamma_k} \cdots \right| + \left| \sum_{k:\Gamma_k \cap 2Q_j = \varnothing} \int_{\Gamma_k} \cdots \right|$$

$$\leq C \sum_{k:\Gamma_k \cap 2Q_j \neq \varnothing} h\big(\mathrm{diam}(\Gamma_k \cap 4Q_j)\big) + C\gamma^h\big([E \cup G^c] \cap 2Q_j\big).$$

To estimate the last term we use the (countable) semiadditivity of γ^h and the fact that $\gamma^h(F) \leq Ch(\mathrm{diam}(F))$ for any compact set $F \subset \mathbb{C}$:

$$\gamma^h\big([E \cup G^c] \cap 2Q_j\big) \leq C\gamma^h(E \cap 2Q_j) + C\gamma^h(G^c \cap 2Q_j)$$

$$\leq C\gamma^h(E \cap 2Q_j) + C \sum_{k:\Gamma_k \cap 2Q_j \neq \varnothing} h\big(\mathrm{diam}(\Gamma_k \cap 4Q_j)\big).$$

Therefore, by Lemma 5 again,

$$\left| \int_{\partial G} f(z)\,dz \right| \leq C \sum_{j \in J_2} \sum_{k:\Gamma_k \cap 2Q_j \neq \varnothing} h\big(\mathrm{diam}(\Gamma_k \cap 4Q_j)\big) + C\gamma^h(E). \tag{29}$$

Our next objective consists of showing that the first term on the right-hand side in the preceding inequality is bounded above by $\gamma^h(E)$. To this end, for each $j \in J_2$ and $k \in K$ such that $\Gamma_k \cap 2Q_j \neq \varnothing$, we consider a square $P_k^j \subset 4Q_j$ with side length $\mathrm{diam}(\Gamma_k \cap 4Q_j)/2$ such that $P_k^j \cap \Gamma_k \neq \varnothing$. By Vitali's covering theorem, there exists a subfamily $\{P_n^m\}_{(m,n)\in S}$ of the squares P_k^j such that the squares $2P_n^m$, $(m,n) \in S$, are pairwise disjoint, and any square P_k^j is contained in some square $6P_n^m$, with $(m,n) \in S$. Now we consider the measure

$$\mu := \sum_{(m,n)\in S} \frac{h(\ell(P_n^m))}{\mathcal{H}^2(P_n^m)} \, \mathcal{H}^2_{|P_n^m}.$$

Observe that $\mathrm{supp}(\mu) \subset \Omega$. We claim that μ satisfies

$$\mu(\Omega) \geq C^{-1} \sum_{j\in J_2} \sum_{k:\Gamma_k\cap 2Q_j\neq\varnothing} h\big(\mathrm{diam}(\Gamma_k \cap 4Q_j)\big), \tag{30}$$

$$\mu(B(z,r)) \leq Ch(r) \quad \text{for all } z \in \mathbb{C}, \, r > 0, \tag{31}$$

and also that the Cauchy integral operator $\mathcal{C}_\mu$ is bounded on $L^2(\mu)$, with constants depending only on the norm of the Cauchy integral operator $\mathcal{C}_{\mathcal{H}^1|\partial G}$ on $L^2(\mathcal{H}^1_{|\partial G})$. Observe that using the characterization of γ^h in terms of the L^2 norm of the Cauchy integral operator in (24), from our claims we deduce

$$\sum_{j\in J_2} \sum_{k:\Gamma_k\cap 2Q_j\neq\varnothing} h\big(\mathrm{diam}(\Gamma_k \cap 4Q_j)\big) \leq C\mu(\Omega) \leq C\gamma^h(E),$$

and then, by (29), the theorem follows.

Let us see that (30) holds. Notice that for $j \in J_2$, and k such that $\Gamma_k \cap 2Q_j \neq \varnothing$, we have

$$\ell(P_k^j) = \frac{1}{2}\,\mathrm{diam}(\Gamma_k \cap 4Q_j) \leq C\mathcal{H}^1(\Gamma_k \cap 4Q_j).$$

Taking into account that $h(r)/r$ is non-decreasing, if $2P_k^j \subset 6P_n^m$, we get

$$h\big(\mathrm{diam}(\Gamma_k \cap 4Q_j)\big) \leq Ch(\ell(P_k^j)) \leq Ch(3\ell(P_n^m)) \frac{\ell(P_k^j)}{3\ell(P_n^m)}$$

$$\leq Ch(\ell(P_n^m)) \frac{\mathcal{H}^1(\Gamma_k \cap 4Q_j)}{\ell(P_n^m)}.$$

Therefore,

$$\sum_{j\in J_2} \sum_{k:\Gamma_k\cap 2Q_j\neq\varnothing} h\big(\mathrm{diam}(\Gamma_k \cap 4Q_j)\big) \leq \sum_{(m,n)\in S} \sum_{\substack{j,k:j\in J_2, \\ \Gamma_k\cap 2Q_j\neq\varnothing, \\ 2P_k^j\subset 6P_n^m}} h\big(\mathrm{diam}(\Gamma_k \cap 4Q_j)\big)$$

$$\leq C \sum_{(m,n)\in S} \sum_{\substack{j,k:j\in J_2, \\ 2P_k^j\subset 6P_n^m}} h(\ell(P_n^m)) \frac{\mathcal{H}^1(\Gamma_k \cap 4Q_j)}{\ell(P_n^m)}. \tag{32}$$

To estimate the last sum notice that, by construction, if $2P_k^j \subset 6P_n^m$, then we have

$$\Gamma_k \cap 4Q_j \subset 6P_k^j \subset 18P_n^m.$$

Thus, using the finite overlap of the Whitney squares Q_j, for each fixed $(m,n) \in S$, we obtain

$$\sum_{\substack{j,k:j\in J_2, \\ 2P_k^j \subset 6P_n^m}} \mathcal{H}^1(\Gamma_k \cap 4Q_j) \leq \sum_k \sum_{j\in J_2} \mathcal{H}^1(18P_n^m \cap \Gamma_k \cap 4Q_j)$$

$$\leq C \sum_k \mathcal{H}^1(18P_n^m \cap \Gamma_k) = \mathcal{H}^1(18P_n^m \cap \partial G) \leq C\ell(P_n^m).$$

If we plug this estimate into (32), (30) follows:

$$\sum_{j\in J_2} \sum_{k:\Gamma_k\cap 2Q_j\neq\varnothing} h\big(\mathrm{diam}(\Gamma_k \cap 4Q_j)\big) \leq C \sum_{(m,n)\in S} h(\ell(P_n^m)) = C\mu(\Omega).$$

Now we will prove (31). Observe that it is enough to prove it for $z \in \mathrm{supp}(\mu)$. So take a fixed P_n^m, $(m,n) \in S$, and $z \in P_n^m$. Suppose first that $r \leq \ell(P_n^m)/2$. Since $2P_n^m$ does not intersect any other square $P_{n'}^{m'}$, $(m',n') \in S$, and on $2P_n^m$ μ coincides with the Lebesgue measure times $h(\ell(P_n^m))/\ell(P_n^m)^2$, we have

$$\mu(B(z,r)) \leq C\frac{h(\ell(P_n^m))}{\ell(P_n^m)^2} r^2.$$

From the property (23) of the function h, setting $\lambda = \ell(P_n^m)/(2r)$, (31) follows (for $0 < r \leq \ell(P_n^m)/2$).

Assume now that $r > \ell(P_n^m)/2$. We set $T = \{(p,q) \in S : P_q^p \cap B(z,r) \neq \varnothing\}$. Since $2P_n^m \cap 2P_q^p = \varnothing$ if $(m,n) \neq (p,q)$, we have $r \geq \ell(P_q^p)/2$ for any $(p,q) \in T$. So there exists some constant C_{12} such that

$$\bigcup_{(p,q)\in T} P_q^p \subset B(z,C_{12}r).$$

Recall that

$$\ell(P_q^p) = \frac{1}{2}\mathrm{diam}(\Gamma_q \cap 4Q_p) \leq C\mathcal{H}^1(\Gamma_q \cap 4Q_p),$$

because $\Gamma_q \cap 2Q_p \neq \varnothing$. Then, by the finite overlap of the squares $4Q_p$, $p \in J_2$ (recall that the Q_p's are Whitney squares), we get

$$\sum_{(p,q)\in T} \ell(P_q^p) \leq C \sum_{(p,q)\in T} \mathcal{H}^1(\Gamma_q \cap 4Q_p) \leq C \sum_{(p,q)\in S} \mathcal{H}^1(B(z,C_{12}r)\cap\Gamma_q\cap 4Q_p)$$

$$\leq C\mathcal{H}^1(B(z,C_{12}r) \cap \partial G) \leq Cr.$$

Taking into account that $h(t)/t$ is non-decreasing, we obtain

$$\mu(B(z,r)) \leq \sum_{(p,q)\in T} h(\ell(P_q^p)) = \sum_{(p,q)\in T} \frac{h(\ell(P_q^p))}{\ell(P_q^p)} \ell(P_q^p)$$

$$\leq \frac{h(2r)}{2r} \sum_{(p,q)\in T} \ell(P_q^p) \leq Ch(2r) \leq Ch(r).$$

The $L^2(\mu)$ boundedness of the Cauchy integral operator $\mathcal{C}_\mu$ follows from the L^2 boundedness of the Cauchy integral operator on $L^2(\mathcal{H}^1_{|\partial G})$ by comparison, and using the linear growth of μ (which is a consequence of (31)). We leave the details for the reader.

Remark 7 The implication (b) $\Rightarrow$ (c) of Theorem 1 can be proved using arguments closer to the ones used to prove Theorem 4 than the ones in Subsection 4.2. More precisely, one can obtain a variant of Lemma 6 suitable for L^∞ functions, and use it to estimate $\int_{\partial G} V_{\varphi_j} f(z)\, dz$ for $j \in J_2$. Recall that, instead, in Subsection 4.2 the integral $\int_{\partial G} V_{\varphi_j} f(z)\, dz$ was estimated with the help of some auxiliary function g_j holomorphic outside some disk D_j.

6. Proof of the implication (c) $\Rightarrow$ (a)

We want to prove that if the estimate of the integral (3) holds, then for any set $F \subset \partial G$ we have

$$\gamma(F) \geq C^{-1} \mathcal{H}(F).$$

Given $n \geq 1$, let $\delta > 0$ be such that $\gamma(U_\delta(F)) \leq \gamma(F) + 1/n$. Using Vitali's covering theorem, it is easy to check that there exists a finite family of pairwise disjoint balls $\{B(x_j, r_j)\}_{j \in J}$ with $x_j \in \partial G$, $B(x_j, r_j) \subset U_\delta(F)$, such that

$$\mathcal{H}^1(F) \leq C\mathcal{H}^1\left(\bigcup_{j \in J} B(x_j, r_j) \cap F\right).$$

Moreover, using well-known elementary properties of the one-dimensional Hausdorff measure, since $\mathcal{H}^1(F) < \infty$, we may assume that the radii r_j are small enough so that $\mathcal{H}^1(B(x_j, r_j) \cap F) \leq 3r_j$, and then we have

$$\mathcal{H}^1(F) \leq C \sum_{j \in J} r_j. \tag{33}$$

Given any ε, with $0 < \varepsilon \leq \min_{j \in J} r_j/10$, for each fixed $j \in J$ there are points $a_j^+, b_j^+, a_j^-, b_j^- \in \mathbb{C}$ which satisfy the following properties:

(a) $a_j^+, b_j^+ \in B(x_j, r_j/3) \cap G$ and $a_j^-, b_j^- \in B(x_j, r_j/3) \cap \mathbb{C} \setminus \bar{G}$,
(b) $r_j/3 \leq |a_j^+ - b_j^+|$ and $r_j/3 \leq |a_j^- - b_j^-|$,
(c) $|a_j^+ - a_j^-| \leq \varepsilon$ and $|b_j^+ - b_j^-| \leq \varepsilon$, and
(d) there exists a simple rectifiable arc σ_j^+ contained in $B(x_j, r_j/2) \cap G$ whose endpoints are a_j^+, b_j^+, and another simple rectifiable arc σ_j^- contained in $B(x_j, r_j/2) \cap \mathbb{C} \setminus \bar{G}$ whose endpoints are a_j^-, b_j^-.

Notice that from (33) and the properties (a) and (b), we deduce

$$\mathcal{H}^1(F) \leq C \sum_{j \in J} |a_j^+ - b_j^+| \approx \sum_{j \in J} |a_j^- - b_j^-|. \tag{34}$$

For each $j \in J$, consider the univalued branch of the function

$$f_j^+(z) := \frac{-\overline{(b_j^+ - a_j^+)}^2}{|b_j^+ - a_j^+|^3} \left(z - \frac{a_j^+ + b_j^+}{2} + \sqrt{(z - a_j^+)(z - b_j^+)} \right),$$

which is holomorphic on $\mathbb{C} \setminus \sigma_j^+$ and vanishes at ∞. Let also f_j^- be the univalued branch, holomorphic on $\mathbb{C} \setminus \sigma_j^-$, of the analogous function f_j^-, vanishing at ∞, defined like f_j^+, interchanging a_j^+, b_j^+ by a_j^-, b_j^-. Then we have

$$\|f_j^+\|_\infty = \|f_j^-\|_\infty \leq C_{13}$$

and

$$\int_{\partial G} f_j^+(z)\, dz = (f_j^+)'(\infty) = |a_j^+ - b_j^+|, \qquad \int_{\partial G} f_j^-(z)\, dz = 0. \tag{35}$$

We set

$$f := \sum_{j \in J} (f_j^+ - f_j^-). \tag{36}$$

It is easy to check that

$$\|f_j^+ - f_j^-\|_{\infty, \mathbb{C} \setminus B(x_j, r_j)} = C(\varepsilon) \to 0 \quad \text{as } \varepsilon \to 0,$$

Thus, we deduce

$$\|f\|_\infty \leq 2C_{13} + C(\varepsilon) \cdot \#J.$$

If we choose ε such that $C(\varepsilon) \leq 1/\#J$, we have $\|f\|_\infty \leq 2C_{13} + 1$, and then using (34), (35), and (3), we get

$$\mathcal{H}^1(F) \leq C \sum_{j \in J} |a_j^+ - b_j^+| = \int_{\partial G} \sum_{j \in J} f_j^+(z)\, dz = \int_{\partial G} f(z)\, dz$$

$$\leq C\gamma\left(\bigcup_{j \in J} \sigma_j^+ \right) \leq C\gamma(U_\delta(F)) \leq C\left(\gamma(F) + \frac{1}{n} \right).$$

Since this estimate holds for any $n \geq 1$, we are done.

7. Proof of the implication (d) $\Rightarrow$ (a)

The proof is analogous to the one of the implication (c) $\Rightarrow$ (a). We only have to change the function f defined in (36) by a continuous function $\tilde{f}$ which coincides with f outside some small neighborhood of $\bigcup_{j \in J} (\sigma_j^+ \cup \sigma_j^-)$, so that $f = \tilde{f}$ on ∂G, and then we apply the estimate of the integral (4) for $\tilde{f}$.

References

[Dd]   G. David, *Opérateurs intégraux singuliers sur certaines courbes du plan complexe*, Ann. Sci. École Norm. Sup. (4) **17** (1984), no. 1, 157–189.

[De]   A.M. Davie, *Analytic capacity and approximation problems*, Trans. Amer. Math. Soc. **171** (1972), 409–444.

[Ga]   T. Gamelin, *Uniform Algebras,* Prentice Hall, Englewood Cliffs N.J., 1969.

[Me1]   M.S. Melnikov, *Estimate of the Cauchy integral over an analytic curve,* (Russian) Mat. Sb. **71(113)** (1966), 503-514. Amer. Math. Soc. Translation **80** (2) (1969), 243–256.

[Me2]   M.S. Melnikov, *Analytic capacity: discrete approach and curvature of a measure,* Sbornik: Mathematics **186** (6) (1995), 827–846.

[MV]   M.S. Melnikov and J. Verdera, *A geometric proof of the L^2 boundedness of the Cauchy integral on Lipschitz graphs,* Internat. Math. Res. Notices (1995), 325–331.

[NTV1]   F. Nazarov, S. Treil and A. Volberg, *Weak type estimates and Cotlar inequalities for Calderón-Zygmund operators in nonhomogeneous spaces,* Int. Math. Res. Not. **9** (1998), 463–487.

[NTV2]   F. Nazarov, S. Treil and A. Volberg, *Accretive system Tb-theorems on nonhomogeneous spaces,* Duke Math. J. **113** (2002), 259–312.

[Pa]   P. V. Paramonov, *Some new criteria for the uniform approximability of functions by rational fractions,* (Russian) Mat. Sb. 186 (1995), no. 9, 97–112. Translation in Sb. Math. 186 (1995), no. 9, 1325–1340.

[To1]   X. Tolsa, *L^2-boundedness of the Cauchy integral operator for continuous measures,* Duke Math. J. **98** (2) (1999), 269–304.

[To2]   X. Tolsa, *A proof of the weak $(1,1)$ inequality for singular integrals with non doubling measures based on a Calderón-Zygmund decomposition,* Publ. Mat. **45:1** (2001), 163–174.

[To3]   X. Tolsa, *On the analytic capacity γ_+,* Indiana Univ. Math. J. **51** (2) (2002), 317–344.

[To4]   X. Tolsa, *Painlevé's problem and the semiadditivity of analytic capacity,* Acta Math. **190** (2003), 105–149.

[To5]   X. Tolsa, *The semiadditivity of continuous analytic capacity and the inner boundary conjecture.* Amer. J. Math. **126** (2004), 523–567.

[Ve]   J. Verdera, *On the $T(1)$-theorem for the Cauchy integral.* Ark. Mat. **38** (2000), 183–199.

[Vi1]   A. G. Vitushkin, *Estimate of the Cauchy integral,* (Russian) Mat. Sb. **71(113)** (1966), 515–534.

[Vi2]   A. G. Vitushkin, *The analytic capacity of sets in problems of approximation theory,* Uspekhi Mat. Nauk. **22** (6) (1967), 141–199 (Russian); in Russian Math. Surveys **22** (1967), 139–200.

Mark Melnikov
Departament de Matemàtiques
Universitat Autònoma de Barcelona, Spain
e-mail: melnikov@mat.uab.es

Xavier Tolsa
ICREA and Departament de Matemàtiques
Universitat Autònoma de Barcelona, Spain
e-mail: xtolsa@mat.uab.es

Operator Theory:
Advances and Applications, Vol. 158, 177–189

Quasinormal Families
of Meromorphic Functions II

Xuecheng Pang, Shahar Nevo and Lawrence Zalcman

Abstract. Let $\mathcal{F}$ be a quasinormal family of meromorphic functions on D, all of whose zeros are multiple, and let φ be a holomorphic function univalent on D. Suppose that for any $f \in \mathcal{F}$, $f'(z) \neq \varphi'(z)$ for $z \in D$. Then $\mathcal{F}$ is quasinormal of order 1 on D. Moreover, if there exists a compact set $K \subset D$ such that each $f \in \mathcal{F}$ vanishes at two distinct points of K, then $\mathcal{F}$ is normal on D.

Mathematics Subject Classification (2000). 30D45.

1. Introduction

In this paper, we are concerned with the order of quasinormality of families of meromorphic functions on plane domains, all of whose zeros are multiple.

Recall that a family $\mathcal{F}$ of functions meromorphic on a plane domain $D \subset \mathbb{C}$ is said to be quasinormal on D [2] if from each sequence $\{f_n\} \subset \mathcal{F}$ one can extract a subsequence $\{f_{n_k}\}$ which converges locally uniformly with respect to the spherical metric on $D \setminus E$, where the set E (which may depend on $\{f_{n_k}\}$) has no accumulation point in D. If E can always be chosen to satisfy $|E| \leq \nu$, $\mathcal{F}$ is said to quasinormal of order ν on D. Thus a family is quasinormal of order 0 on D if and only if it is normal on D. The family $\mathcal{F}$ is said to (quasi)normal at $z_0 \in D$ if it is (quasi)normal on some neighborhood of z_0; thus $\mathcal{F}$ is quasinormal on D if and only if it is quasinormal at each point $z \in D$. On the other hand, $\mathcal{F}$ fails to be quasinormal of order ν on D precisely when there exist points $z_1, z_2, \ldots, z_{\nu+1}$ in D and a sequence $\{f_n\} \subset \mathcal{F}$ such that no subsequence of $\{f_n\}$ is normal at z_j, $j = 1, 2, \ldots, \nu + 1$.

Our point of departure is the following classical result of Gu [4].

Theorem A. *Let $\mathcal{F}$ be a family of functions meromorphic on D, and let $k \geq 1$ be an integer. If for each $f \in \mathcal{F}$ and $z \in D$, $f(z) \neq 0$ and $f^{(k)}(z) \neq 1$, then $\mathcal{F}$ is normal on D.*

The authors were supported by the German-Israeli Foundation for Scientific Research and Development, G.I.F. Grant No. G-643-117.6/1999. The first author was also supported by the NNSF of China Approved No. 10271122.

Theorem A has been generalized in a number of different directions; cf., for instance, [1], [5], [9], [10]. In the present work, we continue our study [7] of the situation in which the condition $f \neq 0$ is replaced by the assumption that all zeros of f are multiple and $\mathcal{F}$ is assumed to be quasinormal on D. Our main result in [7] was

Theorem B. *Let $\mathcal{F}$ be a quasinormal family of meromorphic functions on D, all of whose zeros are multiple. If for any $f \in \mathcal{F}$, $f'(z) \neq 1$ for $z \in D$, then $\mathcal{F}$ is quasinormal of order 1 on D.*

That $\mathcal{F}$ need not be normal on D is shown by the following example.

Example 1.1. Let $D = \{z : |z| < 1\}$ and $\mathcal{F} = \{f_\alpha\}$, where

$$f_\alpha(z) = \frac{(z + \alpha)^2}{z + 2\alpha} = z + \frac{\alpha^2}{(z + 2\alpha)}, \quad \alpha \in \mathbb{C} \setminus \{0\}.$$

Then all zeros of f_α are multiple and $f_\alpha'(z) \neq 1$. However, f_α takes on the values 0 and ∞ in any fixed neighborhood of 0 if α is sufficiently small, so $\mathcal{F}$ fails to be normal at 0.

In certain generalizations of Gu's Theorem, the requirement that $f'(z) \neq 1$ can be weakened to $f'(z) \neq \psi(z)$, where $\psi(z)$ is some fixed analytic function on D [5], [9], which in some cases may be required not to vanish on D. Theorem B does not admit such an extension.

Example 1.2. Consider the family $\mathcal{F} = \{f_n\}$ on $D = \{z : |z| < 1\}$, where

$$f_n(z) = \frac{\left(z - \frac{n+2}{2n}\right)^2}{z - 1/2}.$$

Then $\mathcal{F}$ fails to be normal at $z = 1/2$ but is quasinormal of order 1 on D. Let $\varphi(z) = e^{(z+1)/(z-1)}$. Then $\varphi(D) \subset D$; $\varphi'(z) \neq 0$ on D; and, for each $w \in D \setminus \{0\}$, $\varphi^{-1}(w)$ consists of countably many points of D accumulating at $z = 1$. Consider the family $\tilde{\mathcal{F}} = \{F_n\}$ on D, where $F_n = f_n \circ \varphi$. Then $\tilde{\mathcal{F}}$ is a quasinormal family of meromorphic functions on D, all of whose zeros are multiple. Also, for any $F \in \tilde{\mathcal{F}}$, $F'(z) = f'(\varphi(z))\varphi'(z) \neq \varphi'(z)$ since $f'(z) \neq 1$ for any $f \in \mathcal{F}$. However, $\tilde{\mathcal{F}}$ is not quasinormal of any finite order on D as no subsequence of $\tilde{\mathcal{F}}$ is normal at any point of $\varphi^{-1}(1/2)$.

On the other hand, we do have the following

Theorem. *Let $\mathcal{F}$ be a quasinormal family of meromorphic functions on D, all of whose zeros are multiple, and let φ be a holomorphic function univalent on D. Suppose that for any $f \in \mathcal{F}$, $f'(z) \neq \varphi'(z)$ for $z \in D$. Then $\mathcal{F}$ is quasinormal of order 1 on D. Moreover, if there exists a compact set $K \subset D$ such that each $f \in \mathcal{F}$ vanishes at two distinct points of K, then $\mathcal{F}$ is normal on D.*

Acknowledgment. This work was done while the first author (X.P.) held a research position at Bar-Ilan University. He thanks the Mathematics Department of that institution and his collaborators for their warm hospitality.

2. Notation and preliminary results

Let us set some notation. We denote by Δ the open unit disc in $\mathbb{C}$. For $z_0 \in \mathbb{C}$ and $r > 0$, $\Delta(z_0, r) = \{z : |z - z_0| < r\}$ and $\Delta'(z_0, r) = \{z : 0 < |z - z_0| < r\}$. We write $f_n \xrightarrow{\chi} f$ on D to indicate that the sequence $\{f_n\}$ converges to f in the spherical metric uniformly on compact subsets of D and $f_n \implies f$ on D if the convergence is in the Euclidean metric.

We require the following known results.

Lemma 2.1. *Let $\mathcal{F}$ be a family of functions meromorphic on Δ, all of whose zeros have multiplicity at least k, and suppose that there exists $A \geq 1$ such that $|f^{(k)}(z)| \leq A$ whenever $f(z) = 0$. Then if $\mathcal{F}$ is not normal at z_0, there exist, for each $0 \leq \alpha \leq k$,*

  a) *points $z_n \in \Delta$, $z_n \longrightarrow z_0$;*
  b) *functions $f_n \in \mathcal{F}$; and*
  c) *positive numbers $\rho_n \longrightarrow 0$*

such that $\rho_n^{-\alpha} f_n(z_n + \rho_n \zeta) = g_n(\zeta) \xrightarrow{\chi} g(\zeta)$ on $\mathbb{C}$, where g is a nonconstant meromorphic function on $\mathbb{C}$, all of whose zeros have multiplicity at least k, such that $g^{\#}(\zeta) \leq g^{\#}(0) = kA + 1$. In particular, g has order at most 2.

Here, as usual, $g^{\#}(\zeta) = |g'(\zeta)|/(1 + |g(\zeta)|^2)$ is the spherical derivative.

This is the local version of [8, Lemma 2] (cf. [5, Lemma 1], [11, pp. 216–217]). The proof consists of a simple change of variable in the result cited from [8]; cf. [6, pp. 299–300].

Lemma 2.2. *Let f be a meromorphic function on $\mathbb{C}$ such that $f(z) \neq 0$ and $f'(z) \neq c$ on $\mathbb{C}$, where $c \neq 0$. Then f is constant*

This is a special case of Hayman's alternative; cf. [3, Theorem 3].

Lemma 2.3. *Let $\mathcal{F}$ be a family of functions meromorphic on Δ, all of whose zeros and poles are multiple. If for each $f \in \mathcal{F}$, $f'(z) \neq 1$, $z \in D$, then $\mathcal{F}$ is normal on D.*

This is the case $n = 2$, $k = 1$ of Theorem 5 in [10].

Lemma 2.4. *Let f be a nonconstant meromorphic function of finite order on $\mathbb{C}$, all of whose zeros are multiple. If $f'(z) \neq c$ on $\mathbb{C}$ for $c \neq 0$, then*

$$f(z) = \frac{c(z - a)^2}{z - b}$$

for some a and b ($\neq a$) in $\mathbb{C}$.

This follows from Lemma 6 (with $j = 1$ and $k = 2$) and Lemma 8 (with $k = 1$) of [10].

3. Auxiliary lemmas

The proof of the theorem proceeds by a number of intermediate results. The first of these is a slight extension of Theorem A.

Lemma 3.1. *Let $\{f_n\}$ be a sequence of meromorphic functions on Δ and $\{\psi_n\}$ a sequence of holomorphic functions on Δ such that $\psi_n \Longrightarrow \psi$, where $\psi(z) \neq 0$ on Δ. If for each n, $f_n(z) \neq 0$ and $f'_n(z) \neq \psi_n(z)$ for $z \in \Delta$, then $\{f_n\}$ is normal on Δ.*

Proof. Suppose not. Then by Lemma 2.1, there exist points $z_n \to z_0 \in \Delta$, numbers $\rho_n \to 0^+$ and a subsequence of $\{f_n\}$ (which, renumbering, we continue to denote $\{f_n\}$) such that

$$g_n(\zeta) = \frac{f_n(z_n + \rho_n\zeta)}{\rho_n} \xrightarrow{\chi} g(\zeta),$$

where g is a nonconstant meromorphic function on $\mathbb{C}$. Clearly, $g_n(\zeta) \neq 0$ and $g'_n(\zeta) = f'_n(z_n + \rho_n\zeta) \neq \psi_n(z_n + \rho_n\zeta) \Longrightarrow \psi(z_0) \neq 0$. Since g is nonconstant, it follows from Hurwitz' Theorem that $g(\zeta) \neq 0$ on $\mathbb{C}$ and then that $g'(\zeta) \neq \psi_n(z_0)$. By Lemma 2.2, this implies that g is constant, a contradiction.

Lemma 3.2. *Let $\{a_k\}$ be a sequence in Δ which has no accumulation points in Δ and let $\{\psi_n\}$ be a sequence of holomorphic functions on Δ such that $\psi_n \Longrightarrow \psi$ on Δ, where $\psi(z) \neq 0, \infty$ on Δ. Let $\{f_n\}$ be a sequence of functions meromorphic on Δ, all of whose zeros are multiple, such that $f'_n(z) \neq \psi_n(z)$ for all n and all $z \in \Delta$. Suppose that*

(a) *no subsequence of $\{f_n\}$ is normal at a_1;*
(b) *there exists $\delta > 0$ such that each f_n has a single (multiple) zero on $\Delta(a_1, \delta)$; and*
(c) *$f_n \xrightarrow{\chi} f$ on $\Delta \setminus \{a_k\}_{k=1}^{\infty}$.*

Then

(d) *there exists $\eta_0 > 0$ such that for each $0 < \eta < \eta_0$, f_n has a single simple pole on $\Delta(a_1, \eta)$ for all sufficiently large n; and*
(e) *$f(z) = \int_{a_1}^{z} \psi(\zeta)d\zeta$.*

Remark. Since no subsequence of $\{f_n\}$ is normal at a_1, a_1 is the unique accumulation point of zeros in (b) as well as poles in (d).

Proof. It suffices to prove that each subsequence of $\{f_n\}$ has a subsequence which satisfies (d) and (e). So suppose we have a subsequence of $\{f_n\}$, which (to avoid complication in notation) we again call $\{f_n\}$.

Since $\{f_n\}$ is not normal at a_1, it follows from Lemma 2.1 that we can extract a subsequence (which, renumbering, we continue to call $\{f_n\}$), points $z_n \longrightarrow a_1$, and positive numbers $\rho_n \longrightarrow 0$ such that

$$g_n(\zeta) = \frac{f_n(z_n + \rho_n\zeta)}{\rho_n} \xrightarrow{\chi} g(\zeta), \tag{1}$$

where g is a nonconstant meromorphic function of finite order on $\mathbb{C}$, all of whose zeros are multiple. Since $g_n'(\zeta) = f_n'(z_n + \rho_n\zeta) \neq \psi_n(z_n + \rho_n\zeta) \Rightarrow \psi(a_1)$ and $g_n' \Longrightarrow g'$ on the complement of the poles of g, either $g' \neq \psi(a_1)$ or $g' \equiv \psi(a_1)$, by Hurwitz' Theorem. In the latter case, $g(\zeta) = \psi(a_1)\zeta + c$, which does not have multiple zeros. Thus $g'(\zeta) \neq \psi(a_1)$ on $\mathbb{C}$; so by Lemma 2.4,

$$g(\zeta) = \psi(a_1)\frac{(\zeta - a)^2}{(\zeta - b)} \tag{2}$$

for distinct complex numbers a and b. It now follows from the argument principle that there exist sequences $\xi_n \longrightarrow a$ and $\eta_n \longrightarrow b$ such that, for sufficiently large n, $g_n(\xi_n) = 0$ and $g_n(\eta_n) = \infty$. Thus, writing $z_{n,0} = z_n + \rho_n\xi_n$, $z_{n,1} = z_n + \rho_n\eta_n$, we have $z_{n,j} \longrightarrow a_1$ $(j = 0, 1)$, $f_n(z_{n,0}) = 0$ and $f_n(z_{n,1}) = \infty$.

Let us now assume that (d) has been shown to hold. It follows from Lemma 2.3 that the pole of f_n at $z_{n,1}$ is simple. The limit function f from (c) is either meromorphic on $\Delta \setminus \{a_k\}_{k=1}^{\infty}$ or identically infinite there. Suppose first that it is meromorphic on $\Delta \setminus \{a_k\}_{k=1}^{\infty}$. There exists $\delta_0 > 0$ such that f has no poles on $\Gamma = \{z : |z - a_1| = \delta_0\}$ and f_n' converges uniformly to f' on Γ. We claim that $f' \equiv \psi$ on $\Delta'(a_1, \delta_0)$. Indeed, otherwise by Hurwitz' Theorem, $f'(z) \neq \psi(z)$ on $\Delta'(a_1, \delta_0)$. Now $1/(f_n' - \psi_n)$ is analytic on $\Delta(a_1, \delta_0)$ and converges uniformly on Γ to $1/(f' - \psi)$. By the maximum principle, $1/(f_n' - \psi_n)$ converges uniformly on $\Delta(a_1, \delta_0)$, so $\{f_n'\}$ is normal at a_1. However, since $f_n'(z_{n,0}) = 0$ and $f_n'(z_{n,1}) = \infty$ and $z_{n,j} \longrightarrow a_1$ $(j = 0, 1)$, $\{f_n'\}$ is not equicontinuous at a_1, a contradiction.

Thus f has no poles on $\Delta'(a_1, \delta_0)$ and $f_n' \Longrightarrow \psi$ on $\Delta'(a_1, \delta_0)$. Hence for any $z, z_0 \in \Delta'(a_1, \delta_0)$

$$f_n(z) - f_n(z_0) = \int_{z_0}^{z} f_n'(\zeta)\, d\zeta \longrightarrow \int_{z_0}^{z} \psi(\zeta)d\zeta = \Psi(z) - \Psi(z_0),$$

where $\Psi' = \psi$. Taking a subsequence if necessary, we may suppose that $f_n(z_0) - \Psi(z_0) \longrightarrow \alpha$. We claim that $\alpha = -\Psi(a_1)$. For otherwise, taking $r < \delta_0$ such that

$$\max_{|z-a_1|=r} |\Psi(z) - \Psi(a_1)| < |\alpha + \Psi(a_1)|,$$

we have, for large n,

$$\frac{1}{2\pi i} \int_{|z-a_1|=r} \frac{f_n'(z)}{f_n(z)}\, dz$$

$$= \frac{1}{2\pi i} \int_{|z-a_1|=r} \frac{dz}{\Psi(z) - \Psi(a_1) + [f_n(z_0) - \Psi(z_0) + \Psi(a_1)]} = 0.$$

However, by the argument principle, the left-hand side is the number of zeros minus the number of poles (counting multiplicities) of f_n in $\Delta(a_1, r)$, which for large n is at least $2 - 1 = 1$. It follows that $f(z) = \Psi(z) - \Psi(a_1)$.

Suppose now that $f \equiv \infty$ on $\Delta \setminus \{a_k\}_{k=1}^{\infty}$. Let

$$F_n(z) = f_n(z)\frac{z - z_{n,1}}{(z - z_{n,0})^2}.$$

By (b), $F_n(z) \neq 0$ on $\Delta(a_1, \delta)$. Applying the maximum principle to the sequence $\{1/F_n\}$ of analytic functions, we see that $F_n \Longrightarrow \infty$ on $\Delta(a_1, \delta)$. We have

$$\frac{f_n(z_n + \rho_n \zeta)}{\rho_n} = \frac{F_n(z_n + \rho_n \zeta)}{\rho_n} \frac{(\rho_n \zeta + z_n - z_{n,0})^2}{(\rho_n \zeta + z_n - z_{n,1})}$$

$$\tag{3}$$

$$= F_n(z_n + \rho_n \zeta) \frac{(\zeta - \xi_n)^2}{\zeta - \eta_n}.$$

It follows from (1), (2), and (3) that $F_n(z_n + \rho_n \zeta) \longrightarrow \psi(a_1)$, which contradicts $F_n \Longrightarrow \infty$ near a_1. Thus the possibility $f \equiv \infty$ may be ruled out.

We have shown that when (d) obtains, (e) does as well. Now let us show that (d) must hold. Suppose not. Then, taking a subsequence and renumbering, we may assume that on any neighborhood of a_1, f_n has at least two poles for sufficiently large n. Keeping the notation established above, let $z_{n,2} \neq z_{n,1}$ be such that $f_n(z_{n,2}) = \infty$ and f_n has no poles in $\Delta'(z_{n,1}, |z_{n,1} - z_{n,2}|)$. Write $z_{n,2} = z_n + \rho_n \eta_n^*$. Then $z_{n,2} \longrightarrow a_1$ but $\eta_n^* \longrightarrow \infty$ since the right-hand side of (2) has but a single simple pole. Set

$$G_n(\zeta) = \frac{f_n(z_{n,1} + (z_{n,2} - z_{n,1})\zeta)}{z_{n,2} - z_{n,1}}.$$

Since $z_{n,2} - z_{n,1} \longrightarrow 0$, $G_n(\zeta)$ is defined for any $\zeta \in \mathbb{C}$ if n is sufficiently large; and $G_n'(\zeta) \neq 1$. Now $G_n(1) = \infty$. Also,

$$G_n(0) = \infty, \qquad G_n\left(\frac{z_{n,0} - z_{n,1}}{z_{n,2} - z_{n,1}}\right) = 0$$

and

$$\frac{z_{n,0} - z_{n,1}}{z_{n,2} - z_{n,1}} = \frac{\xi_n - \eta_n}{\eta_n^* - \eta_n} \longrightarrow 0,$$

so $\{G_n\}$ is not normal at 0. On the other hand, for n sufficiently large, G_n has only a single zero (which tends to 0 as $n \longrightarrow \infty$) on any compact subset of $\mathbb{C}$. Since $G_n'(\zeta) \neq \psi(z_{n,1} + (z_{n,2} - z_{n,1})\zeta) \Longrightarrow \psi(a_1)$, it follows from Lemma 3.1 that $\{G_n\}$ is normal on $\mathbb{C} \setminus \{0\}$. Taking a subsequence and renumbering, we may assume that $G_n \xrightarrow{\chi} G$ on $\mathbb{C} \setminus \{0\}$. Since G has only a single pole on Δ, conditions (a), (b), (c), and (d) hold for the sequence $\{G_n\}$ (defined, say, on $\Delta(0, 2)$) with $a_1 = 0$ and $\delta = 1$. Thus, by the first part of the proof, $G(\zeta) = \Psi(\zeta) - \Psi(a_1)$. But this contradicts $G(1) = \infty$. This completes the proof of Lemma 3.1.

Definition. Let $z_1, z_2 \in \mathbb{C}$ and put $\tilde{z} = (z_1 + z_2)/2$. We say that (z_1, z_2) is a nontrivial pair of zeros of f if

(i) $f(z_1) = f(z_2) = 0$ and
(ii) there exists z_3 such that $|z_3 - \tilde{z}| < |z_1 - z_2|$ and $|f'(z_3)| > 1$.

Note that (ii) is equivalent to

(ii') there exists z^* such that $|z^*| < 1$ and $|h'(z^*)| > 1$, where

$$h(z) = \frac{f(\tilde{z} + (z_1 - z_2)z)}{z_1 - z_2}.$$

Since $|h'(z)| \geq h^{\#}(z)$, it suffices to have $h^{\#}(z^*) > 1$ in (ii').

Our next result deals with the situation in which the functions f_n have more than a single zero in each neighborhood of a point of non-normality.

Lemma 3.3. *Let $\{f_n\}$ be a sequence of functions meromorphic on Δ, all of whose zeros are multiple; and let $\{\psi_n\}$ be a sequence of holomorphic functions on Δ such that $\psi_n \implies \psi$ on Δ, where $\psi(z) \neq 0, \infty$ on Δ. Suppose that $f_n'(z) \neq \psi_n(z)$ for all n and all $z \in \Delta$. Suppose further that*

(a) *no subsequence of $\{f_n\}$ is normal at z_0; and*
(b) *for each $\delta > 0$, f_n has at least two distinct zeros on $\Delta(z_0, \delta)$ for sufficiently large n.*

Then for each $\delta > 0$, f_n has a nontrivial pair (a_n, c_n) of zeros on $\Delta(z_0, \delta)$ for sufficiently large n, and

$$\left\{ \frac{f_n(d_n + (a_n - c_n)\zeta)}{a_n - c_n} \right\}$$

is not normal on Δ. Here $d_n = (a_n + c_n)/2$.

Proof. As in the proof of the previous lemma, it follows from (a) and Lemmas 2.1 and 2.4 that for each subsequence of $\{f_n\}$ there exists a (sub)subsequence (which, renumbering, we continue to denote by $\{f_n\}$), points $z_n \longrightarrow z_0$, numbers $\rho_n \longrightarrow 0^+$, and distinct $a, b \in \mathbb{C}$ such that

$$g_n(\zeta) = \frac{f_n(z_n + \rho_n\zeta)}{\rho_n} \overset{\chi}{\implies} g(\zeta) = \psi(z_0)\frac{(\zeta - a)^2}{\zeta - b} \quad \text{on} \quad \mathbb{C}. \tag{4}$$

Thus there exist $\xi_n \longrightarrow a$, $\eta_n \longrightarrow b$ so that $a_n = z_n + \rho_n\xi_n \longrightarrow z_0$, $b_n = z_n + \rho_n\eta_n \longrightarrow z_0$ and $g_n(\xi_n) = f_n(a_n) = 0$, $g_n(\eta_n) = f_n(b_n) = \infty$ for n sufficiently large.

By assumption, there also exist $c_n \neq a_n$, $c_n \longrightarrow z_0$, such that $f_n(c_n) = 0$. Thus $c_n = z_n + \rho_n\xi_n^*$ and $\xi_n^* \longrightarrow \infty$ by (4). Setting $d_n = (a_n + c_n)/2$, we see that the function

$$h_n(\zeta) = \frac{f_n(d_n + (a_n - c_n)\zeta)}{a_n - c_n}$$

is defined for any $\zeta \in \mathbb{C}$ if n is sufficiently large. We claim that $\{h_n\}$ is not normal at $\zeta = 1/2$. Indeed, we have

$$\frac{a_n - d_n}{a_n - c_n} \longrightarrow \frac{1}{2}, \qquad \frac{b_n - d_n}{a_n - c_n} \longrightarrow \frac{1}{2},$$

$$h_n\left(\frac{a_n - d_n}{a_n - c_n}\right) = f_n(a_n) = 0, \qquad h_n\left(\frac{b_n - d_n}{a_n - c_n}\right) = f_n(b_n) = \infty,$$

so $\{h_n\}$ fails to be equicontinuous in a neighborhood of $1/2$. It follows from Marty's Theorem that

$$\lim_{n \longrightarrow \infty} \sup_{|\zeta - \frac{1}{2}| \leq \frac{1}{4}} h_n^{\#}(\zeta) = \infty.$$

Thus (a_n, c_n) is a nontrivial pair of zeros of f_n for n sufficiently large.

Lemma 3.4. *Let $\{f_n\}$ be a sequence of functions meromorphic on Δ, all of whose zeros are multiple; and let $\{\psi_n\}$ be a sequence of holomorphic functions on Δ such that $\psi_n \Longrightarrow \psi$ on Δ, where $\psi(z) \neq 0$ on Δ. Suppose that $f_n'(z) \neq \psi_n(z)$ for all n and all $z \in \Delta$. Suppose further that*

(a) *there exist $d \in \Delta$, $a_n \longrightarrow d$, $c_n \longrightarrow d$, and $z_0 \in \mathbb{C}$ such that for every $\delta > 0$,*

$$h_n(z) = \frac{f_n(d_n + (a_n - c_n)z)}{a_n - c_n}$$

has at least two distinct zeros on $\Delta(z_0, \delta)$ for sufficiently large n, where $d_n = (a_n + c_n)/2$; and

(b) *no subsequence of $\{h_n\}$ is normal at z_0.*

Then for n sufficiently large, f_n has a nontrivial pair of zeros $(z_{n,1}^*, z_{n,2}^*)$ such that $z_{n,j}^* \longrightarrow d$ $(j = 1, 2)$ and $|z_{n,1}^* - z_{n,2}^*| < |a_n - c_n|$.

Proof. As before, it follows from Lemmas 2.1 and 2.4 that to each subsequence of $\{h_n\}$ there corresponds a subsequence (which we continue to write as $\{h_n\}$), $z_n \longrightarrow z_0$, and $\rho_n \longrightarrow 0^+$ such that

$$g_n(\zeta) = \frac{h_n(z_n + \rho_n \zeta)}{\rho_n} \stackrel{\chi}{\Longrightarrow} \psi(z_0) \frac{(\zeta - a)^2}{\zeta - b} \quad \text{on} \quad \mathbb{C}.$$

Thus there exist $\xi_{n,0} \longrightarrow b$, $\xi_{n,1} \longrightarrow a$ so that $z_{n,j} = z_n + \rho_n \xi_{n,j} \longrightarrow z_0$ $(j = 0, 1)$ and $g_n(\xi_{n,0}) = h_n(z_{n,0}) = \infty$, $g_n(\xi_{n,1}) = h_n(z_{n,1}) = 0$. By (a), there exist $z_{n,2} \longrightarrow z_0$, $z_{n,2} \neq z_{n,1}$, such that $h_n(z_{n,2}) = 0$. Setting $z_{n,2} = z_n + \rho_n \xi_{n,2}$, we have $\xi_{n,2} \longrightarrow \infty$. Now put

$$z_{n,j}^* = d_n + (a_n - c_n)z_n + \rho_n(a_n - c_n)\xi_{n,j} \quad j = 0, 1, 2.$$

Clearly $z_{n,j}^* \longrightarrow d$, $j = 0, 1, 2$. Define

$$G_n(\zeta) = \frac{f_n\left(\frac{z_{n,1}^* + z_{n,2}^*}{2} + (z_{n,1}^* - z_{n,2}^*)\zeta\right)}{z_{n,1}^* - z_{n,2}^*}.$$

Then $\{G_n\}$ is not normal at $\zeta = 1/2$. Indeed,

$$G_n\left(\frac{2\xi_{n,0} - \xi_{n,1} - \xi_{n,2}}{2(\xi_{n,1} - \xi_{n,2})}\right) = \infty, \qquad G_n(1/2) = 0.$$

Since $(2\xi_{n,0} - \xi_{n,1} - \xi_{n,2})/2(\xi_{n,1} - \xi_{n,2}) \longrightarrow 1/2$, $\{G_n\}$ is not equicontinuous at $\zeta = 1/2$. As before, it follows from Marty's Theorem that $(z_{n,1}^*, z_{n,2}^*)$ is a nontrivial pair of zeros of f_n. Now $|z_{n,1}^* - z_{n,2}^*| = |a_n - c_n|\,|z_{n,1} - z_{n,2}|$; therefore, since $z_{n,j} \longrightarrow z_0$ $(j = 1, 2)$, we have $|z_{n,1}^* - z_{n,2}^*| < |a_n - c_n|$ for large enough n, as required.

Lemma 3.5. *Let $\{f_n\}$ be a sequence of functions meromorphic on Δ, all of whose zeros are multiple, such that $f_n'(z) \neq 1$ for all n and all $z \in \Delta$. Suppose that*

(a) *$\{f_n\}$ is normal on $\Delta'(0,1)$, but no subsequence of $\{f_n\}$ is normal at 0; and*
(b) *there exists $\delta > 0$ such that f_n has a single (multiple) zero on $\Delta(0,\delta)$ for all sufficiently large n.*

Then there exists a subsequence of $\{f_n\}$ (which we continue to call $\{f_n\}$) such that for any $a \in \mathbb{C}$, $f_n - a$ has at most two zeros (counting multiplicity) on $\Delta(0,1/2)$.

Proof. Taking a subsequence and renumbering, we may assume that $f_n \overset{\chi}{\Longrightarrow} f$ on $\Delta'(0,1)$. By Lemma 3.2, $f(z) = z$. Suppose that $|a| \leq 2/3$. Taking Γ to be the circle $\{|z| = 3/4\}$ traversed once in the positive direction, we have

$$\frac{1}{2\pi i} \int_\Gamma \frac{f_n'(z)}{f_n(z) - a}\, dz \longrightarrow \frac{1}{2\pi i} \int_\Gamma \frac{1}{z - a}\, dz = 1.$$

However, the left-hand side is the number of a-points of f_n minus the number of poles of f_n inside Γ, counting multiplicities. By Lemma 3.2, there exists $0 < \delta < 3/4$ such that f_n has a single simple pole on $\Delta(0,\delta)$ for n sufficiently large. Since f_n converges uniformly to z on $\{z : \delta \leq |z| \leq 3/4\}$, there exists N_1 such that if $n \geq N_1$ f_n has a single simple pole in $\Delta(0,3/4)$. Hence for $n \geq N_1$, f_n takes on the value a (counting multiplicities) exactly twice on $\Delta(0,3/4)$.

Suppose now that $|a| > 2/3$. Let Γ' be the circle $\{|z| = 5/9\}$ traversed in the positive direction. Then

$$\frac{1}{2\pi i} \int_{\Gamma'} \frac{f_n'(z)}{f_n(z) - a}\, dz \longrightarrow \frac{1}{2\pi i} \int_{\Gamma'} \frac{1}{z - a} = 0,$$

so the number of a-points minus the number of poles of f_n (counting multiplicity) inside Γ' is 0 for large n. It follows as before that there exists N_2 such that f_n takes on the value a exactly once (counting multiplicities) on $\Delta(0,5/9)$ if $n \geq N_2$. Dropping the elements f_n with $n < \max(N_1, N_2)$ and renumbering, we obtain the desired sequence.

Lemma 3.6. *Let f be a meromorphic function on $\mathbb{C}$, all of whose zeros are multiple, such that $f'(z) \neq 1$, $z \in \mathbb{C}$. Then either*

(i) *f is rational; or*
(ii) *there exist nontrivial pairs (a_n, c_n) of zeros of f such that $|a_n - c_n| \longrightarrow 0$ and a sequence of functions*

$$h_n(\zeta) = \frac{f(d_n + (a_n - c_n)\zeta)}{a_n - c_n}$$

which is not normal on Δ; here $d_n = (a_n + c_n)/2$.

Proof. Suppose f is not rational. Then by Lemma 2.4, f has infinite order, so there exist $z_n \to \infty$ and $\varepsilon_n \to 0$ such that

$$S(\Delta(z_n, \varepsilon_n), f) = \frac{1}{\pi} \iint_{|z - z_n| \leq \varepsilon_n} [f^\#(z)]^2\, dx\, dy \longrightarrow \infty. \tag{5}$$

Indeed, otherwise there would exist $\varepsilon > 0$ and $M > 0$ such that $S(\Delta(\zeta, \varepsilon), f) \leq M$ for all $\zeta \in \mathbb{C}$. From this follows

$$S(r) = \frac{1}{\pi} \iint_{|z| < r} [f^\#(z)]^2 \, dx dy = O(r^2),$$

so that (cf. [11, p. 217]) f would have order at most 2, a contradiction. In particular, there exist $z_n^* \in \Delta(z_n, \varepsilon_n)$ such that $f^\#(z_n^*) \longrightarrow \infty$. Let $f_n(z) = f(z + z_n^*)$. Then no subsequence of $\{f_n\}$ is normal at 0.

Suppose there exists $\delta > 0$ such that f_n has only a single (multiple) zero ξ_n on $\Delta(0, \delta)$. Since no subsequence of $\{f_n\}$ is normal at 0, $\xi_n \longrightarrow 0$ by Theorem A. Thus, again by Theorem A, $\{f_n\}$ is normal on $\Delta'(0, \delta)$. It follows from Lemma 3.5 that there exist $n_1 < n_2 < \cdots$ such that for any $a \in \mathbb{C}$, $f_{n_k} - a$ has at most two zeros (counting multiplicity) on $\Delta(0, \delta/2)$. Thus, for large enough k,

$$S(\Delta(z_{n_k}, \varepsilon_{n_k}), f) \leq S(\Delta(0, \delta/2), f_{n_k}) \leq 2$$

which contradicts (5).

Thus, for each $\delta > 0$, f_n has at least two distinct zeros on $\Delta(0, \delta)$ for sufficiently large n. The result now follows immediately from Lemma 3.3.

Lemma 3.7. *Let $\{f_n\}$ be a sequence of functions meromorphic on Δ, all of whose zeros are multiple, and let ψ be a non-vanishing holomorphic function on Δ. Suppose that*

(a) *$\{f_n\}$ is quasinormal on Δ;*
(b) *$f_n'(z) \neq \psi(z)$ for $z \in \Delta$ and $n = 1, 2, 3, \ldots$;*
(c) *no subsequence of $\{f_n\}$ is normal at 0.*

Then there exists $\delta > 0$ such that f_n has only a single (multiple) zero on $\Delta(0, \delta)$ for sufficiently large n.

Proof. For otherwise, we may suppose that for any $\delta > 0$, f_n has at least two distinct zeros on $\Delta(0, \delta)$ for sufficiently large n. By Lemma 3.4, f_n has a nontrivial pair of zeros in $\Delta(0, \delta)$ for n large enough. Therefore, some subsequence of $\{f_n\}$ (which, as usual, we continue to call $\{f_n\}$) has a nontrivial pair of zeros (z_n, w_n) such that $|z_n| < 1/n$, $|w_n| < 1/n$. There exists $\delta' > 0$ such that $f_n \overset{\chi}{\Longrightarrow} f$ on $\Delta'(0, \delta')$. We claim that $f \not\equiv 0$. Otherwise, $f_n' - \psi \Rightarrow -\psi$ on $\Delta'(0, \delta')$. But no subsequence of $\{f_n'\}$ is normal at 0. Since $1/(f_n' - \psi)$ is holomorphic, $f_n' \Rightarrow \psi$ on $\Delta'(0, \delta')$, a contradiction.

Thus there exist $\delta_0 > 0$ and $1 < s < 2$ such that f does not vanish on $\{\delta_0 \leq |z| \leq s\delta_0\}$. For $1/n < \delta_0$, let (a_n, c_n) be a nontrivial pair of zeros of f_n in $\Delta(0, \delta_0)$ whose distance is minimal. Clearly, $a_n - c_n \longrightarrow 0$. Set $d_n = (a_n + c_n)/2$. Then $d_n \in \Delta(0, \delta_0)$; and, passing to a subsequence, we may assume that $d_n \longrightarrow a$, so $|a| \leq \delta_0$. Since f and f_n have no zeros on $\{z : \delta_0 \leq |z| \leq s\delta_0\}$ if n is large enough, (a_n, c_n) is a nontrivial pair of zeros of f_n on $\Delta(0, s\delta_0)$ whose distance is minimal.

Set

$$h_n(\zeta) = \frac{f_n(d_n + (a_n - c_n)\zeta)}{a_n - c_n}.$$

Then for each $\zeta \in \mathbb{C}$, $h_n(\zeta)$ is defined if n is sufficiently large. Clearly, all zeros of h_n are multiple and $h'_n(\zeta) \neq \psi(d_n + (a_n - c_n)\zeta)$. We claim that no subsequence of $\{h_n\}$ is normal on $\mathbb{C}$. Otherwise, taking a subsequence and renumbering, we would have $h_n \overset{\chi}{\Longrightarrow} h$ on $\mathbb{C}$. Since (a_n, c_n) is a nontrivial pair of zeros of f_n, $h_n(\pm 1/2) = h'_n(\pm 1/2) = 0$, and $\sup_{\Delta} |h'_n(z)| > 1$. It follows easily that $h'(\zeta) \neq \psi(a)$ on $\mathbb{C}$ and that h is nonconstant. Since all zeros of h are multiple, Lemma 2.4 shows that h must be transcendental. It then follows from Lemma 3.6 that there exist infinitely many nontrivial pairs (ξ_k, η_k) of zeros of h such that $\xi_k \longrightarrow \infty$ and $\xi_k - \eta_k \longrightarrow 0$, and z_k^* with $|z_k^* - (\xi_k + \eta_k)/2| < |\xi_k - \eta_k|$ and $h^\#(z_k^*) \longrightarrow \infty$.

Fix k such that $h^\#(z_k^*) \geq 2$ and $|\xi_k - \eta_k| < 1$. Then there exist $\xi_{n,k} \longrightarrow \xi_k$ and $\eta_{n,k} \longrightarrow \eta_k$ such that for n sufficiently large, $h_n(\xi_{n,k}) = h_n(\eta_{n,k}) = 0$ and $|z_k^* - (\xi_{n,k} + \eta_{n,k})/2| < |\xi_{n,k} - \eta_{n,k}|$. Put

$$\xi_{n,k}^* = d_n + (a_n - c_n)\xi_{n,k}, \quad \eta_{n,k}^* = d_n + (a_n - c_n)\eta_{n,k}, \quad z_{n,k}^* = d_n + (a_n - c_n)z_k^*.$$

Then

$$\left| z_{n,k}^* - \frac{\xi_{n,k}^* + \eta_{n,k}^*}{2} \right| = |a_n - c_n| \left| z_k^* - \frac{\xi_{n,k} + \eta_{n,k}}{2} \right|$$
$$< |a_n - c_n|\,|\xi_{n,k} - \eta_{n,k}| = |\xi_{n,k}^* - \eta_{n,k}^*|,$$

where $\xi_{n,k}^* \longrightarrow a$, $\eta_{n,k}^* \longrightarrow a$ and $|a| < s\delta_0$; also, for n sufficiently large, $|f'_n(z_{n,k}^*)| = |h'_n(z_k^*)| \geq h_n^\#(z_k^*) > 1$. We conclude that $(\xi_{n,k}^*, \eta_{n,k}^*)$ is a nontrivial pair of zeros of f_n on $\Delta(0, s\delta_0)$. However,

$$|\xi_{n,k}^* - \eta_{n,k}^*| = |a_n - c_n|\,|\xi_{n,k} - \eta_{n,k}| < |a_n - c_n|$$

if n is sufficiently large. This contradicts the fact that (a_n, c_n) is a nontrivial pair of zeros of f_n in $\Delta(0, s\delta_0)$ whose distance is minimal.

Thus no subsequence of $\{h_n\}$ is normal on $\mathbb{C}$. Let E be the set on which $\{h_n\}$ is not normal. Suppose that for each $\zeta \in E$, there is a neighborhood on which h_n has only a single (multiple) zero for sufficiently large n. Then by Lemma 3.1, $\{h_n\}$ is quasinormal at each point of E and hence on all of $\mathbb{C}$. Let $\zeta_0 \in E$. Taking a subsequence, we may assume that no subsequence of $\{h_n\}$ is normal at ζ_0 and that $\{h_n\}$ converges locally spherically uniformly on $\mathbb{C} \setminus E_0$, where $E_0 \subset E$ is a discrete set containing ζ_0. By Lemma 3.2, $h_n \overset{\chi}{\Longrightarrow} \psi(a)(\zeta - \zeta_0)$ on $\mathbb{C} \setminus E_0$. Taking additional subsequences and diagonalizing, we may assume that no subsequence of $\{h_n\}$ is normal at any point of E_0. We claim that $E_0 = \{\zeta_0\}$. Indeed, otherwise there exists $\zeta_1 \in E_0$, $\zeta_1 \neq \zeta_0$; then, as before, it follows from Lemma 3.2 that $h_n(\zeta) \overset{\chi}{\Longrightarrow} \psi(a)(\zeta - \zeta_0)$ on $\mathbb{C} \setminus E_0$, so that $\zeta_1 = \zeta_0$, $E_0 = \{\zeta_0\}$, and $h_n(\zeta) \overset{\chi}{\Longrightarrow} \psi(a)(\zeta - \zeta_0)$ on $\mathbb{C} \setminus \{\zeta_0\}$. But this contradicts $h(\pm 1/2) = 0$.

Thus there exists $\zeta_0 \in E$ such that for each $\delta > 0$, there exists a subsequence of $\{h_n\}$ (which we continue to call $\{h_n\}$) such that each h_n has at least two distinct zeros in $\Delta(\zeta_0, \delta)$ for sufficiently large n. Then by Lemma 3.4, for n sufficiently large, f_n has a nontrivial pair of zeros $(w^*_{n,1}, w^*_{n,2})$ such that $w^*_{n,j} \longrightarrow a$ $(j = 1, 2)$ and $|w^*_{n,1} - w^*_{n,2}| < |a_n - c_n|$. This contradicts the fact that (a_n, c_n) is a nontrivial pair of zeros of f_n in $\Delta(0, s\delta_0)$ whose distance is minimal.

4. Proof of the theorem

We may assume that $D = \Delta$. Suppose that $\mathcal{F}$ is quasinormal on Δ but not quasinormal of order 1 there. Then there exists a sequence $\{a^*_k\} \subset \Delta$ with no accumulation point in Δ such that $a^*_1 \neq a^*_2$ and a sequence $\{f_n\} \subset \mathcal{F}$ such that $f_n \overset{\chi}{\Rightarrow} f$ on $\Delta \setminus \{a^*_k\}$ but no subsequence of $\{f_n\}$ is normal at a^*_1 or a^*_2. By Lemma 3.7, there exists $\delta > 0$ such that f_n has a single (multiple) zero on $\Delta(a^*_j, \delta)$ for all sufficiently large n and $j = 1, 2$. By Lemma 3.2,

$$f(z) = \int_{a^*_j}^{z} \varphi'(\zeta)d\zeta = \varphi(z) - \varphi(a^*_j)$$

for $z \in \Delta \setminus \{a_k\}$. Since φ is univalent, $a^*_1 = a^*_2$, a contradiction.

To prove the second assertion of the theorem, suppose that $\mathcal{F}$ is quasinormal, but not normal, on Δ. By the previous paragraph, $\mathcal{F}$ is quasinormal of order 1 on Δ, so there exist $z_0 \in \Delta$ and $\{f_n\} \subset \mathcal{F}$ such that $f_n \overset{\chi}{\Rightarrow} f$ on $\Delta \setminus \{z_0\}$ but no subsequence of $\{f_n\}$ is normal at z_0. As before, it follows from Lemma 3.7 that there exists $\delta > 0$ such that f_n has a single (multiple) zero on $\Delta(z_0, \delta)$ for all large n and from Lemma 3.2 that $f_n(z) \overset{\chi}{\Rightarrow} f(z) = \int_{z_0}^{z} \varphi'(\zeta)d\zeta$ on $\Delta \setminus \{z_0\}$. In fact, since φ is analytic on Δ, the convergence is uniform on compact subsets of $\Delta \setminus \{z_0\}$. Since all zeros of f_n are multiple and $f_n \Rightarrow f$ on $\Delta \setminus \{z_0\}$, any zero of f must also be multiple. It follows that for any compact subset $K \subset \Delta$, at most finitely many f_n can vanish more than once on K. For otherwise, there would exist $z_1 \in K$, $z_1 \neq z_0$, such that $f(z_1) = f'(z_1) = 0$, which contradicts $f'(z_1) = \varphi'(z_1) \neq 0$. This completes the proof.

References

[1] Walter Bergweiler and J.K. Langley, *Multiplicities in Hayman's alternative*, J. Aust. Math. Soc. **78** (2005), 37–57.

[2] Chi-Tai Chuang, *Normal Families of Meromorphic Functions*, World Scientific, 1993.

[3] W.K. Hayman, *Picard values of meromorphic functions and their derivatives*, Ann. of Math. (2) **70** (1959), 9–42.

[4] Ku Yongxing, *Un critère de normalité des familles de fonctions méromorphes*, Sci. Sinica Special Issue **1** (1979), 267–274 (Chinese).

[5] Shahar Nevo, *On theorems of Yang and Schwick*, Complex Variables **46** (2001), 315–321.

[6] Shahar Nevo, *Applications of Zalcman's lemma to Q_m-normal families*, Analysis **21** (2001), 289–325.

[7] Xuecheng Pang, Shahar Nevo, and Lawrence Zalcman, *Quasinormal families of meromorphic functions*, Rev. Mat. Iberoamericana **21** (2005), 249–262.

[8] Xuecheng Pang and Lawrence Zalcman, *Normal families and shared values*, Bull. London Math. Soc **32** (2000), 325–331.

[9] Xuecheng Pang and Lawrence Zalcman, *Normal families of meromorphic functions with multiple zeros and poles*, Israel J. Math. **136** (2003), 1–9.

[10] Yufei Wang and Mingliang Fang, *Picard values and normal families of meromorphic functions with multiple zeros*, Acta Math. Sinica N.S. **14** (1998), 17–26.

[11] Lawrence Zalcman, *Normal families: new perspectives*, Bull. Amer. Math. Soc. **35** (1998), 215–230.

Xuecheng Pang
Department of Mathematics
East China Normal University
Shanghai 200062, P. R. China
e-mail: xcpang@math.ecnu.edu.cn

Shahar Nevo and Lawrence Zalcman
Department of Mathematics
Bar-Ilan University, 52900 Ramat-Gan, Israel
e-mail: nevosh@macs.biu.ac.il
e-mail: zalcman@macs.biu.ac.il

Operator Theory:
Advances and Applications, Vol. 158, 191–211

The Backward Shift on H^p

William T. Ross

In memory of Semen Yakovlevich Khavinson.

Mathematics Subject Classification (2000). Primary 30H05; Secondary 47B38.

Keywords. H^p spaces, backward shift operator, invariant subspaces, pseudo-continuation.

1. Introduction

In this semi-expository paper, we examine the backward shift operator

$$Bf := \frac{f - f(0)}{z}$$

on the classical Hardy space H^p. Though there are many aspects of this operator worthy of study [20], we will focus on the description of its invariant subspaces by which we mean the closed linear manifolds $\mathcal{E} \subset H^p$ for which $B\mathcal{E} \subset \mathcal{E}$. When $1 < p < \infty$, a seminal paper of Douglas, Shapiro, and Shields [8] describes these invariant subspaces by using the important concept of a pseudocontinuation developed earlier by Shapiro [26]. When $p = 1$, the description is the same [1] except that in the proof, one must be mindful of some technical considerations involving the functions of bounded mean oscillation.

The $p \geqslant 1$ case involves heavy use of duality and especially the Hahn-Banach separation theorem where one gets at $\mathcal{E}$ by first looking at $\mathcal{E}^{\perp}$, the annihilator of $\mathcal{E}$, and then returning to $\mathcal{E}$ by $^{\perp}(\mathcal{E}^{\perp})$. On the other hand, when $0 < p < 1$, H^p is no longer locally convex and the Hahn-Banach separation theorem fails [12]. In fact, as we shall see in § 4, there are invariant subspaces $\mathcal{E} \neq H^p$, $0 < p < 1$, for which $^{\perp}(\mathcal{E}^{\perp}) = H^p$. Despite these difficulties, an ingenious *tour de force* approach of Aleksandrov [1] (see also [6]), using such tools as distribution theory and the atomic decomposition theorem, characterizes these invariant subspaces.

The first several sections of this paper are a leisurely, non-technical, treatment of the Douglas-Shapiro-Shields and Aleksandrov results. In § 5, we focus on some new results, based on techniques in [4], which give an alternative description of certain invariant subspaces of H^p. As a consequence, we eventually wind up characterizing the weakly closed invariant subspaces of H^p. In § 6, we make some remarks about the invariant subspaces of the standard Bergman spaces L_a^p when $0 < p < 1$.

2. Preliminaries

We begin with some basic definitions and well-known results about the Hardy spaces H^p. A detailed treatment can be found in [11]. For $0 < p < \infty$, let H^p denote the space of analytic functions f on the open unit disk $\mathbb{D} = \{z \in \mathbb{C} : |z| < 1\}$ whose L^p integral means

$$\int_0^{2\pi} |f(re^{i\theta})|^p \frac{d\theta}{2\pi}$$

are uniformly bounded for $r \in (0, 1)$. These means increase as $r \nearrow 1$ and we define

$$\|f\|_p := \lim_{r \to 1^-} \left(\int_0^{2\pi} |f(re^{i\theta})|^p \frac{d\theta}{2\pi} \right)^{1/p} .$$

For almost every (with respect to Lebesgue measure on the unit circle $\mathbb{T} := \partial\mathbb{D}$) $e^{i\theta}$, the radial limit

$$\lim_{r \to 1^-} f(re^{i\theta})$$

exists and we denote its value by $f(e^{i\theta})$, or perhaps $f^*(e^{i\theta})$ when we want to emphasize this almost everywhere defined boundary function. Moreover,

$$\|f\|_p = \left(\int_0^{2\pi} |f^*(e^{i\theta})|^p \frac{d\theta}{2\pi} \right)^{1/p} .$$

One can show that $f \in H^p$ satisfies the pointwise estimate

$$|f(z)| \leqslant 2^{1/p} \|f\|_p (1 - |z|)^{-1/p}, \quad z \in \mathbb{D}.$$

As a result, for $1 \leqslant p < \infty$, the quantity $\|f\|_p$ defines a norm that makes H^p a Banach space while for $0 < p < 1$, $\|f - g\|_p^p$ defines a translation invariant metric that makes H^p a complete metric space. In either case, $f(re^{i\theta}) \to f^*(e^{i\theta})$ almost everywhere and in the norm (metric) of L^p. When $p = \infty$, H^∞ will denote the bounded analytic functions on $\mathbb{D}$ with the sup-norm $\|f\|_\infty := \sup\{|f(z)| : z \in \mathbb{D}\}$.

Since $f \to f^*$ is an isometry of H^p to L^p, one can regard H^p as a closed subspace of L^p. In fact, at least when $1 \leqslant p < \infty$, we can think of H^p in the following way

$$H^p = \{f \in L^p : \widehat{f}(n) = 0 \text{ for all } n < 0\},$$

where $\widehat{f}(n)$ is the nth Fourier coefficient of f. This follows from the F. and M. Riesz theorem [11, p. 41].

Every function $f \in H^p$ can be factored as $f = \phi\Theta$, where $\phi \in H^\infty$ with $|\phi^*(e^{i\theta})| = 1$ almost everywhere (such functions are called 'inner functions') and

$\Theta \in H^p$ has no zeros on $\mathbb{D}$ and satisfies

$$\log |\Theta(0)| = \int_0^{2\pi} \log |\Theta^*(e^{i\theta})| \frac{d\theta}{2\pi}$$

(such functions are called 'outer functions'). Moreover, except for a unimodular constant, this factorization is unique.

Identifying the dual, $(H^p)^*$, of H^p with a space of analytic functions on $\mathbb{D}$ is often, but not always, the key to understanding the structure of its invariant subspaces. The dual pairing between H^p and $(H^p)^*$, as a space of analytic functions, is the following 'Cauchy pairing'. For analytic functions

$$f(z) = \sum_{n=0}^{\infty} a_n z^n \text{ and } g(z) = \sum_{n=0}^{\infty} b_n z^n$$

on the disk $\mathbb{D}$, define

$$\langle f, g \rangle := \lim_{r \to 1^-} \sum_{n=0}^{\infty} a_n \overline{b_n} r^n, \tag{2.1}$$

whenever this limit exists. A simple computation with power series shows that if $\langle f, g \rangle$ exists, then

$$\langle f, g \rangle = \lim_{r \to 1^-} \int_0^{2\pi} f(re^{i\theta}) \overline{g}(re^{i\theta}) \frac{d\theta}{2\pi}.$$

For $1 < p < \infty$, the dual of H^p can be identified with H^q, where q is the conjugate index to p, and this comes somewhat easily. Notice that for $f \in H^p$ and $g \in H^q$, we have $f_r \to f$ in H^p and $g_r \to g$ in H^q, where $f_r(z) := f(rz)$. Thus, by Hölder's inequality,

$$\langle f, g \rangle = \int_0^{2\pi} f\overline{g} \frac{d\theta}{2\pi}.$$

Certainly, the linear functional $f \to \langle f, g \rangle$ is continuous on H^p for fixed $g \in H^q$. On the other hand, if $\ell \in (H^p)^*$, the Hahn-Banach extension theorem says that

$$\ell(f) = \int_0^{2\pi} f\overline{g} \frac{d\theta}{2\pi} \tag{2.2}$$

for some $g \in L^q$. Using the continuity of the Riesz projection operator $P : L^q \to H^q$

$$P\left(\sum_{n=-\infty}^{\infty} \widehat{g}(n) e^{in\theta} \right) = \sum_{n=0}^{\infty} \widehat{g}(n) e^{in\theta}$$

and the identities

$$\int_0^{2\pi} f\overline{g} \frac{d\theta}{2\pi} = \int_0^{2\pi} f\overline{Pg} \frac{d\theta}{2\pi} = \langle f, Pg \rangle,$$

one can replace, in (2.2) and hence (2.1), the above $g \in L^q$ with a unique function in H^q. A little technical detail shows that norm of the linear functional $f \to \langle f, g \rangle$ is equivalent to the H^q norm of g. Thus $(H^p)^*$ can be identified with H^q via the dual pairing in (2.1).

194 W.T. Ross

When $p = 1$, the above analysis breaks down. Certainly if $\ell \in (H^1)^*$, then

$$\ell(f) = \int_0^{2\pi} f\bar{g}\,\frac{d\theta}{2\pi}$$

for some $g \in L^\infty$. However, when one tries to imitate the above analysis and replace g with Pg in the above integral, there are problems. For one, $P(L^\infty) = BMOA \supsetneq H^\infty$, where $BMOA$ are the analytic functions of bounded mean oscillation[1]. Secondly, there are $f \in H^1$ and $g \in BMOA$, for which $f\bar{g} \notin L^1$. These technical problems are not insurmountable since, for $f \in H^1$ and $g \in BMOA$, the quantity $\langle f, g \rangle$ (as in (2.1)) does indeed exist and $f \to \langle f, g \rangle$ defines a continuous linear functional on H^1. In fact, these are all the linear functionals on H^1. Another technical detail says that the norm of $f \to \langle f, g \rangle$ is equivalent to the $BMOA$ norm of g. In summary, we can identify the dual of H^1 with $BMOA$ via the dual pairing in (2.1). See [13, Ch. 6] for more details on all this.

When $0 < p < 1$, surprisingly, there are non-trivial bounded linear functionals on H^p. Surprisingly since when $0 < p < 1$, $(L^p)^* = (0)$ [7]. The theorem here is one of Duren, Romberg, and Shields [12] and says that if ℓ is a bounded linear functional on H^p, then there is a unique g belonging O_p, a subspace of the disk algebra, so that $\ell(f) = \langle f, g \rangle$. Conversely, for $g \in O_p$, $f \to \langle f, g \rangle$ defines an element of $(H^p)^*$. The space O_β, for $\beta > 0$, is the set of analytic functions g on the disk for which

$$\|g\|_\beta := \sup_{|z|<1} |g^{[1/\beta]}(z)|(1 - |z|) < \infty,$$

where, if $g(z) = \sum a_n z^n$,

$$g^{[\alpha]}(z) := \sum_{n=0}^{\infty} \frac{\Gamma(n + 1 + \alpha)}{\Gamma(n + 1)} a_n z^n$$

is the fractional derivative of g of order α. The classes O_β can be equivalently characterized as Lipschitz or Zygmund spaces. For example, if $1/2 < p < 1$, then O_p is the space of analytic functions on $\mathbb{D}$ which have continuous extensions to $\mathbb{D}^-$ and such that

$$\sup_{\theta \neq t} \frac{|g(e^{i\theta}) - g(e^{it})|}{|\theta - t|^{1/p-1}} < \infty.$$

For other p's, one requires the derivatives (depending on p) of g to have certain smoothness on $\mathbb{T}$. In general, the smaller the p, the more derivatives of g that need to satisfy a Lipschitz or Zygmund condition on $\mathbb{T}$ in order for g to belong to O_p. One can show that the norm of the functional $f \to \langle f, g \rangle$ is equivalent to the O_p

---

[1] $g \in L^1$ is of bounded mean oscillation BMO if

$$\|g\| = \|g\|_{L^1} + \sup_I \frac{1}{|I|} \int_I |g - g_I|\,d\theta < \infty,$$

where $g_I = |I|^{-1} \int_I g\,d\theta$ and $|I|$ is the length of an arc $I \subset \mathbb{T}$. $BMOA := BMO \cap H^1$.

norm of g [2]. Thus we identify $(H^p)^*$ with O_p when $0 < p < 1$ via (2.1). Again, consult [12] for the details.

3. The backward shift on H^p for $1 \leqslant p < \infty$

If $1 < p < \infty$, notice that the backward shift B on H^p is the Banach space adjoint of the forward shift operator $Sf = zf$ on H^q, that is to say

$$\langle Bf, g \rangle = \langle f, Sg \rangle, \quad f \in H^p, \quad g \in H^q.$$

Thus if $\mathcal{E} \subsetneq H^p$ is an invariant subspace for B, then

$$\mathcal{E}^\perp := \{ g \in H^q : \langle f, g \rangle = 0 \ \forall f \in \mathcal{E} \},$$

the 'annihilator' of $\mathcal{E}$, is an S-invariant subspace of H^q. A celebrated theorem of Beurling [11, p. 114] says that $\mathcal{E}^\perp = \phi H^q$ for some non-constant inner function ϕ. By the Hahn-Banach separation theorem,

$$\mathcal{E} = {}^\perp(\mathcal{E}^\perp) = {}^\perp(\phi H^q),$$

where for $A \subset H^q$, ${}^\perp A := \{ f \in H^p : \langle f, g \rangle = 0, \forall g \in A \}$ is the 'pre-annihilator' of A. So the problem of describing $\mathcal{E}$ is reduced to characterizing, in some function-theoretic way, this pre-annihilator ${}^\perp(\phi H^q)$.

The function theoretic tool, the concept of a pseudocontinuation, used here was developed by Shapiro in some earlier work [26] and we now take a few moments to point out some basic facts about pseudocontinuations. Suppose that h is a meromorphic function on $\mathbb{D}$ and H is a meromorphic on $\mathbb{D}_e$. There is no *a priori* reason why the non-tangential limits of h (from $\mathbb{D}$) and H (from $\mathbb{D}_e$) need to exist. But *if* they do, *and* they are equal almost everywhere, we say that H is a 'pseudocontinuation' of h. Two representative examples of functions with a pseudocontinuation are the following.

Example 3.1.

1. If h is an inner function, then

$$H(z) = \frac{1}{\overline{h(1/\overline{z})}}$$

is a pseudocontinuation of h. This follows from that fact that $h^* \overline{h^*} = 1$ almost everywhere. Also notice, for example, that if h is a Blaschke product whose zeros accumulate on all of the circle, then h, although a pseudocontinuable function, will not have an analytic continuation across any point of the unit circle.

---

[2] Technically $\|g\|_\beta$ is only a semi-norm on O_β. One can make this a true norm by adding in $|g(0)| + |g'(0)| + \cdots + |g^{(\lfloor 1/\beta \rfloor)}|$, where $\lfloor x \rfloor$ is the greatest integer less than x.

2. Another example of a pseudocontinuation is when h is a Cauchy integral

$$h(z) := \int \frac{1}{1 - e^{-i\theta}z} d\mu(e^{i\theta}),$$

where μ is a finite Borel measure on $\mathbb{T}$ that is singular with respect to Lebesgue measure. If H is the above Cauchy integral but with $z \in \mathbb{D}_e$, one can show that h and H are H^p functions (for $0 < p < 1$) on their respective domains [11, p. 39][3] and so have finite non-tangential limits almost everywhere. Notice that

$$h(z) - h(1/\bar{z}) = \int_0^{2\pi} P_z(e^{i\theta})d\mu(e^{i\theta}),$$

where $P_z(e^{i\theta})$ is the Poisson kernel. Using a classical theorem of Fatou [11, p. 39], which says that

$$\int_0^{2\pi} P_z(e^{i\theta})d\mu(e^{i\theta}) \to 2\pi\mu'(e^{i\theta})$$

for almost every $e^{i\theta}$ as $z \to e^{i\theta}$, and the fact that μ is singular (and so $\mu' = 0$ almost everywhere), one can show that the non-tangential limits of h and H are equal almost everywhere.

Let us make a few general comments about pseudocontinuations. The first is that they are unique. Indeed, if H_1 and H_2 are two pseudocontinuations of h, then $H_1 - H_2$ is a meromorphic function on $\mathbb{D}_e$ that has zero non-tangential limits almost everywhere. A classical theorem of Privalov [16, p. 62] says that any meromorphic function that has zero non-tangential limits on a subset of $\mathbb{T}$ with positive Lebesgue measure must be identically zero. Hence h can have only one pseudocontinuation. Here is why we use *non-tangential limits* rather than *radial limits* in the definition of a pseudocontinuation. If radial limits were used, then pseudocontinuations would not be unique. Indeed, there are non-trivial analytic functions on $\mathbb{D}$ which have radial limits equal to zero almost everywhere [5] - thus the zero function would be a pseudocontinuation without the original function being the zero function. Certainly, when we are talking about H^p functions this cannot happen since the non-tangential limits exist almost everywhere anyway. But in general, we need to make this important distinction.

Another consequence of Privalov's uniqueness theorem is that if h has an analytic continuation to a neighborhood U of $e^{i\theta}$ and a pseudocontinuation H, then $h = H$ on $U \cap \mathbb{D}_e$, that is to say, pseudocontinuation is compatible with the classical notion of analytic continuation.

The point at infinity is important. The function $h(z) = e^z$ certainly has an analytic continuation across $\mathbb{T}$. However, it does not have a pseudocontinuation as we have defined it above since $H(z) = e^z$ has an essential singularity at infinity.

---

[3]The Hardy space of the extended exterior disk $\mathbb{D}_e := \{z \in \widehat{\mathbb{C}} : 1 < |z| \leqslant \infty\}$ is defined by $H^p(\mathbb{D}_e) := \{f(1/z) : f \in H^p\}$. Note that if $f \in H^p$, then $\overline{f}(e^{i\theta})$ is the boundary function for a function belonging to $H^p(\mathbb{D}_e)$, the function being $\overline{f}(1/\bar{z})$.

The interested reader is invited to consult [23] for a more detailed discussion of pseudocontinuations.

The function theoretic description of $^{\perp}(\phi H^q)$ is the following well-known theorem.

Proposition 3.2 (Douglas-Shapiro-Shields). *Let ϕ be an inner function and $1 < p < \infty$. For $f \in H^p$, the following are equivalent:*

1. $f \in {}^{\perp}(\phi H^q)$
2. $f^* \in H^p \cap \phi \overline{H_0^p}$, where $H_0^p = \{f \in H^p : f(0) = 0\}$.
3. *The meromorphic function f/ϕ on $\mathbb{D}$ has a pseudocontinuation to a function $\widetilde{f_\phi} \in H^p(\mathbb{D}_e)$ with $\widetilde{f_\phi}(\infty) = 0$.*

It is important to note that the space

$$H^p \cap \phi \overline{H_0^p} = \{f \in H^p : f^* = \phi^* \overline{h^*}, h \in H_0^p\}^4 \tag{3.3}$$

must be understood as a space of functions on the circle and not on the disk. For fixed $1 < p < \infty$ and inner function ϕ, we let $\mathcal{E}^p(\phi)$ be the collection of H^p functions that satisfy one of the equivalent conditions in Proposition 3.2. Since $\mathcal{E}^p(\phi)$ is an annihilating subspace, it is closed in H^p. It also follows from the above argument that $\mathcal{E}^p(\phi)$ is invariant. Combining this with what was said above, we have the following summary theorem.

Theorem 3.4 (Douglas-Shapiro-Shields). *For $1 < p < \infty$, a subspace $\mathcal{E} \subsetneq H^p$, is invariant if and only if $\mathcal{E} = \mathcal{E}^p(\phi)$ for some inner function ϕ.*

Before proceeding to the $p = 1$ case, we mention a few other items of interest. Using a Morera type argument, one can show that every $f \in \mathcal{E}^p(\phi)$ has an analytic continuation to the set

$$\widehat{\mathbb{C}} \setminus \{1/\overline{z} : z \in \sigma(\phi)\}, \quad \text{where} \quad \sigma(\phi) := \left\{ z \in \mathbb{D}^- : \liminf_{\lambda \to z} |\phi(\lambda)| = 0 \right\}^5.$$

Note that ϕ has an analytic continuation to $\widehat{\mathbb{C}} \setminus \{1/\overline{z} : z \in \sigma(\phi)\}$ [13, p. 75–76]. Furthermore, by the compatibility of pseudocontinuation with analytic continuation, the analytic continuation of f/ϕ to $\mathbb{D}_e \setminus \{1/\overline{z} : z \in \sigma(\phi)\}$ must be equal to $\widetilde{f_\phi}$, the pseudocontinuation of f/ϕ.

Another interesting item is that if $f = B\phi$, then

$$[f]_{H^p} := \bigvee \{B^n f : n = 0, 1, 2, \ldots\} = \mathcal{E}^p(\phi),$$

where $\bigvee$ is the closed linear span in the H^p norm. This says that $\mathcal{E}^p(\phi)$ is a 'cyclic invariant subspace' generated by $f = B\phi$. While we are mentioning cyclic vectors, there is a celebrated result that determines exactly when a particular $f \in H^p$ is 'cyclic', that is to say $[f]_{H^p} = H^p$.

---

[4] Recall from the preliminaries that $f^*(e^{i\theta}) = \lim_{r \to 1-} f(re^{i\theta})$ almost everywhere.

[5] A basic fact about inner functions is that if $\phi = b s_\mu$, where b is a Blaschke product and s_μ is a singular inner function with associated positive singular measure μ on $\mathbb{T}$ (all inner functions can be factored this way), then $\sigma(\phi)$ is the closure of the zeros of b together with the support of μ.

Theorem 3.5 (Douglas-Shapiro-Shields). *For $1 \leqslant p < \infty$, a vector $f \in H^p$ is non-cyclic for the backward shift if and only if f has a pseudocontinuation of bounded type, i.e., there is a meromorphic function $\tilde{f}$ on $\mathbb{D}_e$, that can be written as a quotient of two bounded analytic functions on $\mathbb{D}_e$, such that $\tilde{f}$ is a pseudocontinuation of f.*

Though this theorem is both necessary and sufficient, the hypothesis (having a pseudocontinuation of bounded type) is not something easily tested. There are some obvious examples of cyclic vectors like

$$e^{1/(z-2)} \quad \text{and} \quad e^z$$

which are not meromorphic on $\mathbb{D}_e$, and

$$f = \sum_{n=1}^{\infty} \frac{2^{-n}}{z - (1 + 1/n)}$$

which has a pseudocontinuation, but not of bounded type (too many poles). Notice that we are using the uniqueness of pseudocontinuations here and the fact that if a function has an analytic continuation across a point of the circle, then the analytic continuation must agree with its pseudocontinuation. Along these lines, the vector $\sqrt{1-z}$ is a cyclic vector since its pseudocontinuation, which must be $\sqrt{1-z}$, can not have a branch cut. Less obvious examples of cyclic vectors are H^p functions given by Hadamard gap series [27] such as

$$f(z) = \sum_{n=0}^{\infty} 2^{-n} z^{2^n} \quad \text{or Fabry gap series [2]} \quad f(z) = \sum_{n=0}^{\infty} 2^{-n} z^{n^2}.$$

Actually, both of these gap series have the following stronger pathological property: There exists no $1 < R < \infty$ and no meromorphic function $\tilde{f}$ on $\{z : 1 < |z| < R\}$ such that the nontangential limits of $\tilde{f}$ and f agree almost everywhere. See also [23] for further details and other pathological examples of this type.

The $p = 1$ case is a bit pesky and poses some technical challenges that were overcome by Aleksandrov (see [1] or [6, p. 101]). If $\mathcal{E} \subset H^1$ is invariant, then $\mathcal{E}^{\perp}$ is an S-invariant subspace of $BMOA$, closed in the weak-* topology $BMOA$ inherits by being the dual of H^1. However, the description of these S-invariant subspaces is not as simple as $\phi BMOA$ (ϕ inner), as in Beurling's theorem for H^p. In fact, $\phi BMOA$ may not even be a subset of $BMOA$. That is to say, ϕ may not be a 'multiplier' of $BMOA$ [30]. The second technical challenge is that the dual pairing between H^1 and $BMOA$ is

$$\lim_{r \to 1^-} \int_0^{2\pi} f \overline{g_r} \frac{d\theta}{2\pi} \tag{3.6}$$

and not simply

$$\int_0^{2\pi} f \overline{g} \frac{d\theta}{2\pi}. \tag{3.7}$$

This may not seem like a major difference but the proof of the Douglas-Shapiro-Shields theorem makes use of the F. and M. Riesz theorem [11, p. 40–41]

$$\int_{[0,2\pi]} e^{in\theta}\,d\mu(\theta) = 0 \quad n = 0,1,2,\dots \quad \Leftrightarrow \quad d\mu(\theta) = f(e^{i\theta})\frac{d\theta}{2\pi}, \quad f \in H^1, \quad (3.8)$$

for which we need to write the dual pairing $\langle f, g \rangle$ as an integral, as in (3.7), and not as a limit of integrals, as in (3.6). Nevertheless, one can show that $\mathcal{E} \cap H^2$ is not equal to H^2, is closed in the norm of H^2, and is invariant and hence takes the form $\mathcal{E}^2(\phi)$ (Theorem 3.4). Using the $(H^1, BMOA)$ duality, one can show that $\mathcal{E}^2(\phi)$ is dense in $\mathcal{E}^1(\phi)$. Here $\mathcal{E}^1(\phi) := H^1 \cap \phi\overline{H_0^1}$, or equivalently, the space of H^1 functions f such that f/ϕ has a pseudocontinuation to a function $\widetilde{f_\phi} \in H^1(\mathbb{D}_e)$ that vanishes at infinity. Thus $\mathcal{E}^1(\phi) \subset \mathcal{E}$. The other inclusion is also a bit tricky but nevertheless true. The summary theorem here is the following.

Theorem 3.9 (Aleksandrov).

1. *A subspace $\mathcal{E} \subsetneq H^1$ is invariant if and only if $\mathcal{E} = \mathcal{E}^1(\phi)$ for some inner function ϕ.*
2. *If $f = B\phi$, then $[f]_{H^1} = \mathcal{E}^1(\phi)$, that is to say $\mathcal{E}^1(\phi)$ is cyclic.*
3. *A vector $f \in H^1$ is non-cyclic, i.e., $[f]_{H^1} \neq H^1$, if and only if f has a pseudocontinuation of bounded type.*

4. The backward shift on H^p for $0 < p < 1$

Characterizing the invariant subspaces of H^p when $0 < p < 1$ poses special challenges. For example, H^p $(0 < p < 1)$, with its metric topology, is no longer locally convex and the Hahn-Banach separation theorem, a key tool in understanding the $p \geqslant 1$ case, fails[6].

Example 4.1. For each $\theta \in [0, 2\pi]$, the function $(1 - e^{-i\theta}z)^{-1}$ belongs to H^p for all $0 < p < 1$ and so we can consider the following subspace of H^p:

$$\mathcal{E} := \bigvee \left\{ \frac{1}{1 - e^{-i\theta}z} : 0 \leqslant \theta < 2\pi \right\}.$$

When z is on the unit circle, we have

$$\frac{1}{1 - e^{-i\theta}z} = \frac{\overline{z}}{\overline{z} - e^{-i\theta}} \in \overline{H_0^p}$$

and so $\mathcal{E} \subset H^p \cap \overline{H_0^p} \neq H^p$. As an aside, one can show that indeed $\mathcal{E} = H^p \cap \overline{H_0^p}$ (see [1] or [6, p. 116]). Again we remind the reader that $H^p \cap \overline{H_0^p}$ is a space

---

[6]The Hahn-Banach extension theorem also fails in H^p $(0 < p < 1)$. Indeed, there is a closed subspace A of H^p and a continuous linear functional on A which cannot be extended continuously to all of H^p [12].

of functions on the unit circle (see (3.3)). We claim that $\mathcal{E}^{\perp} = (0)$. Indeed, if $g = \sum_n b_n z^n \in O_p = (H^p)^*$ belongs to $\mathcal{E}^{\perp}$, then for all θ,

$$0 = \left\langle \frac{1}{1 - e^{-i\theta} z}, g \right\rangle = \lim_{r \to 1} \sum_{n=0}^{\infty} e^{-in\theta} \overline{b_n} r^n = \lim_{r \to 1^-} \overline{g}(re^{i\theta}) = \overline{g}(e^{i\theta}),$$

making g the zero function.

Example 4.1 shows that describing an invariant subspace $\mathcal{E}$ of H^p ($0 < p < 1$) by first examining $\mathcal{E}^{\perp}$ and then returning to $\mathcal{E}$ via the Hahn-Banach separation theorem $\mathcal{E} = {}^{\perp}(\mathcal{E}^{\perp})$ is of no use here. In the above example, $\mathcal{E} \neq H^p$, but ${}^{\perp}(\mathcal{E}^{\perp}) = {}^{\perp}(0) = H^p$. As it turns out though, the invariant subspaces of H^p ($0 < p < 1$) can be characterized but the description is not the same as before (namely $\mathcal{E}^p(\phi)$ spaces) and the proof is much more difficult, involving many advanced tools in analysis. This complicated but beautiful characterization was accomplished by Aleksandrov [1] and we spend a few moments stating his result.

With the use of duality out, one must discover what functions belong to a given invariant subspace almost by hand. Given $0 < p < 1$ and an invariant subspace $\mathcal{E} \subset H^p$, we notice that $\mathcal{E} \cap H^2$ is a closed (in the H^2 norm) invariant subspace of H^2 which, by the Douglas-Shapiro-Shields theorem (Theorem 3.4), equals $\mathcal{E}^2(\phi)$ for some inner function ϕ. If $\mathcal{E} \cap H^2 = (0)$, which can indeed be the case by Example 4.1, we take the ϕ to be the constant function $\phi = 1$. This makes sense since $\mathcal{E}^2(1) = H^2 \cap \overline{H_0^2} = (0)$ (F. and M. Riesz theorem – (3.8)).

Let $F \subset \mathbb{T}$ be the following set

$$F := \left\{ e^{i\theta} \in \mathbb{T} : \frac{1}{1 - e^{-i\theta} z} \in \mathcal{E} \right\}.$$

One can show that F is a closed subset of $\mathbb{T}$ and that $\sigma(\phi) \cap \mathbb{T} \subset F$. Also consider a function

$$k : F \to \mathbb{N} \cap [1, n_p],$$

where

$$n_p := \max\{n \in \mathbb{N} \cap [1, 1/p)\}, \tag{4.2}$$

defined by

$$k(e^{i\theta}) := \max \left\{ j \in \mathbb{N} \cap [1, n_p] : \frac{1}{(1 - e^{-i\theta} z)^j} \in \mathcal{E} \right\}.$$

Note that a simple integral calculation shows that $(1 - e^{-i\theta} z)^{-j} \in H^p$ for all $j \in \mathbb{N} \cap [1, n_p]$. One can show that if F_0 is the set of isolated points of F, then $k(e^{i\theta}) = n_p$ whenever $e^{i\theta} \in (F \setminus F_0) \cup (\sigma(\phi) \cap \mathbb{T})$.

With these three parameters ϕ, F, k, form the space $\mathcal{E}^p(\phi, F, k)$ of functions $f \in H^p$ such that

1. $f^* \in H^p \cap \phi \overline{H_0^p}$, or equivalently f/ϕ has a pseudocontinuation to a function $\widetilde{f_\phi} \in H^p(\mathbb{D}_e)$ that vanishes at infinity;
2. f has an analytic continuation to a neighborhood of $\mathbb{T} \setminus F$;
3. At each $e^{i\theta} \in F_0 \setminus \sigma(\phi)$, f has a pole of order at most $k(e^{i\theta})$.

Before moving on, let us give a non-trivial example of a function belonging to $\mathcal{E}^p(\phi, F, k)$. This example will become important later on.

Example 4.3. Suppose $F \subset \mathbb{T}$ is a closed set of Lebesgue measure zero. We assume, as usual, that $\sigma(\phi) \cap \mathbb{T} \subset F$. Consider the function

$$(K\mu)(z) := \int \frac{d\mu(e^{i\theta})}{1 - e^{-i\theta}z},$$

where μ is a finite Borel measure on $\mathbb{T}$ whose support is exactly F. As mentioned earlier in Example 3.1, $K\mu|\mathbb{D} \in H^p$ and $K\mu|\mathbb{D}_e \in H^p(\mathbb{D}_e)$ and moreover, since μ is singular with respect to Lebesgue measure, these two functions are pseudocontinuations of each other. Furthermore, $K\mu$ has an analytic continuation across $\mathbb{T} \setminus F$ and at each isolated point of F, $K\mu$ has a pole of order one. Finally, note from Example 3.1, that the inner function ϕ has a pseudocontinuation

$$\widetilde{\phi}(z) = \frac{1}{\overline{\phi(1/\overline{z})}}, \quad z \in \mathbb{D}_e$$

and so $K\mu/\phi$ has a pseudocontinuation $K\mu/\widetilde{\phi}$ which belongs to $H^p(\mathbb{D}_e)$ and vanishes at infinity. Thus $K\mu \in \mathcal{E}^p(\phi, F, k)$, at least when F has Lebesgue measure zero.

Though somewhat involved to prove, one can show that $\mathcal{E}^p(\phi, F, k)$ is a non-trivial closed invariant subspace of H^p (invariance is clear, closed is what is difficult to prove). Furthermore,

$$\mathcal{E} \subset \mathcal{E}^p(\phi, F, k).$$

To obtain the reverse inclusion, Aleksandrov defines the space

$$e^p(\phi, F, k) := \mathcal{E}^2(\phi) \bigvee \left\{ \frac{1}{(1 - e^{-i\theta}z)^j} : e^{i\theta} \in F; j = 1, 2, \ldots, k(e^{i\theta}) \right\}. \tag{4.4}$$

From the very definition of the parameters ϕ, F and k, it follows that

$$e^p(\phi, F, k) \subset \mathcal{E}.$$

What is very difficult to prove here is that

$$e^p(\phi, F, k) = \mathcal{E}^p(\phi, F, k). \tag{4.5}$$

Aleksandrov's proof of this fact is quite involved and uses, among other tricks, distribution theory and the Coifman atomic decomposition theorem for H^p. To summarize, we have the following.

Theorem 4.6 (Aleksandrov). *For fixed $0 < p < 1$ and parameters ϕ, F, and k above, the space $\mathcal{E}^p(\phi, F, k)$ is an invariant subspace of H^p. Moreover, every proper invariant subspace of H^p is of the form $\mathcal{E}^p(\phi, F, k)$.*

We close this section with a few remarks. The characterization of the cyclic vectors remains the same: f is non-cyclic if and only if f has a pseudocontinuation of bounded type. One can also show, as in the H^p case when $p \geqslant 1$ but with a more complicated vector, that $\mathcal{E}^p(\phi, F, k)$ is a cyclic subspace (i.e., generated by one vector). Later on (Theorem 5.6) we will give an alternative characterization of $\mathcal{E}^p(\phi, F, k)$. The curious reader might be wondering why the parameters F and k are not needed in the $1 \leqslant p < \infty$ case. Notice that $(1 - e^{-i\theta}z)^{-j} \notin H^1$ for any $\theta \in [0, 2\pi)$ and $j \in \mathbb{N}$.

5. A closer look at Aleksandrov's theorem

Aleksandrov's theorem says that when $0 < p < 1$, a non-trivial invariant subspace of H^p takes the form $\mathcal{E}^p(\phi, F, k)$ (as described in §4). In this section, we show that under certain natural conditions, there is an alternative description of $\mathcal{E}^p(\phi, F, k)$. To do this, we will characterize the weakly closed invariant subspaces of H^p.

Let us say a few words about the weak topology on H^p $(0 < p < 1)$. The reader can refer to [12] for further details and examples. Recall from § 2 that $(H^p)^*$ can be identified (with equivalent norm) with a Lipschitz or Zygmund space O_p by means of the pairing

$$\langle f, g \rangle = \lim_{r \to 1^-} \sum_{n=0}^{\infty} a_n \overline{b_n} r^n.$$

A set $U \subset H^p$ is 'weakly open' if given any $f_0 \in U$, there is an $\varepsilon > 0$ and $g_1, \ldots, g_n \in O_p$ so that

$$\bigcap_{j=1}^{n} \{f \in H^p : |\langle f - f_0, g_j \rangle| < \varepsilon\} \subset U.$$

Since the family of semi-norms

$$\{\rho_g(f) := |\langle f, g \rangle| : g \in O_p\}$$

on H^p separates points, standard functional analysis says that (H^p, wk) $(H^p$ endowed with the weak topology) is a locally convex topological vector space [24, p. 64]. Furthermore, $(H^p, wk)^* = O_p$. As a consequence, a linear manifold $E \subset H^p$ is weakly closed if and only if it satisfies the Hahn-Banach separation property: If $f \notin E$, there is a $g \in O_p$ so that $g \perp E$ but $\langle f, g \rangle = 1$, i.e., each point not in E can be separated from E by a bounded linear functional [24, p. 60]. Viewing this another way,

$$\mathrm{clos}_{(H^p, wk)} E = {}^{\perp}(E^{\perp}), \tag{5.1}$$

where, for $C \subset O_p$, ${}^{\perp}C := \{f \in H^p : \langle f, c \rangle = 0 \;\; \forall c \in C\}$ is the pre-annihilator of C. Finally notice that if E is weakly closed then E is closed in the metric topology.

There is a containing Banach space B^p of H^p namely, the weighted Bergman space of analytic functions f on $\mathbb{D}$ for which the quantity

$$\|f\|_{B^p} := \int_0^{2\pi} \int_0^1 |f(re^{i\theta})|(1 - r)^{1/p - 2} dr \, \frac{d\theta}{2\pi}$$

is finite. Certain standard facts about B^p are that H^p is a dense subset of B^p and

$$\|f\|_{B^p} \leqslant A_p \|f\|_{H^p}, \quad f \in H^p,$$

that is to say, the containment $H^p \subset B^p$ is continuous. Moreover, B^p is a Banach space and $(B^p)^*$ can be identified (with equivalent norm) with the space O_p via the dual pairing in (2.1), i.e.,

$$\langle f, g \rangle = \lim_{r \to 1^-} \sum_{n=0}^{\infty} a_n \overline{b_n} r^n.$$

Thus B^p and H^p have the same continuous linear functionals. Using this fact along with the Hahn-Banach separation theorem, applied to the Banach space B^p, one can show that if E is a linear manifold in H^p, then

$$\mathrm{clos}_{(H^p, wk)} E = \left(\mathrm{clos}_{B^p} E \right) \cap H^p.$$

See [12, Lemma 8] for details.

For the rest of this section we will assume that $1/2 < p < 1$. For other values of p, most of the results are still true but the notation becomes cumbersome since the description of O_p changes very much with p. That being said, we fix $1/2 < p < 1$, an inner function ϕ, and a closed set $F \subset \mathbb{T}$. Without loss of generality, we assume that $\sigma(\phi) \cap \mathbb{T} \subset F$. Define

$$\mathcal{I}(\phi, F) := \{ g \in O_p : g \in \phi H^\infty, g|F = 0 \}^7.$$

One can easily observe that $\mathcal{I}(\phi, F)$ is an ideal of O_p. What is more difficult to prove is that when O_p is endowed with the weak-* topology it naturally inherits by being the dual of B^p, then $\mathcal{I}(\phi, F)$ is weak-* closed. In fact, every non-zero weak-* closed ideal of O_p is of the form $\mathcal{I}(\phi, F)$. There is a direct proof of this result (with an equivalent weak-* topology on O_p) in [21]. Another, perhaps more indirect, proof is found in [4, Thm. 3.2][8] Also, $\mathcal{I}(\phi, F) \neq (0)$ if and only if

$$\int_0^{2\pi} \log \mathrm{dist}(e^{i\theta}, \sigma(\phi) \cup F) \frac{d\theta}{2\pi} > -\infty \tag{5.2}$$

(see [32]). In fact, if (5.2) holds, then there is a $g \in A^\infty$ ($g^{(k)}$ has a continuous extension to $\mathbb{D}^-$ for all k) such that $g \in \mathcal{I}(\phi, F) \setminus (0)$ and g generates $\mathcal{I}(\phi, F)$ in the sense that the smallest weak-* closed ideal containing g is $\mathcal{I}(\phi, F)$. In this case, ϕ_g, the inner part of g, must be ϕ and $g^{-1}(\{0\}) \cap \mathbb{T}$ must be F [15].

It is worth repeating here that we are assuming, to avoid technical details, that $1/2 < p < 1$. In this case, $n_p = 1$ (see (4.2)) and so for ϕ, F, k as before,

$$\mathcal{E}^p(\phi, F, k) = \mathcal{E}^p(\phi, F, 1).$$

Proposition 5.3. $\mathcal{E}^p(\phi, F, 1)^\perp = \mathcal{I}(\phi, F).$

---

[7] Recall that functions in O_p have a continuous extension to $\mathbb{D}^-$ and so the notation $g|F = 0$ makes sense.

[8] The characterization of the ideals of functions 'smooth up to the boundary' has been well worked over [15, 17, 18, 19, 25, 31].

Proof. Let $g \in \mathcal{I}(\phi, F)$. Then $g \in \phi H^\infty$ and so g annihilates $\mathcal{E}^2(\phi)$ (being the annihilator of ϕH^2 in H^2). Also, for $e^{i\theta} \in F$,

$$\left\langle \frac{1}{1 - e^{-i\theta}z}, g \right\rangle = \overline{g}(e^{i\theta}) = 0.$$

Recalling the definition of $e^p(\phi, F, 1)$ from (4.4), we see that g annihilates $e^p(\phi, F, 1)$ and hence, by Aleksandrov's approximation (4.5), $\mathcal{E}^p(\phi, F, 1)$. For the other direction, suppose $g \in O_p$ annihilates $\mathcal{E}^p(\phi, F, 1)$. Then g annihilates $\mathcal{E}^2(\phi)$ as well as $(1 - e^{-i\theta}z)^{-1}$ for all $e^{i\theta} \in F$. It follows now that $g \in \mathcal{I}(\phi, F)$. $\square$

This proposition yields the following corollary.

Corollary 5.4. *The following are equivalent.*

1. $\mathcal{E}^p(\phi, F, 1)$ *is weakly closed.*
2. *Condition* (5.2) *is satisfied.*
3. $\mathcal{E}^p(\phi, F, 1)$ *is not weakly dense.*

Before getting into the proof, let us set some notation. For $f \in H^p$, let $[f]$ denote the linear span of $\{B^n f : n = 0, 1, \ldots\}$, $[f]_{H^p}$ the closure of $[f]$ in the H^p metric, and $[f]_w$ denote the weak-closure of $[f]$. From the definitions of the metric and weak topologies follow the inclusions

$$[f] \subset [f]_{H^p} \subset [f]_w. \tag{5.5}$$

Proof of Corollary 5.4. We will show that $(1) \Leftrightarrow (2)$ and $(2) \Leftrightarrow (3)$. If $\mathcal{E}^p(\phi, F, 1)$ is weakly closed, it is not weakly dense and so by Proposition 5.3,

$$(0) \neq \mathcal{E}^p(\phi, F, 1)^\perp = \mathcal{I}(\phi, F).$$

Since $\mathcal{I}(\phi, F) \neq (0)$, then (5.2) must be satisfied. So $(1) \Rightarrow (2)$.

For the other direction, we assume (5.2) is satisfied. We will show that $\mathcal{E}^p(\phi, F, 1)$ is weakly closed by showing it has the Hahn-Banach separation property. Let $f_0 \in H^p \setminus (0)$ satisfy $\langle f_0, g \rangle = 0$ for all $g \in \mathcal{E}^p(\phi, F, 1)^\perp = \mathcal{I}(\phi, F)$. We will show that $f_0 \in \mathcal{E}^p(\phi, F, 1)$.

Since $\mathcal{I}(\phi, F)$ is an ideal, then $z^n \mathcal{I}(\phi, F) \subset \mathcal{I}(\phi, F)$ and, by using the identity

$$\langle B^n f_0, g \rangle = \langle f_0, z^n g \rangle = 0, \quad n = 0, 1, 2, \ldots, \quad g \in \mathcal{I}(\phi, F),$$

we see that

$$\langle f, g \rangle = 0 \text{ for all } g \in \mathcal{I}(\phi, F) \text{ and } f \in [f_0].$$

But since $[f_0]^\perp \neq (0)$ (since $\mathcal{I}(\phi, F) \neq (0)$), then $[f_0]_w \neq H^p$ and hence, by (5.5), $[f_0]_{H^p} \neq H^p$.

It follows now, by Aleksandrov's theorem (Theorem 4.6), that

$$[f_0]_{H^p} = \mathcal{E}^p(\psi, H, 1) = e^p(\psi, H, 1) = \mathcal{E}^2(\psi) \bigvee \left\{ \frac{1}{1 - e^{-i\theta}z} : e^{i\theta} \in H \right\},$$

where ψ is inner and H is a closed subset of $\mathbb{T}$. We assume, as always, that $\sigma(\psi) \cap \mathbb{T} \subset H$. Let $g_1 \in \mathcal{I}(\phi, F)$ so that ϕ_{g_1} (the inner part of g_1) is equal to ϕ

and $g_1^{-1}(\{0\}) \cap \mathbb{T} = F$. This is possible since we are assuming (5.2) and so we can invoke a result in [15] (the ideals are singly generated).

Since $g_1 \perp [f_0]_{H^p}$, then $g_1 \perp \mathcal{E}^2(\psi)$ and so $g_1 \in \psi H^\infty$. It follows now, since $\phi_{g_1} = \phi$, that ψ divides ϕ and so $\mathcal{E}^2(\psi) \subset \mathcal{E}^2(\phi)$. Notice again that $g_1 \perp [f_0]_{H^p}$ and so

$$g_1 \perp \bigvee \left\{ \frac{1}{1 - e^{i\theta}z} : e^{i\theta} \in H \right\}.$$

This means that

$$\left\langle \frac{1}{1 - e^{-i\theta}z}, g_1 \right\rangle = \overline{g_1}(e^{i\theta}) = 0, \quad e^{i\theta} \in H.$$

Since $g^{-1}(\{0\}) \cap \mathbb{T} = F$, then $H \subset F$ and so, again using Aleksandrov's approximation theorem $\mathcal{E}^p(\phi, F, 1) = e^p(\phi, F, 1)$ and $\mathcal{E}^p(\psi, H, 1) = e^p(\psi, H, 1)$,

$$f_0 \in [f_0]_{H^p} = \mathcal{E}^p(\psi, H, 1) \subset \mathcal{E}^p(\phi, F, 1).$$

Thus $\mathcal{E}^p(\phi, F, 1)$ satisfies the Hahn-Banach separation property and hence is weakly closed. Hence $(2) \Rightarrow (1)$.

Finally, from (5.1) and Proposition 5.3, notice that for any ϕ and F, the weak closure of $\mathcal{E}^p(\phi, F, 1)$ is

$$^\perp(\mathcal{E}^p(\phi, F, 1)^\perp) = {}^\perp\mathcal{I}(\phi, F).$$

Thus $\mathcal{E}^p(\phi, F, 1)$ is not weakly dense if and only if (5.2) is satisfied. Hence $(2) \Leftrightarrow 3)$.
$\square$

The following is our alternative description of $\mathcal{E}^p(\phi, F, 1)$. The theorem and proof is very similar to a result for weighted Bergman spaces in [4] but, for the sake of completeness, and since there are enough differences, we include it anyway.

Theorem 5.6. *If* (5.2) *is satisfied, then* $\mathcal{E}^p(\phi, F, 1)$ *is the space of functions* $f \in H^p$ *such that*

    1. $fg \in H^1$

    2. f/ϕ *has a pseudocontinuation to an* $\widetilde{f_\phi} \in H^p(\mathbb{D}_e)$ *with* $\widetilde{f_\phi}(\infty) = 0$,

where $g \in A^\infty$ *with* $\phi_g = \phi$ *and* $g^{-1}(\{0\}) \cap \mathbb{T} = F$.

Proof. Since $\mathcal{I}_g$ (the weak-* closed ideal generated by g) is equal to $\mathcal{I}(\phi, F)$, then, by the equality $\mathcal{E}^p(\phi, F, 1) = {}^\perp\mathcal{I}(\phi, F)$, we need to show that an $f \in H^p$ satisfies the two hypotheses of the theorem if and only if $f \in {}^\perp\mathcal{I}_g$.

Let $\phi\Theta = g$ be the inner-outer factorization of g. If $f \in H^p$ satisfies the two hypotheses of the theorem, then for almost every θ,

$$(f\overline{g})(e^{i\theta}) = \widetilde{f_\phi}(e^{i\theta})\overline{\Theta}(e^{i\theta}).$$

The right-hand side of the above equation is the boundary function for

$$\widetilde{f_\phi}(z)\overline{\Theta}(1/\overline{z})$$

which belongs to $H^p(\mathbb{D}_e)$. Moreover, by the assumption that $fg \in H^1$, this boundary function belongs to L^1 and so, by a classical theorem of Smirnov [11, p. 28],

206   W.T. Ross

$(f\bar{g})(e^{i\theta})$ is the boundary function for a function belonging to $H^1(\mathbb{D}_e)$. Hence, by the F. and M. Riesz theorem (3.8),

$$\int_0^{2\pi} (f\bar{g})(e^{i\theta})e^{-in\theta}\frac{d\theta}{2\pi} = 0, \quad n = 0, 1, 2, \dots$$

By our dual pairing between H^p and O_p, and the fact that $fg \in H^1$, we conclude that

$$\langle f, z^n g\rangle = \int_0^{2\pi} (f\bar{g})(e^{i\theta})e^{-in\theta}\frac{d\theta}{2\pi} = 0, \quad n = 0, 1, 2, \dots{}^9$$

This shows that f annihilates the weak-* closed S-invariant subspace of O_p containing g. One can prove (see [4, Thm. 3.2]) that any weak-* closed S-invariant subspace of O_p is an ideal and so $f \in {}^{\perp}\mathcal{I}_g$ (the weak-* closed ideal generated by g).

Conversely, suppose $f \in {}^{\perp}\mathcal{I}_g$, or equivalently $f \in \mathcal{E}^p(\phi, F, 1)$. By the definition of $\mathcal{E}^p(\phi, F, 1)$, f satisfies the second (pseudocontinuation) condition of the theorem and so we just need to show that fg belongs to H^1. To this end note that for any integer $n \geqslant 1$,

$$\langle f, g\rangle = n! \int f\overline{(z^{n+1}g)^{(n+1)}}(1 - |z|^2)^n\frac{dA}{\pi},$$

where dA is area measure on the disk $\mathbb{D}$. For ease in notation, let

$$g_n := \overline{(z^{n+1}g)^{(n+1)}}.$$

We also assume that $n > 1/p$ so that $fg_n(1 - |z|^2)^n$ is bounded on $\mathbb{D}$. This is possible since g_n is bounded on $\mathbb{D}$, since we are assuming that $g \in A^\infty$, and all H^p functions f satisfy the growth estimate $|f(z)| \leqslant C_f(1 - |z|)^{-1/p}$ (recall this from § 2).

With this fixed n, let $\lambda \in \mathbb{D}$ and note, using the definition of $\mathcal{E}^p(\phi, F, 1)$, that

$$\frac{f - f(\lambda)}{z - \lambda} \in \mathcal{E}^p(\phi, F, 1)$$

and so, since g annihilates $\mathcal{E}^p(\phi, F, 1)$,

$$0 = \langle\frac{f - f(\lambda)}{z - \lambda}, g\rangle = n! \int \frac{f - f(\lambda)}{z - \lambda}g_n(1 - |z|^2)^n\frac{dA}{\pi}. \tag{5.7}$$

Let

$$G(\lambda) := n! \int \frac{g_n(1 - |z|^2)^n}{z - \lambda}\frac{dA}{\pi}, \qquad H(\lambda) := n! \int \frac{fg_n(1 - |z|^2)^n}{z - \lambda}\frac{dA}{\pi}.$$

Elementary facts about Cauchy transforms of bounded functions on the plane [33, p. 40] show that G and H are continuous functions on $\mathbb{C}$ and satisfy the Lipschitz-type condition

$$|G(\lambda_1) - G(\lambda_2)| \leqslant C_G|\lambda_1 - \lambda_2| \log\frac{1}{|\lambda_1 - \lambda_2|}. \tag{5.8}$$

---

[9] Note, by the dominated convergence theorem and the fact that $fg \in H^1$ and so $f_r g_r \to fg$ as $r \to 1$, that $f_r\bar{g}_r - f\bar{g} = (f_r g_r - fg)\bar{g}_r/g_r + fg(\bar{g}_r/g_r - \bar{g}/g)$ converges to zero as $r \to 1$.

Furthermore, by (5.7),

$$f(\lambda)G(\lambda) = H(\lambda), \quad \lambda \in \mathbb{D}. \tag{5.9}$$

A computation with power series shows that for $0 < r < 1$

$$G(e^{i\theta}/r) = re^{-i\theta}\overline{g}(re^{i\theta})$$

and so for $r > 1/2$,

$$|(fg)(re^{i\theta})| \leqslant C\left[|f(re^{i\theta})G(re^{i\theta})| + |f(re^{i\theta})|\,|G(re^{i\theta}) - G(e^{i\theta}/r)|\right]. \tag{5.10}$$

By (5.9), the first term on the right-hand side of the above is equal to $|H(re^{i\theta})|$ which is uniformly bounded in r and θ. For the second term, notice from (5.8) that

$$\left|G(re^{i\theta}) - G(e^{i\theta}/r)\right| \leqslant C_G(1-r)\log\frac{1}{1-r}$$

and so, by (5.10),

$$|(fg)(re^{i\theta})| \leqslant C_1 + C_2|f(re^{i\theta})|(1-r)\log\frac{1}{1-r}. \tag{5.11}$$

Since $H^p \subset B^p$ then, for any $0 < r < 1$,

$$
\begin{aligned}
\|f\|_{B^p} &\geqslant \int_r^1 (1-s)^{1/p-2}\int_0^{2\pi} |f(se^{i\theta})|\frac{d\theta}{2\pi}\,ds \\
&\geqslant \int_r^1 (1-s)^{1/p-2}\int_0^{2\pi} |f(re^{i\theta})|\frac{d\theta}{2\pi}\,ds \\
&\geqslant (1/p-1)^{-1}(1-r)^{1/p-1}\int_0^{2\pi} |f(re^{i\theta})|\frac{d\theta}{2\pi}
\end{aligned}
$$

and so

$$\int_0^{2\pi} |f(re^{i\theta})|\frac{d\theta}{2\pi} \leqslant C_p(1-r)^{1-1/p}.$$

Combining this with (5.11) along with the fact that $1/2 < p < 1$ (and so

$$(1-r)^{2-1/p}\log\frac{1}{1-r}$$

is bounded in r), we conclude that

$$\int_0^{2\pi} |(fg)(re^{i\theta})|\frac{d\theta}{2\pi}$$

is uniformly bounded in r. Hence $fg \in H^1$. $\qquad\square$

For other p, not in $(1/2, 1)$, the above theorem is still true, though the proof is more technical since the resulting ideal $\mathcal{E}^p(\phi, F, k)^\perp$ will involve the zeros of the *derivatives* of g on the circle. The proof presented here needs to be changed slightly and for this we refer the reader to [4] where there is a similar result for the invariant subspaces of B^p. Notice that since every weakly closed invariant subspace is also closed in the metric of H^p, we have shown the following corollary.

Corollary 5.12. *For $0 < p < 1$ and ϕ, F, k satisfying the condition*

$$\int_0^{2\pi} \log dist(e^{i\theta}, \sigma(\phi) \cup F)\frac{d\theta}{2\pi} > -\infty, \tag{5.13}$$

$\mathcal{E}^p(\phi, F, k)$ *is a non-trivial weakly closed invariant subspace of H^p. Conversely, every non-trivial weakly closed invariant subspace of H^p takes the form $\mathcal{E}^p(\phi, F, k)$ for some ϕ, F, k satisfying* (5.13).

6. The Bergman spaces L_a^p, $0 < p < 1$

We end with some remarks about the invariant subspaces of the Bergman spaces[10] L_a^p $(0 < p < \infty)$ of analytic functions f on $\mathbb{D}$ for which

$$\|f\|_p := \left(\int_{\mathbb{D}} |f|^p dA\right)^{1/p} < \infty.$$

The quantity $\|f - g\|_p$ defines a norm when $1 \leqslant p < \infty$ while $\|f - g\|_p^p$ defines a translation invariant metric when $0 < p < 1$. In either case, one can use the pointwise estimate

$$|f(z)| \leqslant \pi^{-1/p}\|f\|_p(1 - |z|)^{-2/p}, \quad z \in \mathbb{D}$$

to show that L_a^p is an F-space [9, p. 51]. For $f \in L_a^p$, routine integral estimates show that $Bf \in L_a^p$. Using the above pointwise estimate, one proves that the graph of B is closed and so, by the closed graph theorem (which is valid in an F-space [9, p. 57]), B is continuous on L_a^p.

When $1 \leqslant p < \infty$, one can make heavy use of duality to show that if $\mathcal{E}$ is a non-trivial invariant subspace of L_a^p, then every $f \in \mathcal{E}$ has a pseudocontinuation of bounded type. Moreover, when $1 \leqslant p < 2$, there is a complete description of $\mathcal{E}$ [3, 4, 22]. We pause for a moment to remark that in order for $f \in \mathcal{E}$ to have a pseudocontinuation, it must first have non-tangential limits almost everywhere on $\mathbb{T}$. This is automatic for H^p but not for L_a^p. There are indeed examples of functions in L_a^p (or in any of the weighted Bergman spaces such as B^p) which do not even have radial limits almost everywhere [11, p. 86]. Such poorly behaved functions do not belong to non-trivial invariant subspaces of L_a^p.

When $0 < p < 1$, can we say anything about the invariant subspaces of L_a^p? In this case, L_a^p is not locally convex and so, as in H^p $(0 < p < 1)$, duality is of little use. J. Shapiro [28, 29] showed, assuming $0 < p < 1$ as we will do from now on, that $L_a^p \subset B^{p/2}$ and this containment is continuous. Moreover, L_a^p and $B^{p/2}$ have the same set of continuous linear functionals (via the same 'Cauchy duality' as in (2.1)), namely $O_{p/2}$. As in the H^p case, there is a corresponding weak topology on L_a^p induced by $O_{p/2}$ and if A is a linear manifold in L_a^p, then

$$\operatorname{clos}_{(L_a^p, wk)} A = \left(\operatorname{clos}_{B^{p/2}} A\right) \cap L_a^p. \tag{6.1}$$

---

[10]Two nice references about Bergman spaces are the following books [10, 14].

If $\mathcal{E}$ is a norm-closed invariant subspace of $B^{p/2}$, then $\mathcal{E}^{\perp}$ is an S-invariant subspace of $O_{p/2}$ and hence a weak-* closed ideal which, as before (see also [4, Thm. 3.2]), is of the form $\mathcal{I}_g$ (the weak-* closed ideal generated by g) for some $g \in A^{\infty}$. By the Hahn-Banach theorem, which is applicable here since $B^{p/2}$ is a Banach space, we have $\mathcal{E} = {}^{\perp}\mathcal{I}_g$. One can prove [4] that $f \in B^{p/2}$ belongs to ${}^{\perp}\mathcal{I}_g$ if and only if (i) $fg \in H^1$; (ii) f/ϕ_g (where ϕ_g is the inner part of g) has a pseudocontinuation $\widetilde{f_{\phi_g}} \in N^+(\mathbb{D}_e)$ [11] which vanishes at infinity. Combining this with (6.1) we have the following result.

Theorem 6.2. *Let* $0 < p < 1$ *and* $\mathcal{E}$ *be a non-trivial weakly closed invariant subspace of* L_a^p. *Then there is a* $g \in A^{\infty}$ *such that* $\mathcal{E}$ *is the set of* $f \in L_a^p$ *such that*

1. $fg \in H^1$.
2. f/ϕ_g *has a pseudocontinuation* $\widetilde{f_{\phi_g}} \in N^+(\mathbb{D}_e)$ *which vanishes at infinity.*

Certainly if $\mathcal{E}$ is a weakly closed invariant subspace of L_a^p, then $\mathcal{E}$ is closed in the metric of L_a^p. Is every closed invariant subspace weakly closed? In H^p, this is not the case (see Corollary 5.12). Though we do not have a proof, we conjecture that every closed invariant subspace of L_a^p is indeed weakly closed. In H^p, the space $\mathcal{E} = \bigvee\{(1 - e^{-i\theta}z)^{-1} : 0 \leqslant \theta < 2\pi\}$, where $\bigvee$ is the closed linear span in the metric topology of H^p, is a proper closed invariant subspace that is weakly dense. This same example, with the linear span in H^p replaced by the closed linear span in the metric topology of L_a^p, is certainly weakly dense. However, since $L_a^1 \subset L_a^p$ with continuous inclusion, and since the linear span of $\{(1 - e^{-i\theta}z)^{-1} : 0 \leqslant \theta < 2\pi\}$ is norm dense in L_a^1, we see that $\mathcal{E}$ is dense in the metric topology of L_a^p. We end with the following open question.

Question 6.3. For $0 < p < 1$, what are the closed (in the metric topology) invariant subspaces of L_a^p?

References

[1] A.B. Aleksandrov, *Invariant subspaces of the backward shift operator in the space H^p ($p \in (0, 1)$)*, Zap. Nauchn. Sem. Leningrad. Otdel. Mat. Inst. Steklov. (LOMI) **92** (1979), 7–29, 318, Investigations on linear operators and the theory of functions, IX. MR **81h:**46018

[2] ______, *Gap series and pseudocontinuations. An arithmetic approach*, Algebra i Analiz **9** (1997), no. 1, 3–31. MR **98d:**30006

[3] A. Aleman, S. Richter, and W.T. Ross, *Pseudocontinuations and the backward shift*, Indiana Univ. Math. J. **47** (1998), no. 1, 223–276. MR **2000i:**47009

[4] A. Aleman and W.T. Ross, *The backward shift on weighted Bergman spaces*, Michigan Math. J. **43** (1996), no. 2, 291–319. MR **97i:**47053

[5] F. Bagemihl and W. Seidel, *Some boundary properties of analytic functions*, Math. Z. **61** (1954), 186–199. MR 16,460d

---

[11] $N^+(\mathbb{D}_e)$ is the Smirnov class (see [11, p. 25] for a definition).

[6] J.A. Cima and W.T. Ross, *The backward shift on the Hardy space*, American Mathematical Society, Providence, RI, 2000. MR **2002f**:47068

[7] M.M. Day, *The spaces L^p with $0 < p < 1$*, Bull. Amer. Math. Soc. **46** (1940), 816–823. MR 2,102b

[8] R.G. Douglas, H.S. Shapiro, and A.L. Shields, *Cyclic vectors and invariant subspaces for the backward shift operator*, Ann. Inst. Fourier (Grenoble) **20** (1970), no. fasc. 1, 37–76. MR 42 #5088

[9] N. Dunford and J. Schwartz, *Linear Operators. I. General Theory*, With the assistance of W. G. Bade and R. G. Bartle. Pure and Applied Mathematics, Vol. 7, Interscience Publishers, Inc., New York, 1958. MR 22 #8302

[10] P. Duren and A. Schuster, *Bergman spaces*, American Mathematical Society, Providence, RI, 2004.

[11] P.L. Duren, *Theory of H^p spaces*, Academic Press, New York, 1970. MR 42 #3552

[12] P.L. Duren, B.W. Romberg, and A.L. Shields, *Linear functionals on H^p spaces with $0 < p < 1$*, J. Reine Angew. Math. **238** (1969), 32–60. MR 41 #4217

[13] J.B. Garnett, *Bounded analytic functions*, Academic Press Inc., New York, 1981. MR **83g**:30037

[14] H. Hedenmalm, B. Korenblum, and K. Zhu, *Theory of Bergman spaces*, Graduate Texts in Mathematics, vol. 199, Springer-Verlag, New York, 2000. MR **2001c**:46043

[15] S.V. Hruščev, *Every ideal of the ring C_A^n is principal*, Zap. Naučn. Sem. Leningrad. Otdel. Mat. Inst. Steklov (LOMI) **65** (1976), 149–160, 206, Investigations on linear operators and the theory of functions, VII. MR 58 #30282

[16] P. Koosis, *Introduction to H_p spaces*, second ed., Cambridge University Press, Cambridge, 1998. MR **2000b**:30052

[17] B.I. Kornblum, *Closed ideals of the ring A^n*, Funkcional. Anal. i Priložen. **6** (1972), no. 3, 38–52. MR 48 #2776

[18] A. Matheson, *Closed ideals in rings of analytic functions satisfying a Lipschitz condition*, Banach spaces of analytic functions (Proc. Pelczynski Conf., Kent State Univ., Kent, Ohio, 1976), Springer, Berlin, 1977, pp. 67–72. Lecture Notes in Math., Vol. 604. MR 57 #3864

[19] ______, *Cyclic vectors for invariant subspaces in some classes of analytic functions*, Illinois J. Math. **36** (1992), no. 1, 136–144. MR **93e**:30074

[20] N.K. Nikol'skiĭ, *Treatise on the shift operator*, Springer-Verlag, Berlin, 1986. MR **87i**:47042

[21] T. Pedersen, *Ideals in big Lipschitz algebras of analytic functions*, to appear, Studia Math.

[22] S. Richter and C. Sundberg, *Invariant subspaces of the Dirichlet shift and pseudocontinuations*, Trans. Amer. Math. Soc. **341** (1994), no. 2, 863–879. MR **94d**:47026

[23] W.T. Ross and H.S. Shapiro, *Generalized analytic continuation*, University Lecture Series, vol. 25, American Mathematical Society, Providence, RI, 2002. MR **2003h**:30003

[24] W. Rudin, *Functional analysis*, second ed., McGraw-Hill Inc., New York, 1991. MR **92k**:46001

[25] F.A. Shamoyan, *Closed ideals in algebras of functions that are analytic in the disk and smooth up to its boundary*, Mat. Sb. **184** (1993), no. 8, 109–136. MR **94m**:46087

[26] H.S. Shapiro, *Generalized analytic continuation*, Symposia on Theoretical Physics and Mathematics, Vol. 8 (Symposium, Madras, 1967), Plenum, New York, 1968, pp. 151–163. MR 39 #2953

[27] ______, *Functions nowhere continuable in a generalized sense*, Publ. Ramanujan Inst. No. **1** (1968/1969), 179–182. MR 42 #1982

[28] J. Shapiro, *Linear functionals on non-locally convex spaces*, Ph.D. thesis, University of Michigan, 1969.

[29] ______, *Mackey topologies, reproducing kernels, and diagonal maps on the Hardy and Bergman spaces*, Duke Math. J. **43** (1976), no. 1, 187–202. MR 58 #17806

[30] D. Stegenga, *Bounded Toeplitz operators on H^1 and applications of the duality between H^1 and the functions of bounded mean oscillation*, Amer. J. Math. **98** (1976), no. 3, 573–589. MR 54 #8340

[31] B.A. Taylor and D.L. Williams, *Ideals in rings of analytic functions with smooth boundary values*, Canad. J. Math. **22** (1970), 1266–1283. MR 42 #7905

[32] ______, *Zeros of Lipschitz functions analytic in the unit disc*, Michigan Math. J. **18** (1971), 129–139. MR 44 #409

[33] I.N. Vekua, *Generalized analytic functions*, Pergamon Press, London, 1962. MR 27 #321

William T. Ross
Department of Mathematics and Computer Science
University of Richmond
Richmond
Virginia 23173, USA
e-mail: `wross@richmond.edu`

Operator Theory:
Advances and Applications, Vol. 158, 213–222

On Some Classes of q-parametric Positive linear Operators

Victor S. Videnskii

To the Memory of my Friend S.Ya. Havinson

Abstract. Recently, G.M. Phillips [6], [7] has introduced the q-parametric Bernstein polynomials denoted by $B_n(f, x, q)$, $0 < q \leq 1$. In [6] the investigation of the properties of these polynomials is based on some generalization of divided differences and on the Newton interpolation formula. A. Il'inskii and S. Ostrovska [4] have considered linear operators $B_\infty(f, x, q)$ which were derived by means of $B_n(f, x, q)$ with the help of an informal passing to the limit for $n \to \infty$. In this paper we give another more natural proof of the main results of [6], [4]. Moreover, for the Voronovskaya's asymptotic formula we obtain the estimate of the remainder term. We also consider the modification of these operators in order to improve the degree of approximation of twice differentiable functions.

Mathematics Subject Classification (2000). 41A10, 41A25.

Keywords. approximation, q-Bernstein polynomials, central moments, inequalities.

1. Introduction

The classical Bernstein polynomial of degree n for $f \in C[0, 1]$ is given by

$$B_n(f, x) = \sum_{k=0}^{n} f\left(\frac{k}{n}\right) p_{nk}(x), \tag{1.1}$$

$$p_{nk}(x) = \binom{n}{k} x^k (1 - x)^{n-k}. \tag{1.2}$$

This construction is due to S. Bernstein [1]. These polynomials possess many remarkable properties and are the leading special case in the theory of approximation by positive linear operators (p.l.o.) (cf., e.g., [8, 9, 10]).

In 1997 G. M. Phillips [6] generalized the Bernstein polynomials in the following way. Let $0 < q \leq 1$ and denote by

$$[n]_q = [n] = 1 + q + \cdots + q^{n-1} \text{ for } n \in \mathbb{N}; \quad [0] = 0; \tag{1.3}$$

$$[n]! = [1][2]\ldots[n] \text{ for } n \in \mathbb{N}; \quad [0]! = 1; \tag{1.4}$$

$$\binom{n}{k}_q = \frac{[n]!}{[k]![n-k]!}, \quad k = 0, 1, \ldots, n; \tag{1.5}$$

if $k < 0$ or $k > n$ then $\binom{n}{k}_q = 0$. The notation (1.5) is due to Gauss [3, p. 16]. The q-Bernstein polynomials G.M. Phillips defined by formulas

$$B_n(f, x, q) = \sum_{k=0}^{n} f\left(\frac{[k]}{[n]}\right) p_{nk}(x, q), \tag{1.6}$$

$$p_{nk}(x, q) = \binom{n}{k}_q x^k (1-x)(1-xq)\ldots(1-xq^{n-k-1}). \tag{1.7}$$

For the special case $q = 1$ we have $B_n(f, x, 1) = B_n(f, x)$. As $p_n(x, q) \geq 0$ on $[0, 1]$ therefore $B_n(f, x, q)$ is p.l.o.

2. Main identities

From (1.3)–(1.5) we have

$$[k] + q^k[n-k] = [n], \tag{2.1}$$

$$\binom{n-1}{k-1}_q + q^k \binom{n-1}{k}_q = \binom{n}{k}_q, \tag{2.2}$$

$$\frac{[k]}{[n]}\binom{n}{k}_q = \binom{n-1}{k-1}_q, \tag{2.3}$$

$$\frac{[n-k]}{[n]}\binom{n}{k}_q = \binom{n-1}{k}_q. \tag{2.4}$$

In order to consider the approximation properties of p.l.o. $B_n(f, x, q)$ we shall derive the following identities:

$$B_n(1, x, q) = 1, \tag{2.5}$$

$$B_n(t, x, q) = x, \tag{2.6}$$

$$B_n(t^2, x, q) = x^2 + \frac{x(1-x)}{[n]}. \tag{2.7}$$

These formulas are the generalization of those for classical Bernstein polynomials. Our proofs are based on (2.1)–(2.4) and are essentially different from the proofs in [6].

3. Proofs of relations (2.5)–(2.7)

We shall obtain (2.5) by induction. The case $n = 1$ is immediate. In fact,

$$B_1(1, x, q) = p_{10}(x, q) + p_{11}(x, q) = (1 - x) + x = 1.$$

Assume that (2.5) is true for $n - 1$, in other words $B_{n-1}(1, x, q) = 1$. Using (2.2) we get

$$p_{nk}(x, q) = \left\{ \binom{n-1}{k-1}_q + q^k \binom{n-1}{k}_q \right\} x^k (1 - x) \dots (1 - xq^{n-k-1})$$
$$= x p_{n-1,k-1}(x, q) + (1 - x) p_{n-1,k}(qx, q).$$

Then because of the induction hypothesis we have

$$B_n(1, x, q) = \sum_{k=0}^{n} p_{nk}(x, q) = x B_{n-1}(1, x, q) + (1 - x) B_{n-1}(1, qx, q) = 1.$$

The equality (2.6) is readily derived from (2.5) with the help of (2.3). Indeed,

$$B_n(t, x, q) = \sum_{k=0}^{n} \frac{[k]}{[n]} p_{nk}(x, q) = x \sum_{k=1}^{n} p_{n-1,k-1}(x, q) = x.$$

We shall prove now the following recurrence formula

$$B_n(t^{m+1}, x, q) = B_n(t^m, x, q) - \frac{[n-1]^m}{[n]^m}(1 - x) B_{n-1}(t^m, xq, q). \qquad (3.1)$$

In the special case $m = 1$ we shall obtain (2.7) from (3.1). We write explicitly

$$B_n(t^{m+1}, x, q) = \sum_{k=0}^{n} \frac{[k]^{m+1}}{[n]^{m+1}} \binom{n}{k}_q x^k (1 - x)(1 - xq) \dots (1 - xq^{n-k-1}) \qquad (3.2)$$

and transform the first factor using (2.1) and (2.4) in the following manner:

$$\frac{[k]^{m+1}}{[n]^{m+1}} \binom{n}{k}_q = \frac{[k]^m}{[n]^m} \left(1 - q^k \frac{[n-k]}{[n]} \right) \binom{n}{k}_q$$
$$= \frac{[k]^m}{[n]^m} \binom{n}{k}_q - \frac{[n-1]^m}{[n]^m} \frac{[k]^m}{[n-1]^m} \binom{n-1}{k}_q q^k. \qquad (3.3)$$

Finally, if we substitute (3.3) in (3.2) we get (3.1). In fact, according to (2.6) and (3.1) we find for $m = 1$

$$B_n(t^2, x, q) = x - \frac{[n-1]}{[n]}(1 - x)xq = x^2 + \frac{x(1 - x)}{[n]}.$$

We conclude from (3.1) that for $m \le n$

$$\deg B_n(t^m, x, q) = m. \qquad (3.4)$$

Possibly the recurrence formula (3.1) is new even for the classical Bernstein polynomials. Another recurrence relation for $B_n(t^m, x)$ is stated in [9, p. 28].

4. Approximation theorem

If $0 < q < 1$ then by (2.7)

$$\lim_{n \to \infty} \{B_n(t^2, x, q) - x^2\} = (1 - q)x(1 - x).$$

Consequently, the polynomials $B_n(t^2, x, q)$ do not converge to x^2 in $(0, 1)$ as $n \to \infty$. In what follows it is assumed that $0 < q_n < 1$. It is easy to see that if $q_n \to 1$ for $n \to \infty$ then

$$\lim_{n \to \infty} [n]_{q_n} = +\infty.$$

In fact, there exists for every q_0, $0 < q_0 < 1$, such a number n_0 that $[n_0]_{q_0} > 2^{-1}(1 - q_0)^{-1}$. But for $n > n_0$ such that $q_n > q_0$ we have $[n]_{q_n} > [n_0]_{q_0}$. Therefore, if $q_n \to 1$ then

$$\lim_{n \to \infty} B_n(t^2, x, q_n) = x^2.$$

It should be noted that as $[n]_{q_n} < n$ then

$$B_n(t^2, x, q_n) - x^2 > B_n(t^2, x) - x^2 \qquad (0 < x < 1).$$

Consequently, the degree of approximation of t^2 by $B_n(t^2, x, q_n)$ for arbitrary rate of convergence of q_n to the unity is worse than the degree of approximation of t^2 by $B_n(t^2, x)$.

So-called central moments of p.l.o. play a significant role in the theory of approximation by sequence of p.l.o. For the q-Bernstein polynomials let us denote them by

$$S_{nm}(x, q) = B_n((t - x)^m, x, q). \tag{4.1}$$

For $m = 2$ using (2.5)–(2.7) we find immediately

$$S_{n2}(x, q) = \frac{x(1 - x)}{[n]}. \tag{4.2}$$

As the conditions (2.5) and (4.2) take place then by a well-known general result due to Popoviciu (see [8, p. 13]) we obtain:

Theorem 4.1. *For any $f \in C[0, 1]$ the following inequality holds*

$$|B_n(f, x, q_n) - f(x)| \leq 2\omega(f, \sqrt{S_{n2}(x, q_n)}) \leq 2\omega(f, 2^{-1}[n]_{q_n}^{-\frac{1}{2}}), \tag{4.3}$$

where $\omega(f, s)$ denotes the modulus of continuity of the function f on the segment $[0, 1]$.

Therefore, the sequence $B_n(f, x, q_n)$ converges uniformly to f for each continuous function f if and only if

$$\lim_{n \to \infty} q_n = 1. \tag{4.4}$$

5. Generalization of Voronovskaya's theorem

In this section we will follow the method due to S. Bernstein [2] (see also [9, §6]). First of all we shall find the explicit formula for $S_{n3}(x, q)$. Using (3.1) and taking into account (2.5)–(2.7) we obtain

$$B_n(t^3, x, q) = x^2 + \frac{x(1-x)}{[n]} - \frac{x(1-x)}{[n]}\left(1 - \frac{1}{[n]}\right)(1 + q^2[n-2]x). \qquad (5.1)$$

On the other hand, if we express t^3 at the following form

$$t^3 = x^3 + 3x^2(t-x) + 3x(t-x)^2 + (t-x)^3$$

and use the q-Bernstein polynomial, then we get

$$B_n(t^3, x, q) = x^3 + 3\frac{x^2(1-x)}{[n]} + S_{n3}(x, q). \qquad (5.2)$$

By (5.1) and (5.2) we finally obtain the explicit expression

$$S_{n3}(x, q) = \frac{x(1-x)(1-Q_n x)}{[n]^2}, \qquad (5.3)$$

where $Q_n = 2 + q(1 - q^{n-1})$. Therefore,

$$B_n(t^3, x, q) = x^3 + 3\frac{x^2(1-x)}{[n]} + \frac{x(1-x)(1-Q_n x)}{[n]^2}. \qquad (5.4)$$

The analogous calculation can be applied to $S_{n4}(x, q)$. Substituting (5.4) in (3.1) and comparing the obtained result with the formula

$$B_n(t^4, x, q) = x^4 + 6\frac{x^3(1-x)}{[n]} + 4\frac{x^2(1-x)(1-Q_n x)}{[n]^2} + S_{n4}(x, q), \qquad (5.5)$$

similar to (5.2), we easily receive the important inequality

$$S_{n4}(x, q) \leq K\frac{x(1-x)}{[n]^2}. \qquad (5.6)$$

Here and further K denotes a positive absolute constant.

Theorem 5.1. *For any $f \in C^{(2)}[0, 1]$ the following inequality holds*

$$\left| B_n(f, x, q_n) - f(x) - \frac{f''(x)}{2}\frac{x(1-x)}{[n]_{q_n}} \right| \leq \frac{Kx(1-x)}{[n]_{q_n}}\omega(f'', [n]_{q_n}^{-\frac{1}{2}}). \qquad (5.7)$$

Let $f \in C^{(2)}[0, 1]$ and $x \in [0, 1]$ is fixed. By Taylor's formula we may write

$$f(t) = f(x) + \frac{f'(x)}{1!}(t-x) + \frac{f''(\xi_t)}{2!}(t-x)^2$$

$$= f(x) + \frac{f'(x)}{1!}(t-x) + \frac{f''(x)}{2!}(t-x)^2 + r_2(f, t, x), \qquad (5.8)$$

$$r_2(f, t, x) = \frac{f''(\xi_t) - f''(x)}{2}(t-x)^2, \qquad (5.9)$$

where ξ_t is situated between x and t, therefore, $|\xi_t - x| < |t - x|$. Applying p.l.o. $B_n(f, x, q)$ to (5.8) we obtain

$$\left| B_n(f, x, q_n) - f(x) - f''(x)\frac{x(1-x)}{2[n]_{q_n}} \right| \leq B_n(|r_2|, x, q_n),$$

because $S_{n1}(x, q) = 0$. For the estimate of the remainder r_2 we shall use the well-known inequality $\omega(f, \lambda\delta) \leq (1 + \lambda^2)\omega(f, \delta)$. We have

$$|f''(\xi_t) - f''(x)| \leq \omega(f'', |\xi_t - x|) \leq \omega(f'', |t - x|)$$

$$\leq \omega(f'', [n]_{q_n}^{-\frac{1}{2}})(1 + [n]_{q_n}(t - x)^2).$$

Hence,

$$B_n(|r_2|, x, q_n) \leq \frac{1}{2}\omega(f'', [n]_{q_n}^{-\frac{1}{2}})(S_{n2}(x, q_n) + [n]_{q_n}S_{n4}(x, q_n))$$

$$\leq \frac{1}{2}\omega(f'', [n]_{q_n}^{-\frac{1}{2}})\frac{x(1-x)}{[n]_{q_n}}(1 + K).$$

The proof is complete. $\qquad\qquad\qquad\qquad\qquad\qquad\qquad\qquad\qquad\qquad$ $\square$

Corollary 5.2. *If $f \in C^{(2)}[0, 1]$ and $q_n \to 1$ as $n \to \infty$, then*

$$\lim_{n \to \infty} [n]_{q_n}\{B_n(f, x, q_n) - f(x)\} = \frac{f''(x)}{2}x(1 - x) \qquad (5.10)$$

uniformly on $[0, 1]$.

It is of interest the fact that for the function $f(t) = t^2$ takes place the exact equality

$$[n]_{q_n}\{B_n(t^2, x, q_n) - x^2\} = \frac{1}{2}(x^2)''x(1 - x)$$

without passing to the limit.

Corollary 5.3. *If the function f is convex on $[0, 1]$ then*

$$B_n(f, x, q) \geq f(x). \qquad (5.11)$$

Evidently, it suffices to verify (5.11) for $f \in C^{(2)}[0, 1]$. Using the first line of (5.8) we get

$$B_n(f, x, q) = f(x) + \frac{1}{2}\sum_{k=0}^{n} f''(\xi_k)\left(\frac{[k]}{[n]} - x\right)^2 p_{nk}(x, q) \geq f(x)$$

as $f''(x) \geq 0$ for every x.

6. Modification of q-Bernstein polynomials

Let $f \in C^{(2)}[0,1]$ and let

$$D_n(f,x,q) = B_n(f,x,q) - \frac{x(1-x)}{2[n]}B_n(f'',x,q) \qquad (6.1)$$

denote the linear operator which represents some modification of p.l.o. $B_n(f,x,q)$. The special case $q = 1$ is due to S. Bernstein [2] (see also [9, §7]). Applying $D_n(f,x,q)$ instead of $B_n(f,x,q)$ to $f \in C^{(2)}[0,1]$ we are able to improve considerably the degree of approximation. In this connection we shall prove the following theorem.

Theorem 6.1. *If* $f \in C^{(2)}[0,1]$ *then*

$$|D_n(f,x,q_n) - f(x)| \le \frac{Kx(1-x)}{[n]_{q_n}}\omega(f'', [n]_{q_n}^{-\frac{1}{2}}). \qquad (6.2)$$

Combining (5.7) and (4.3) we get immediately (6.2). Obviously,

$$|D_n(f,x,q) - f(x)|$$

$$\le \left|B_n(f,x,q) - f(x) - f''(x)\frac{x(1-x)}{2[n]}\right| + \frac{x(1-x)}{2[n]}|f''(x) - B_n(f'',x,q)|.$$

This establishes the statement.

Concerning the modification of the classical Bernstein polynomials for $f \in C^{(\nu)}[0,1]$, $\nu \ge 3$, see [9, §7]. For the further application of our idea of modification to $B_n(f,x,q)$ in the case $f \in C^{(\nu)}[0,1]$, $\nu \ge 3$, we have need of bounds for the central moments $S_{n,2m}(x,q)$ for every natural number m. For our purpose it suffices to show that

$$S_{n,2m}(x,q) \le K_m \frac{x(1-x)}{[n]^m}. \qquad (6.3)$$

These inequalities take place for the classical Bernstein polynomials for all m. If $0 < q < 1$ then (6.3) holds in the special cases $m = 1$ and $m = 2$ according to (4.2) and (5.6). It would be of interest to know if (6.3) is true in general case. (See [10]).

Recently, S. Ostrovska [5] has published very interesting paper concerning the approximation of an analytic function f in the z-plane by $B_n(f,z,q)$ if it is assumed that $q \in (1,\infty)$.

7. P.l.o. introduced by S. Ostrovska and A. Il'inskii

In 2001 A. Il'inskii and S. Ostrovska [4] introduced a certain new set of p.l.o. and denoted it by $B_\infty(f,x,q)$, $0 < q < 1$. These p.l.o. were derived by means of $B_n(f,x,q)$ with the help of an informal passing to the limit for $n \to \infty$. As we shall see the behavior of $B_\infty(f,x,q)$ is very similar to that of $B_n(f,x,q)$. In [4] the investigation is based on some probabilistic considerations as in the original basic note [1] due to S. Bernstein. Our method is different from the authors' one and is similar to the discussion in §§2–6. For a sake of simplicity we shall apply the notation $A_q(f, x)$ instead of $B_\infty(f, x, q)$.

Let $0 < q < 1$; it is valid the following Euler's identity:

$$\sum_{k=0}^{\infty} \frac{x^k \psi(x)}{(1-q)\ldots(1-q^k)} = 1, \qquad 0 \le x < 1, \tag{7.1}$$

where

$$\psi(x) = \prod_{s=0}^{\infty}(1 - xq^s). \tag{7.2}$$

In order to deduce (7.1) we may use the expansion of the function $\varphi(x) = (1/\psi(x))$ in the power series for $0 \le x < 1$.

Obviously, we have

$$\psi(x) = (1-x)\psi(qx), \quad \varphi(qx) = (1-x)\varphi(x). \tag{7.3}$$

Hence,

$$\varphi(0) = 1, \quad \frac{\varphi^{(k)}(0)}{k!} = \frac{\varphi^{(k-1)}(0)}{(k-1)!(1-q^k)} = \frac{1}{(1-q)\ldots(1-q^k)}. \tag{7.4}$$

Thus (7.1) is established and the series in the left-hand side converges uniformly to the unity in $[0, 1-\varepsilon]$. P.l.o. in question $A_q(f, x)$ for $f \in C[0, 1]$ is defined by

$$A_q(f, x) = \sum_{k=0}^{\infty} f(1 - q^k)p_{qk}(x), \tag{7.5}$$

$$p_{qk}(x) = \frac{x^k \psi(x)}{(1-q)\ldots(1-q^k)}. \tag{7.6}$$

On account of (7.1) the equality

$$A_q(1, x) = \sum_{k=0}^{\infty} p_{qk}(x) = 1 \tag{7.7}$$

holds. It is easy to see also that

$$A_q(t, x) = \sum_{k=0}^{\infty}(1 - q^k)p_{qk}(x) = x \sum_{k=1}^{\infty} p_{q,k-1}(x) = x. \tag{7.8}$$

It is not difficult to verify the following recurrence formula

$$A_q(t^{m+1}, x) = A_q(t^m, x) - (1 - x)A_q(t^m, qx). \tag{7.9}$$

Indeed, it suffices to observe that

$$q^k p_{qk}(x) = (1 - x)p_{qk}(qx).$$

Hence,

$$(1 - q^k)^{m+1}p_{qk}(x) = (1 - q^k)^m p_{qk}(x) - (1 - x)(1 - q^k)^m p_{qk}(qx).$$

Let us note that it is possible to obtain (7.7)–(7.9) as limits of (2.5), (2.6) and (3.1) for $n \to \infty$.

It follows from (7.9) that for any positive integer m the analytic function $A_q(t^m, x)$ is a polynomial of exact degree m.

Putting $m = 1$ into (7.9) we get immediately

$$A_q(t^2, x) = x^2 + (1 - q)x(1 - x). \tag{7.10}$$

If we denote a central moment of p.l.o. A_q by

$$S_m(A_q, x) = A_q((t - x)^m, x) \tag{7.11}$$

then by (7.7), (7.8) and (7.10) we have

$$S_0(A_q, x) = 1, \quad S_1(A_q, x) = 0, \quad S_2(A_q, x) = (1 - q)x(1 - x). \tag{7.12}$$

Later we shall establish results analogous to the preceding theorems on approximation by q-Bernstein polynomials.

8. Approximation by p.l.o. A_q

We shall consider an approximation of the function $f \in C[0, 1]$ by $A_q f$ assuming that $0 < q < 1$ and q tends to the unity. Taking into account (7.12) and applying the general Popoviciu's theorem we obtain

Theorem 8.1. *If $f \in C[0, 1]$ then*

$$|A_q(f, x) - f(x)| \leq 2\omega(f, \sqrt{S_2(A_q, x)}) \leq 2\omega(f, \frac{1}{2}\sqrt{1 - q}). \tag{8.1}$$

Consequently, according to (7.10) and (8.1) the sequence $\{A_{q_n} f\}$ converges uniformly to any $f \in C[0, 1]$ if and only if $q_n \to 1$ for $n \to \infty$. In order to establish a theorem of Voronovskaya's type it is enough to estimate the fourth central moment $S_4(A_q, x)$. Now the consideration is easier then in §5. Using (7.9) and (7.11) we get immediately

$$S_3(A_q, x) = (1 - q)^2 x(1 - x)(1 - (q + 2)x). \tag{8.2}$$

Further, applying (7.9) for $m = 4$ and taking into account (8.2) we obtain required inequality

$$S_4(A_q, x) \leq K(1 - q)^2 x(1 - x). \tag{8.3}$$

Theorem 8.2. *If $f \in C^{(2)}[0, 1]$ then*

$$\left| A_q(f, x) - f(x) - \frac{1 - q}{2} f''(x)x(1 - x) \right| \leq K(1 - q)x(1 - x)\omega(f'', \sqrt{1 - q}). \tag{8.4}$$

For the proof of (8.4) a slight variation of the argument of §5 may be made. We have just to apply (8.3) instead of (5.6).

Corollary 8.3. *If $f \in C^{(2)}[0, 1]$ then*

$$\lim_{q \to 1} \frac{A_q(f, x) - f(x)}{1 - q} = \frac{f''(x)}{2}x(1 - x) \tag{8.5}$$

uniformly on $[0, 1]$.

Of course, for $f(t) = t^2$ we have because of (7.10) the exact equality

$$A_q(t^2, x) - x^2 = (1 - q)\frac{(x^2)''}{2}x(1 - x).$$

Corollary 8.4. *If the function f is convex on $[0, 1]$ then $A_q(f, x) \geq f(x)$.*

For $f \in C^{(2)}[0, 1]$ we define the modification of p.l.o. $A_q(f, x)$ by the formula

$$C_q(f, x) = A_q(f, x) - \frac{1 - q}{2}x(1 - x)A_q(f'', x). \tag{8.6}$$

The following statement is similar to Theorem 6.1 and is the immediate consequence of Theorems 8.1 and 8.2.

Theorem 8.5. *If $f \in C^{(2)}[0, 1]$ then*

$$|C_q(f, x) - f(x)| \leq K(1 - q)x(1 - x)\omega(f'', \sqrt{1 - q}). \tag{8.7}$$

References

[1] S.N. Bernstein, *Demonstration du théoreme de Weierstrass fondée sur la calcul des probabilités*, Comm. Soc. Math. Charkow Ser. 2, Vol. 13 (1912), 1–2. *Collected works.* Vol. 1, 1952, 106–107.

[2] S.N. Bernstein, *Complément à l'article de E. Voronovskaya, Détermination de la forme asymptotique de l'approximation de fonction par les polynômes de M. Bernstein*, Dokl. AN USSR, Ser. A, No. 4 (1932), 86–92. *Collected works.* Vol. 2, 1954, 155–158.

[3] C.F. Gauss, *Summatio quarumdam serierum singularium*, Werke. Vol. 2, 9–45.

[4] A. Il'inskii, S. Ostrovska *Convergence of generalized Bernstein polynomials*, J. Approx. Theory, **116** (2002), 100–112.

[5] S. Ostrovska, *q-Bernstein polynomials and their iterates*, J. Approx. Theory, **123** (2003), 232–255.

[6] G.M. Phillips, *Bernstein polynomials based on the q-integers*, Ann. Numer. Math., **4** (1997), 511–518.

[7] G.M. Phillips, *Interpolation and approximation by polynomials*, CMS Books in Math., Vol. 14, Springer Verlag, 2003.

[8] V.S. Videnskii, *Linear positive operators of finite rank*, Leningrad, 1985. (Russian)

[9] V.S. Videnskii, *Bernstein polynomials*, Leningrad, 1990. (Russian)

[10] V.S. Videnskii, *New topics concerning approximation by positive linear operators*, Proceedings of International Conference "Open problems in approximation theory", Bulgaria, 1993; SCT Publishing, Singapore, 1994, 205–212.

Victor S. Videnskii
Russian State Pedagogical University, Dept. of Mathematics
Moika 48, Saint-Petersburg, 191186 Russia

Author's home address: Kazanskaya ul. 50, kv. 27, Saint-Petersburg, 190031, Russia
e-mail: `Ilya@Viden.lek.ru`